ADVANCES IN CHEMICAL PHYSICS

VOLUME LXXVI

SERIES EDITORIAL BOARD

ADVANCES IN CHEMICAL PHYSICS – VOLUME LXXVI

I. Prigogine and Stuart A. Rice – Editors

MOLECULE SURFACE INTERACTIONS

Edited by

K. P. LAWLEY

Department of Chemistry
Edinburgh University

A WILEY–INTERSCIENCE PUBLICATION

JOHN WILEY & SONS

CHICHESTER · NEW YORK · BRISBANE · TORONTO · SINGAPORE

Other Wiley Editorial Offices

John Wiley & Sons, Inc., 605 Third Avenue,
New York, NY 10158-0012, USA

Jacaranda Wiley Ltd, G.P.O. Box 859, Brisbane,
Queensland 4001, Australia

John Wiley & Sons (Canada) Ltd, 22 Worcester Road,
Rexdale, Ontario M9W 1L1, Canada

John Wiley & Sons (SEA) Pte Ltd, 37 Jalan Pemimpin 05-04,
Block B, Union Industrial Building, Singapore 2057

British Library Cataloguing in Publication Data:

Molecule surface interactions.
1. Molecules. Interactions
I. Lawley, K.P. (Kenneth Patrick) II.
Series
539′.6
ISBN 0 471 91782 6

Typeset by Thomson Press (India) Ltd, New Delhi
Printed and bound in Great Britain by Courier International, Tiptree, Essex

CONTRIBUTORS TO VOLUME LXXVI

A. T. Amos, Department of Mathematics, University of Nottingham, Nottingham, NG7 2RD, UK

K. Binder, Institut für Physik, Johannes–Gutenberg–Universität–Mainz, D-6500 Mainz, Postfach 3980, FDR

D. W. Brenner, Naval Research Laboratory, Washington, DC 20375, USA

R. C. Cavanagh, Center for Atomic, Molecular and Optical Physics, National Institute of Standards and Technology, Gaithersburg, MD 20899, USA

G. Comsa, Institut für Grenzflächenforschung und Vakuumphysik, KFR Jülich, P.O. Box 1913, D-5170 Jülich, FDR

S. G. Davison, Department of Physics, Texas A&M University, College Station, Texas 77843–4242, USA

B. J. Garrison, Department of Chemistry, The Pennsylvania State University, University Park, PA 16802, USA

D. W. Goodman, Surface Science Division, Sandia National Laboratories, Albuquerque, NM 87185, USA

K. Kern, Institut für Grenzflächenforschung und Vakuumphysik, KFR Jülich, P. O. Box 1913, D-5170 Jülich, FDR

D. S. King, Center for Atomic, Molecular and Optical Physics, National Institute of Standards and Technology, Gaithersburg, MD 20899, USA

D. P. Landau, Center for Simulational Physics, The University of Georgia, Athens, GA 30602, USA

R. Ryberg, Physics Department, Chalmers University of Technology, S-412 96 Göteborg, Sweden

A. G. Sault, Surface Science Division, Sandia National Laboratories, Albuquerque, NM 87185, USA

K. W. Sulston, Department of Mathematics, University of Nottingham, Nottingham, NG7 2RD, UK

INTRODUCTION

Few of us can any longer keep up with the flood of scientific literature, even in specialized subfields. Any attempt to do more, and be broadly educated with respect to a large domain of science, has the appearance of tilting at windmills. Yet the synthesis of ideas drawn from different subjects into new, powerful, general concepts is as valuable as ever, and the desire to remain educated persists in all scientists. This series, *Advances in Chemical Physics*, is devoted to helping the reader obtain general information about a wide variety of topics in chemical physics, which field we interpret very broadly. Our intent is to have experts present comprehensive analyses of subjects of interest and to encourage the expression of individual points of view. We hope that this approach to the presentation of an overview of a subject will both stimulate new research and serve as a personalized learning text for beginners in a field.

ILYA PRIGOGINE

STUART A. RICE

CONTENTS

Advances in Chemical Physics
Edited by K. P. Lawley

INFRARED SPECTROSCOPY OF MOLECULES ADSORBED ON METAL SURFACES

ROGER RYBERG

Physics Department, Chalmers University of Technology, S-412 96 Göteborg, Sweden

CONTENTS

Abbreviations

EELS	Electron energy loss spectroscopy
FTIRS	Fourier transform infrared spectroscopy
FWHM	Full width at half maximum
IRS	Infrared spectroscopy
LEED	Low energy electron diffraction
NEXAFS	Near edge X-ray absorption fine-structure spectroscopy

NMR	Nuclear magnetic resonance
UHV	Ultrahigh vacuum
UPS	Ultraviolet photoemission spectroscopy
cm^{-1}	Wavenumbers; $8.065 cm^{-1} = 1 meV$, $2000 cm^{-1}$ at a wavelength of $5 \mu m$

1. INTRODUCTION

During the present century infrared spectroscopy (IRS) has been used as one of the main tools for studying molecules in the gas phase. It was therefore quite natural when modern surface physics started to evolve some twenty years ago, that one wanted to use the same approach also for studies of molecules adsorbed on a solid surface. However, when using the same type of spectrometers the experimentalists faced a sensitivity problem, which could not be overcome by simply increasing the number of molecules, as one was restricted to study one monolayer or less. This led to the development of dedicated spectrometers taking advantage of the special boundary conditions set up by the surface and also using some kind of modulation technique to increase the sensitivity. Today, infrared spectroscopy is able to detect submonolayers even for very weakly absorbing vibrational modes with energies above $2000 cm^{-1}$, of rather strongly absorbing modes above $800 cm^{-1}$, but has until now only been able to detect low energy substrate-molecule modes under very special conditions.

Although infrared spectroscopy has suffered from a somewhat limited energy range compared to the competing technique of electron energy loss spectroscopy, due to its inherent high resolution it has had a great success in studies of the fine structure of the vibrational spectra. It is the purpose of this review to summarize the present picture of fundamental processes in adsorption systems as they are revealed by the infrared spectra. This incorporates different kinds of vibrational coupling and the transfer and dissipation of the vibrational energy. We will also discuss how infrared spectroscopy can be used to study surface reaction intermediates and outline its use in high pressure situations. As we will discuss fundamental processes, we will mainly consider studies made on single crystalline metal surfaces under well characterized (often ultrahigh vacuum) conditions and for the same reason we will consider small molecules and frequently carbon monoxide will be our test molecule.

It is not the purpose of this chapter to describe the evolution of the spectroscopy. For such a historical background we refer the reader to the reviews by Hoffmann[1], Hollins and Pritchard[2] and Ueba[3]. Neither do we attempt to give references to all works in the field. This is done in the proceedings of the Conferences on Vibrations at Surfaces, where Darville[4–6] has presented careful and very useful tabulations of all systems investigated by

IRS. References in this chapter are only given to either the most relevant or latest works and the interested reader is referred to these papers for the appropriate background. The literature has been searched until the end of 1987.

2. EXPERIMENTAL ASPECTS

The fact that the molecules are adsorbed on a solid surface gives rise to a number of new effects compared to the gas phase situation. The experimental situation consists of a monolayer of molecules adsorbed on a metal surface, on which we shine infrared radiation and then detect the reflected light. The macroscopic theory for the electromagnetic response of such a system is reproduced in the previous reviews[1-3]. A more microscopic treatment has been given by Persson[7], showing that the integrated infrared absorptance for p-polarized light is given by;

$$\int (I_0 - I)/I_0 \, d\omega = (8\pi^2/hc)(N/A)\mu^2\Omega G(\alpha) \tag{1}$$

where N/A is the number of molecules per unit area, $G(\alpha)$ is a constant depending on the dielectric properties of the substrate and the angle of incidence α and Ω is the vibration frequency. How the dynamical dipole moment μ of the vibrational mode should be interpreted for adsorbed molecules was for a long time debated and we will try to clarify this in section 3.

An important consequence of the presence of the metal surface is the so-called *infrared selection rule.* If the metal is a good conductor the electric field parallel to the surface is screened out and hence it is only the p-component (normal to the surface) of the external field that is able to excite vibrational modes. In other words, it is only possible to excite a vibrational mode that has a nonvanishing component of its dynamical dipole moment normal to the surface. This has the important implication that one can obtain information by infrared spectroscopy about the orientation of a molecule and definitely decide if a mode has its dynamical dipole moment parallel with the surface (and hence is undetectable in the infrared spectra) or not. This strong polarization dependence must also be considered if one wishes to use Eq. (1) as an independent way of determining μ. It is necessary to put a polarizer in the incident beam and use optically passive components (which means polycrystalline windows and mirror optics) to avoid serious errors. With these precautions we have obtained pretty good agreement[8] for the value of μ determined from Eq. (1) and by independent means as will be discussed in section 3.2.

Over the last decade there have been great developments of the infrared spectrometers used in surface science. One has moved from simple, single beam reflectometers to different kinds of modulation setups, from dispersive to multiplex Fourier transform infrared spectrometers and in addition to reflexion also tried direct absorption or emission spectroscopy. One has tested other light sources than blackbodies, like lasers[9] and synchrotron radiation[10]. Richards and Tobin[11] have critically discussed in an excellent way the present status of spectroscopy. They conclude that for a reflexion setup with proper, modern technology and careful design the sensitivity will be limited by the stability of the background. The reason is that in an optical reflexion measurement one is always, in one way or another, looking at a small difference between two large signals, i.e. in the present case with and without molecules adsorbed on the surface. This important conclusion, which is supported by my own practical experience, shows that from a sensitivity point of view there is nothing to gain in using other light sources than blackbody radiators, as they will only increase the source noise and the use of multiplex spectrometers can even be disadvantageous. Therefore, if one attempts to achieve maximum sensitivity and wants to study small molecules with few vibrational modes, the best solution is still a dispersive spectrometer, probably using the wavelength modulation technique which at present has shown the best reported signal/noise[12]. The main efforts should be made in designing an instrument as stable as possible, with special attention to temperature stability of all parts, repeatability of the grating motion, minimizing electronic drift and with special care to positional stability of the sample whilst cooling and heating during the actual experiment. If one is looking for very weak structures it is also important that one has the necessary dynamical response in the signal processing system (the lock-in amplifier and signal averager).

Much of the effort in recent years has been devoted to the development of spectrometers that are able to detect low frequency modes, such as the metal-molecule stretch at energies below 500 cm^{-1}. This has turned out to be a very difficult problem, because the integrated absorptance (1) suffers from an Ω dependence and the dynamical dipole moment μ normally is smaller for these modes than for the intramolecular high frequency ones. If we consider the CO/Cu(100) system and use the dynamical dipole moments of the two stretch modes as obtained from electron energy loss spectroscopy (EELS)[13], we find that the integrated infrared absorptance is about 90 times smaller (assuming constant $G(\alpha)$) for the low frequency mode. In this context we can mention that the energy loss peaks in EELS do not suffer from the Ω dependence in (1), which here makes a difference by a factor of 6, and that many EELS spectrometers enhance low energy modes as the angular distribution for dipole excitation losses narrows[13].

Furthermore, if one uses the wavelength modulation technique, which is strictly not surface sensitive but only enhances sharp structures and as the bandwidth of a grating monochromator decreases with decreasing energy, the low frequency peak will appear broad for reasonable slit widths. The solution to this problem is to use special gratings with very low dispersion, decrease the angle of incidence and use larger samples to enable an increase of the slit width. There exists in the literature no report that it has been possible to detect such a low frequency mode in a reflexion configuration, despite several attempts both with dispersive and multiplex spectrometers[14]. However, Chiang *et al.*[15] in a very nice experiment used a quit different approach with a new design of infrared *emission* spectrometer. The sample is surrounded by liquid nitrogen-cooled baffles and the emitted light is measured by a liquid helium-cooled grating spectrometer. With this technique they were able to detect the very weak absorption of the low frequency stretch mode of CO—Ni[15] and CO—Pt[16].

However, during the last three to four years, I have in several stages improved on my infrared spectrometer along the lines discussed above.

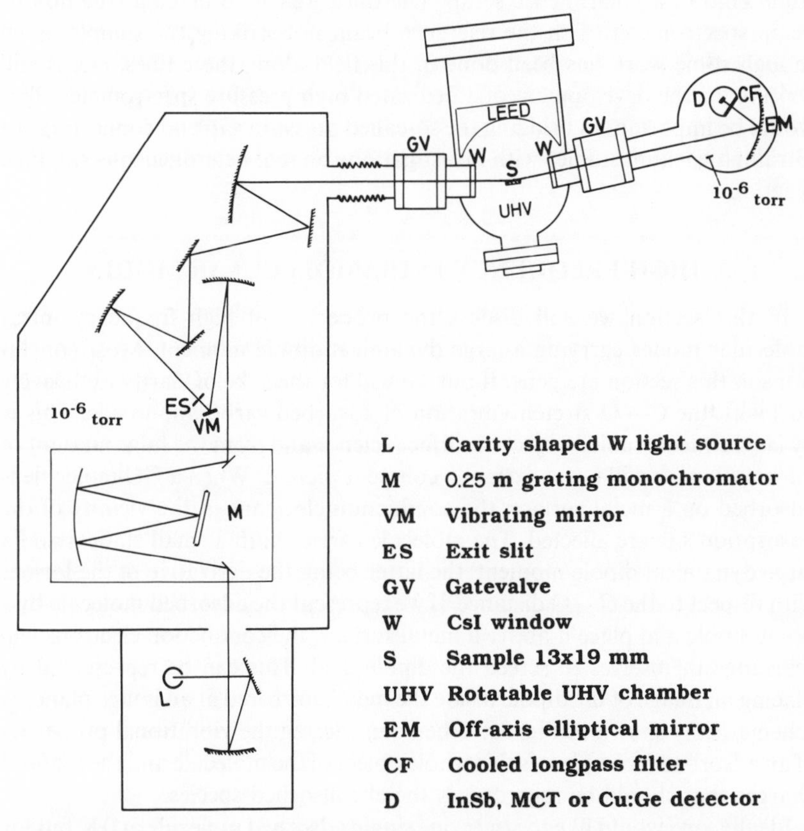

Fig. 1. Schematic drawing of the wavelength modulation spectrometer setup. The spectrometer keeps the throughput ($f/3.6$) of the monochromator through the whole optical path.

Figure 1 shows this latest version of the experimental setup with the wavelength modulation spectrometer, where emphasis has been put on stability of all parts. During the course of writing this paper I was finally able to detect the low frequency metal-molecule stretch of CO/Pt(111), as will briefly be discussed in section 4.

It is often stated that one of the advantages of infrared spectroscopy is the possibility to work under high ambient pressures. This is certainly true if one studies a vibrational mode of a surface intermediate, which is well away in energy from any of the modes of the gasphase molecules. When this is not the case, the strong gasphase absorption with the broad rotational bands will interfere with the spectrum of the adsorbed molecules. The straightforward approach is of course just to take the difference of two spectra in the ordinary way, but this demands high spectrometer stability and dynamical range. The second possibility is to take advantage of the infrared selection rule and make some kind of a polarimetric setup. The third way is to make a true double beam spectrometer, with the reference beam not striking the sample. Even though some work has been done in this field along these lines, one is still waiting for the development of a dedicated high pressure spectrometer. This would be important in bridging the so-called pressure gap and connecting the ultrahigh vacuum studies with investigations on real heterogeneous catalytic reactions.

3. HIGH FREQUENCY INTRAMOLECULAR MODES

In this section we will discuss the properties of high frequency intramolecular modes carrying a large dynamical dipole moment. Most conclusions in this section are general but we will for the sake of clarity exclusively deal with the C—O stretch vibration of adsorbed carbonmonoxide. This is by far the most studied object in surface science and from the huge amount of information we will try to extract a coherent picture. When a CO molecule is adsorbed on a metal surface, the conduction electrons in the vicinity of the adsorption site are affected. The molecule carries both a small statical and a large dynamical dipole moment, the latter being the derivative of the former with respect to the C—O distance. If we represent the adsorbed molecule by a point dipole and place it above a metal surface, the conduction electrons will rearrange themselves to screen the dipole field. This can be represented by placing an image of the dipole inside the metal, mirrored at an image plane, as schematically shown in Fig. 3a. When considering the vibrational properties of an adsorbed molecule it is this whole object of the molecule and the induced charges that should be regarded as the chemisorbed species.

Ideally, one would like to study one single adsorbed molecule at 0 K but for practical reasons one has to study an ensemble of molecules at finite temperatures. Even if it is feasible to cool the sample to very low temperatures,

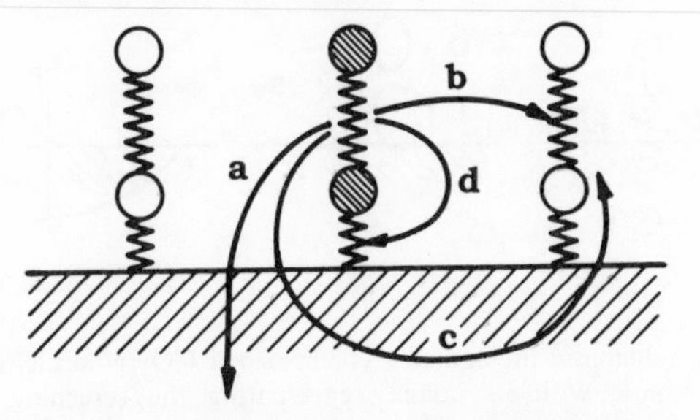

Fig. 2. The different kinds of interactions of the intramolecular high frequency mode that may be seen in the infrared spectra. Interactions with the substrate (a) will both affect the vibration frequency and the lifetime of the vibrational excitation. Molecule–molecule interactions can be direct (b) or mediated via the substrate (c). There may also be a substantial interaction with thermally excited, low frequency modes (d).

studies of very dilute overlayers are seldom meaningful. Even for the best prepared metal surface there is a defect concentration of the order of 1%, so even if one has the sensitivity to record 1/1000 of a monolayer, at very low coverages one will often mainly probe molecules sitting at these defects. Therefore, considering rather dense overlayers we have to take into account the different kinds of interaction that can occur, as indicated in Fig. 2. For a chemisorbed molecule there is of course an interaction with the substrate, which will affect the vibration frequency. The vibrational energy can be dissipated by different kinds of excitations in the metal. We can imagine different kinds of molecule–molecule interactions, direct or mediated via the substrate. The high frequency mode can couple to other, thermally excited low frequency modes. All these kinds of interactions will in principle affect the infrared spectrum in one way or another, so it is essential to sort these things out to be able to give a correct interpretation of the measured data and we will discuss each phenomenon in the following sections.

3.1. Metal–molecule interaction

It is a well known fact that the vibration frequency of the internal stretch mode of CO decreases from its gas phase value of 2143 cm^{-1} on chemisorption onto a metal surface. In general terms, depending on the adsorption site it takes values between 2000–2100 cm^{-1} in the ontop position, 1900–2000 cm^{-1} when bridgebonded and 1800–1900 cm^{-1} for molecules in the hollow site[1–3]. The vibration frequency of a particular system is of course the total effect of the different kinds of interactions sketched in Fig. 2. For example, simply the fact that the oscillator is attached to a more or less rigid substrate so that the

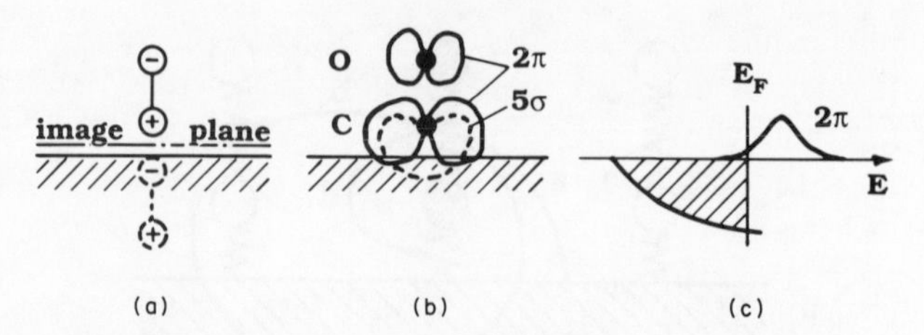

Fig. 3. Schematic picture of a chemisorbed CO molecule. (a) As a point dipole with its image, representing the screening by the conduction electrons. (b) The spatial extension of the two molecular orbitals involved in the chemisorption bond, the highest occupied 5σ and the lowest unoccupied 2π orbital. (c) The density of states of the conduction electrons and the 2π orbital, which by interaction with the metal electrons has broadened into a resonance and shifted down in energy.

carbon atom cannot oscillate freely gives rise to a *mechanical shift*. This has been estimated by a simple masses and springs model[1] to give an increase in frequency by $\sim 50\,\mathrm{cm}^{-1}$.

Furthermore, as mentioned above the screening of the dipole field by the conduction electrons can be represented by an image dipole inside the metal. This complex of the chemisorbed molecule and its image has a vibration frequency different from that of the free molecule. The *electrodynamic interaction between a dipole and its image* has been discussed in many works[17]. The theoretical problem is that the calculated frequency shift is extremely sensitive to the position of the image plane (Fig. 3a). One can with reasonable parameter values obtain a downward frequency shift of the order of 5–$50\,\mathrm{cm}^{-1}$, but the latest work[17] indicates that the shift due to this interaction is rather small.

Even if the distance of the molecule above the image plane varies with adsorption site, most of the observed site dependence of the frequency must come from changes in the *electronic structure*. The present, somewhat revised picture of the CO chemisorption stems from recent cluster calculations by Bagus and coworkers[18]. The lowest unoccupied 2π molecular orbital in free CO is, due to the interaction with the metal electrons, broadened and pulled down in energy close to the Fermi level, as sketched in Fig. 3. The chemisorption bond is formed by transfer of charge from the metal into this 2π resonance, while the 5σ molecular orbital interacts with the conduction band with little net charge transfer. Since the 2π orbital has an antibonding character with respect to the C—O bond, the filling of this resonance lowers the vibration frequency. The 2π resonance has also been observed experimentally a few eV above the Fermi level by inverse photoemission. Its position in energy varies somewhat for different adsorption sites and metal substrates. For some systems the lower energy is found for the more strongly bonded species[19], quite in line with the picture above. However, a critical review of all present data[20] shows that the problem is not yet completely understood.

Within the same cluster calculation approach, attempts have also been made to mimic the vibrational motion of the chemisorbed molecule. If the cluster gives a good representation of the metal, the different interactions above should be included. For the internal stretch mode of CO in the ontop position on Cu(100) Müller and Bagus[21] obtain a downward frequency shift of 40–50 cm^{-1}, depending on the cluster size. This rather close to the observed value[8] of 67 cm^{-1}. Even if this good agreement may be rather fortuitous, they obtain the correct trends of the shifts for CO bonded on different metals. In the ontop position at low coverages IRS data give for Cu(100)[8] 2076 cm^{-1} and for Ni(100)[11] 2020 cm^{-1}, which gives a ratio of 1.03, which is just the same as for the calculated values[21]. Hopefully, larger clusters or complementary calculation techniques in the future will be able to predict the vibration frequencies of molecules chemisorbed at different sites and on different metals.

The situation is quite different for *physisorbed* molecules. In that case, there is no transfer of charge, the mechanical renormalization is weaker due to a much weaker metal–molecule bond and also the image interaction is smaller as the molecule probably is adsorbed further out from the surface. In a recent IRS investigation of CO physisorbed on Al(100)[22] the measured frequency is only shifted down a few cm^{-1} from the gasphase value. However, there is for this system also a short range intermolecular interaction that certainly will affect the vibrational frequency. As yet there exist no theoretical calculations for the van der Waals interaction between a CO molecule and a metal.

3.2. Molecule–molecule interaction

When a dipole is placed above a metal surface, the conduction electrons rearrange themselves to screen the dipole field as discussed above. This charge redistribution is found mainly in the very vicinity of the adsorption site but a small tail extends over several angstroms as so-called Friedel oscillations. When the adsorbed molecule is vibrationally excited, this will also give rise to a time-dependent variation in these oscillations. In principle these could affect the nearest neighbours and give rise to a *short range electronic interaction mediated via the substrate*. Treating the metal in the jellium model (that is neglecting the localized d-electrons), Persson[23] has estimated that the effect on the high frequency intramolecular modes should be negligible for $R > 4/k_F$, where R is the molecule–molecule distance and k_F is the Fermi wavenumber of the metal. This corresponds for copper to a distance of 2.9 Å. CO on close packed metal surfaces forms structures with a nearest neighbour distance larger than 3.5 Å and one has not seen any effect of this interaction in the infrared spectra. The main reason is that it is only in special cases that one can obtain the densely packed overlayers that produce the necessary short intermolecular distance. The situation for atomic adsorbates is different; one has found, in closed packed overlayers of hydrogen on Pd(100)[24] and Ni(100)[25], by EELS a very strong vibrational coupling, which has to be attributed to this electronic interaction through the substrate.

When the chemisorbed molecule is vibrationally excited this influences not only the metal electrons but also the ion cores in the neighbourhood. The vibrating ion cores can then in turn couple to other molecules and give rise to a *short range interaction mediated via the substrate lattice.* However, as Ω is much larger than the highest substrate phonon frequency the effect of this interaction is very small[26], but it can be important for low frequency modes[27].

CO chemisorbed on a metal surface often occupies distinct adsorption sites, giving at certain coverages ordered structures in phase with the substrate. On fcc(100) and (111) surfaces saturation occurs at coverages well below 1, the molecules are chemisorbed in an upright position with the oxygen end pointing outwards and with a molecule–molecule distance larger than 3 Å. However, it is now well established that on the (110) surfaces of Ni, Pd and Pt it is possible to obtain a coverage of 1, that is one adsorbed molecule on every metal surface atom. The surface consists of close packed, well separated rows of surface metal atoms. The most studied system is CO/Ni(110)[28] and at coverages less than 0.8 the molecules have the normal orientation but at higher coverages one obtains a CO–CO separation of only 2.5 Å. At such a short distance there is a *direct short range interaction* between the molecular orbitals of neighbouring molecules, which forces them to tilt away from each other to increase the intermolecular distance. The situation is similar for CO/Pt(110), where an angular resolved ultraviolet photoemission spectroscopy (UPS) study[29] has shown that at a coverage of 1 the molecules are tilted with about 25° away from the surface normal. This is confirmed by an infrared investigation of this system by Hayden *et al.*[30]. They find a decrease in the integrated absorptance for coverages above 0.5, reflecting how the component normal to the surface of the dynamical dipole moment decreases when the molecules tilt away from each other. One could expect that the onset of this molecule–molecule interaction also would show up in the infrared spectrum as a frequency shift and/or a change in the line shape. However, the data show a monotonous coverage-dependent frequency shift solely caused by the dipole–dipole interaction discussed below and the peak has a width of about 20 cm^{-1} at all coverages, probably caused by inhomogeneous broadening[30].

If the interaction with the substrate is weaker the situation becomes different. For a chemisorbed molecule an electronic rearrangement or even charge transfer between the substrate and the molecule has taken place, as discussed in the previous section. Physisorbed molecules on the other hand are only bonded via the fluctuating polarization, that is the van der Waals interaction. The problem when investigating physisorbed or weakly chemisorbed molecules is that as the interaction with the substrate is so weak they do not form any ordered structures and one does not get from the low energy electron diffraction (LEED) pattern any information about the nearest neighbour distance. This was seen in the infrared study on CO physisorbed on an Al(100) surface[22]. As much of the binding energy comes from the molecule–molecule interaction they form close packed clusters on the surface at all coverages and it is not possible to observe the onset of the direct short range interaction. This interaction manifests itself in the infrared spectrum as a relatively large inhomogeneous broadening and asymmetry of the infrared absorption peak (see Fig. 10).

The vibrational interaction that by far gives the strongest influence on the vibrational spectra is the *long range dipole–dipole interaction.* It originates from the fact that the vibrating molecule in Fig. 3a gives rise to a long range dipole field which oscillates in time and which is felt by the other adsorbed molecules. Already from the beginning of the infrared studies of adsorbed molecules one often found an upward frequency shift of 5–50 cm^{-1} as the coverage increased up to a full monolayer[1–3]. It was discussed how much of this shift was due to a vibrational coupling and how much was caused by changes in the electronic environment of the molecules, the 'chemical shift'. Hammaker *et al.*[31] early realized that the two effects could be separated by studies of mixtures of two different isotopes of the molecule. If the study is made at a constant coverage, the chemical environment in the different mixtures is the same, while the vibrational properties are varied. However, it was only after one had obtained enough sensitivity/resolution in IRS that one realized how strong the effect of the interaction could be. Figure 4 shows the infrared spectra of different mixtures of $^{12}C^{16}O/^{12}C^{18}O$ on a Cu(100) surface in the ordered $c(2 \times 2)$ structure[32]. The difference in vibration frequency for the two isotopes in the gas phase is 50 cm^{-1}, so for noninteracting molecules we would expect two absorption peaks, each with a height proportional to the concentration of the corresponding isotope. However, we observe one absorption peak dominant for nearly all concentrations and we find two peaks of equal height not at a 50/50 concentration but for 5% of the light isotope and 95% of the heavy one. Obviously, the overlayer must be regarded as a strongly coupled system and at that time the existing theories of the dipole–dipole interaction were unable to explain such a strong intensity transfer.

In the first treatment of this problem, taking up ideas from molecular crystal work, Hammaker *et al.*[31] considered only the direct interaction between the dipoles. Mahan and Lucas[33] made a substantial improvement by taking into account both the contribution from the images and the electronic polarizability of the molecules. Still they were not able to explain the observed coverage dependent freqency shifts. The reason for this became evident when Persson *et al.*[32,34] developed a new theory for the dipole–dipole interaction based on the old ideas but with several important new concepts. They showed

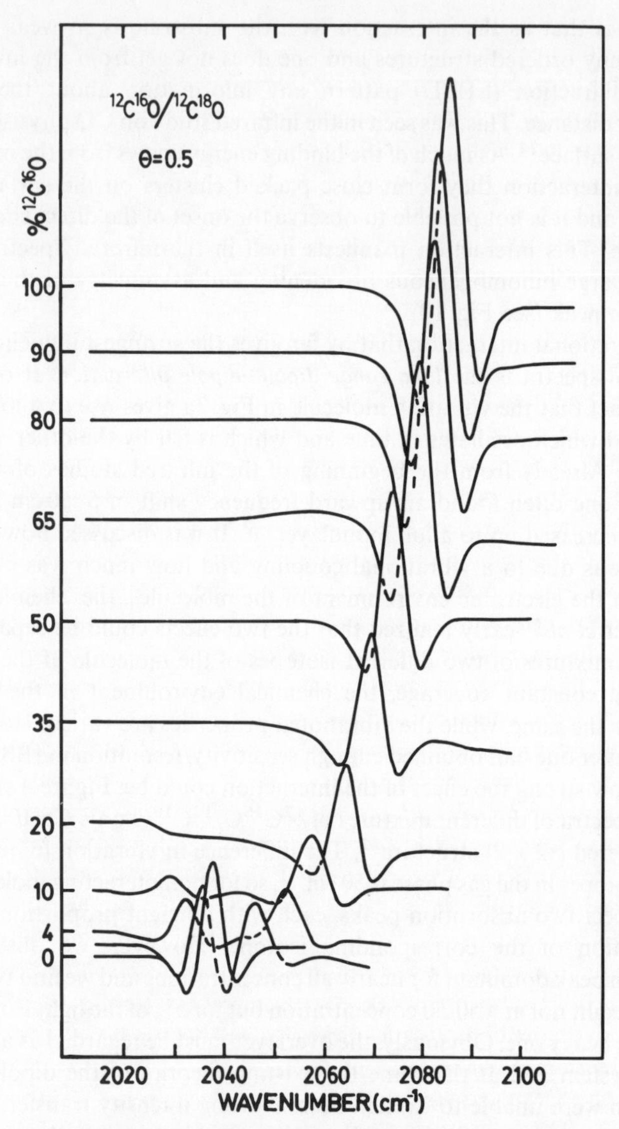

Fig. 4. Infrared derivative spectra for different isotopic mixtures of $^{12}C^{16}O/^{12}C^{18}O$ in the $c(2 \times 2)$ structure at 100 K on Cu(100). (*Reproduced by permission from Persson and Ryberg*[32].)

that the dynamical dipole moments that previously had been deduced from the strength of the IRS absorption or EELS loss peaks were incorrect, because earlier theories had neglected the screening of the external field by the electronic polarizability of the adsorbed molecules. Including this screening it was found that the dynamical dipole moment of chemisorbed CO is twice as large as for the free molecule. Using this new value in the previous theory[33] it was possible to explain the observed frequency shift. This result also solved an old problem as to why the dynamical dipole moment of the C—O stretch mode in carbonyls such as $Ni(CO)_4$ seemed to be about twice as large as for chemisorbed CO. With this new evaluation the values became about the same[32].

It was then possible to describe the behaviour of the frequencies but not the strong, anomalous intensity transfer. However, by using a theoretical approach called the coherent potential approximation it was possible to reproduce also these fine structures of the problem[32]. We ended up with a two-parameter theory, the parameters being the vibrational polarizability α_v (related to the dynamical dipole moment μ by $\alpha_v = 2\mu^2/\hbar\Omega$) and the electronic polarizability α_e of the adsorbed molecule. These quantities are attributed to a vibrationally excited object consisting of the *chemisorbed molecule and its image charge.* It has been argued that the molecule and its image should be treated separately and that the position of the molecule above the image plane could be used as a theoretical parameter[2,35]. However, Persson and Liebsch[34] have shown that it is the whole object to which one should attribute the different measured quantities like the vibrational resonance frequency, the dynamical dipole moment and the electronic polarizability.

A best fit to the experimental data in Fig. 4 gave $\alpha_v = 0.25 \pm 0.02$ Å^3 and $\alpha_e = 2.5 \pm 0.5$ Å^3 (compared to 2.5 Å^3 for the free molecule)[34]. The vibrational polarizability can also be obtained in two independent ways, from the total infrared absorptance using Eq. (1), which gave $\alpha_v = 0.19$ Å^3 [8] and from EELS data, giving $\alpha_v = 0.23$ Å^3 [36], both assuming $\alpha_e = 2.5$ Å^3. These three values are in satisfactory agreement and give a dynamical dipole moment twice as large as for the free molecule. This increase is caused both by a polarization of the chemisorbed molecule and by the charge transfer to the 2π molecular resonance discussed in section 3.1, as charge is oscillating between this resonance and the metal during the vibration. Theoretical cluster calculations on CO/Cu(100)[37] also predict an enhancement in μ of this magnitude.

Using this theory of the dipole–dipole interaction it was then possible to get a good description of the experimentally measured integrated absorptance and vibration frequency as function of coverage. This is shown in Fig. 5 for CO/Ru(001). The solid lines are calculated using

$$(\Omega/\omega_1)^2 = 1 + c\alpha_v U(\mathbf{0})/[1 + c\alpha_e U(\mathbf{0})] \tag{2}$$

$$\int (I_0 - I)/I_0 \, \mathrm{d}\omega \sim c\alpha_v U(\mathbf{0})/[1 + c\alpha_e U(\mathbf{0})]^2 \tag{3}$$

where ω_1 is the vibration frequency when the coverage c goes to zero and $U(\mathbf{0})$ is the dipole sum for the actual adsorbate structure with the wavevector $\mathbf{q} = \mathbf{0}$.

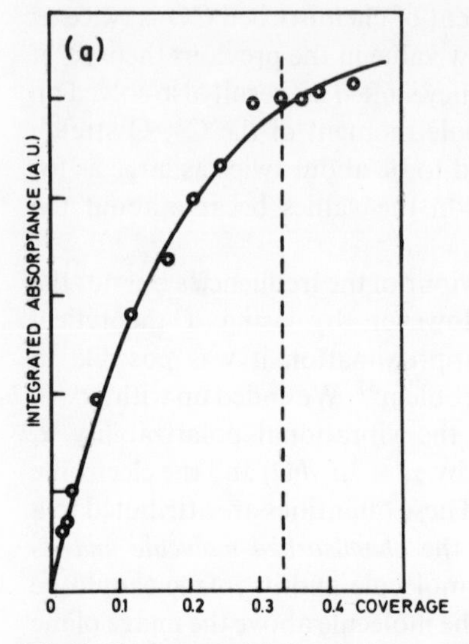

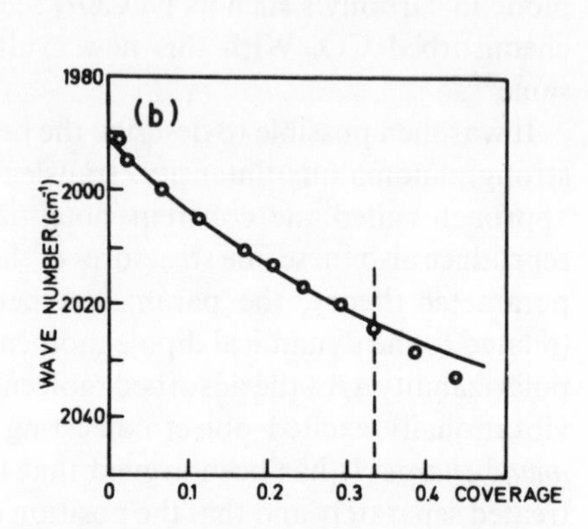

Fig. 5. The integrated infrared absorptance (a) and the vibration frequency (b) as function of converage for CO/Ru(001). The circles are experimental data from Pfnür *et al.*[38] and the solid lines are calculated using Eqs. (2) and (3), giving $\alpha_v = 0.28\,\text{Å}^3$ and $\alpha_e = 2.8\,\text{Å}^3$. (*Reproduced by permission from Persson and Ryberg*[32].)

From the fit one obtains values of α_v and α_e. Note how the electronic polarizability of the adsorbed molecules gives the absorptance a nonlinear coverage dependence. However, there exist several systems that do not follow Eqs. (2) and (3). This can be caused either by a coverage dependent change in the electronic structure, that is an additional chemical shift, or because the system exhibits clustering or the molecules occupy more the one adsorption site, since the theory assumes a random filling of the adsorbate lattice.

For the interpretation of infrared spectra of molecules adsorbed on a metal surface the dipole–dipole interaction have the following important implications:

1. An *ordered* monolayer of molecules having a large dynamical dipole moment must not be regarded as an ensemble of individual oscillators but a strongly coupled system, the vibrational excitations being collective modes (phonons) for which the wavevector **q** is a good quantum number. The dispersion of the mode for CO/Cu(100) in the $c(2 \times 2)$ structure has been measured by off-specular EELS[36], while the infrared radiation of course only excites the $\mathbf{q} = \mathbf{0}$ mode.
2. One must account for the screening of the external field by the electronic polarization of the adsorbed molecules. This screening gives rise to a reduced infrared absorption. Taking it into account in a proper way shows

that the dynamical dipole moment of chemisorbed CO is roughly twice as large as for the free molecule.

3. The anomalous strong transfer of intensity to the high frequency mode that is seen in Fig. 4 arises from the fact that high frequency modes can screen low frequency ones but not vice versa. If this screening effect is overlooked it can give rise to a significant misinterpretation when one tries to use the peak height as a measure of the concentration as is done in gas phase spectroscopy. This is particularly important when one considers two or more modes separated by less than a hundred cm^{-1}. This intensity transfer to the high frequency mode has been shown to give an apparently too low concentration of a surface species[39], too small intensity from molecules chemisorbed in the bridge position[40] and even missing peaks in some structures[8].
4. It has turned out to be important to use the full theory[32] of the dipole–dipole interaction also when calculating other vibrational properties of an adsorbate layer. It was important for the interpretation of the coupling to the low frequency modes[41], the so-called dephasing discussed in the next section. Further, it had drastic effects on the calculated shape of the infrared absorption peak[42], as discussed in section 3.4.

All this work on the dipole–dipole interaction has been made for modes oriented normal to the surface or for the normal component of μ and they predict an upward frequency shift for increasing coverage. Hayden *et al.*[43] suggested that a downward shift could occur for modes oriented parallel to the surface and this idea has also been used to assign modes of H/W(100)[44]. However, it should be clear that the interaction must be much weaker for modes parallel to the surface, as the dipole field in accordance with the infrared selection rule mentioned in section 2 is screened by the metal surface. At least, in a theoretical model this has to be taken into account.

In conclusion, the effects of the dipole–dipole interaction cause the greatest difference between the interpretation of infrared spectra of gas phase and adsorbed molecules and if the mode has a large dynamical dipole moment this interaction is always in operation.

3.3. Interaction with low frequency modes

Consider the possible normal modes of a diatomic molecule oriented with the axis normal to the surface. If we neglect the surface structure, we get starting from the lowest energy as schematically shown in Fig. 6;

(a) a *frustrated translation* consisting mainly of translational but also some rotational motion, because the molecule has its highest binding energy at a certain site;

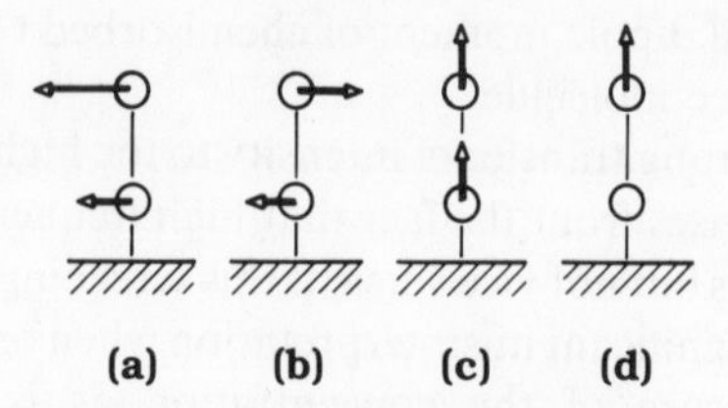

Fig. 6. Schematic representation of the normal modes of an adsorbed diatomic molecule neglecting the surface structure, after Richardson and Bradshaw[45]. In parentheses the experimentally measured values for CO in the ontop position on Pt(111). (a) A frustrated translation ($60\,cm^{-1}$)[46]. (b) A frustrated rotation (not yet detected). (c) The metal–molecule stretch ($460\,cm^{-1}$)[16]. (d) The intramolecular stretch model ($2100\,cm^{-1}$)[14].

(b) a *frustrated rotation* consisting mainly of rotational but also some a translational motion;
(c) a mainly *metal–molecule stretch* mode; and finally
(d) a mainly *intramolecular stretch* mode.

The C—O stretch vibrational mode of a free molecule is almost a perfect harmonic oscillator and if the interaction with the substrate were negligible this would also hold for an adsorbed molecule. However, we know that for a chemisorbed molecule there is a substantial molecule–substrate interaction which manifests itself, for example, as a dependence of the vibration frequency on the binding position on the surface. The trend is that the vibration frequency decreases as we go from ontop to bridge site and from bridge to hollow site, the difference being about $100–200\,cm^{-1}$ in each step. Intuitively we feel that if the whole molecule is making any motion with respect to the surface described by the low frequency modes above, this will affect the high frequency mode and should in principle be seen in the infrared spectrum. Such effects of an anharmonic coupling (which does not exist for a perfect harmonic oscillator) gives rise to vibrational phase relaxation of the C—O stretch mode, a process often called *dephasing*. There is no transfer of vibrational energy but the high frequency mode, after a certain time due to this anharmonic coupling will be out of phase compared with an unperturbed harmonic oscillator. Characteristic for this kind of interaction would be a strong temperature dependence, because the low frequency modes are thermally excited and the effect should vanish at low temperatures. Gadzuk and Luntz[47] have reviewed the different kinds of phase relaxation mechanisms that may be applicable to adsorption systems. On the other hand there exist other temperature dependent effects such as phase transitions, increased surface mobility causing disorder in the overlayer and so on. Experimentally we find for some systems, like CO/Cu(100)[12], almost no temperature dependence at all, while for other systems drastic temperature effects have been reported.

Rather strong temperature variation has been seen in the infrared spectra of CO/Ni(111)[48,49]. Figure 7 shows the peak width and position as a function of substrate temperature for the $c(4 \times 2)$ structure, where all molecules are chemisorbed in the bridge position. As seen in the figure the peak width increases strongly with increasing temperature and there is also a small upward shift of the peak position. We observe that both the peak width and position reaches a constant value for very low temperatures, so it could be possible that the behaviour can be explained in terms of an anharmonic coupling. Persson *et al.*[40,41] developed a new theory for this problem, partly

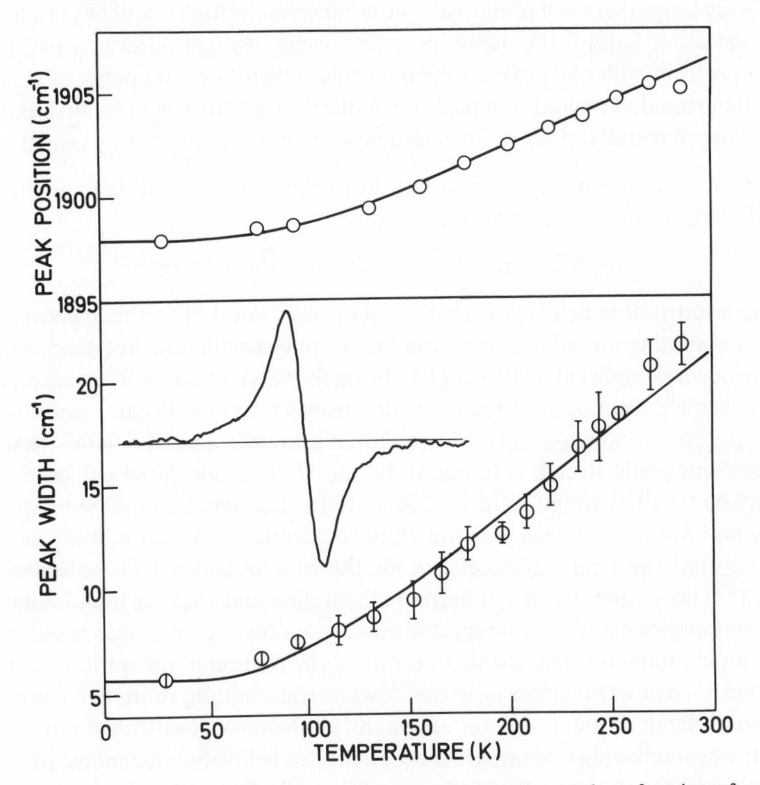

Fig. 7. The position and width of the infrared absorption peak as function of substrate temperature for the $c(4 \times 2)$ structure of CO on Ni(111). All molecules are chemisorbed in the bridge position and the solid lines are calculated within the theory describing the anharmonic coupling to a low frequency mode. (*Reproduced by permission from Persson and Ryberg*[40].)

based on older work[50] but incorporating new important concepts, in particular the lateral dipole–dipole interaction. The theory contains three parameters, which describe the dephasing mechanism:

1. The frequency ω_0 of the low energy mode. This frequency must of course be low enough to give a reasonable population at temperatures where the adsorption system is stable.
2. The coupling strength $\delta\omega$, which is the shift in frequency of the high energy mode when the low frequency mode is in its first excited state. $\delta\omega$ has to be large enough to give a measurable effect. It can take both positive and negative values, e.g. a frustrated translation for ontop bonded molecules gives a negative $\delta\omega$, as the frequency decreases when going away from the ontop position.
3. The damping rate η of the low energy mode. These modes are predominantly damped by excitations of substrate phonons. If the damping rate is very large, there will of course be little effect on the high frequency mode. On the other hand if the damping is very weak, we will observe a two-peak spectrum with one peak for the molecules where the low energy mode is in its ground state and one peak for molecules where it is in its first excited state (naturally, if ω_0 is low enough we can get a number of peaks).

The solid lines in Fig. 7 are a best fit to the experimental data, giving the following values for the parameters above[42]:

$$\omega_0 = 235\,\text{cm}^{-1} \quad \delta\omega = 20\,\text{cm}^{-1} \quad \eta = 20\,\text{cm}^{-1}.$$

The important conclusion is that we get a very good fit to the experimental data assuming an anharmonic coupling to one specific low frequency mode. The normal mode calculation of CO bridgebonded on Ni by Richardson and Bradshaw[45] estimates for the frustrated translation $\omega = 76\,\text{cm}^{-1}$ and for the frustrated rotation $\omega = 184\,\text{cm}^{-1}$, while it is known from EELS data[51] that the metal–molecule stretch is found at $400\,\text{cm}^{-1}$. The calculated values should only be taken as rough estimates, because the frequencies of these frustrated modes have not yet been experimentally determined. However, it seems very likely that the temperature effects for the bridgebonded CO molecules on Ni(111) are caused by this anharmonic coupling and that the high frequency mode couples mainly to one specific mode, possibly the frustrated rotation. In a simple model it was possible to estimate the damping rate η for a mode at $235\,\text{m}^{-1}$ to be of the order of $50\,\text{cm}^{-1}$, while the coupling strength $\delta\omega$ is much more difficult to treat without an extensive chemisorption calculation.

It was interesting to compare these results for bridgebonded molecules with similar data of ontop bonded ones. Figure 8 shows the temperature dependence of CO on Ru(001)[42] in the $(\sqrt{3} \times \sqrt{3})\text{R}30°$ structure, with all molecules in the ontop position. In this case the temperature dependence on the peak width is small while there is a large downward frequency shift. Again

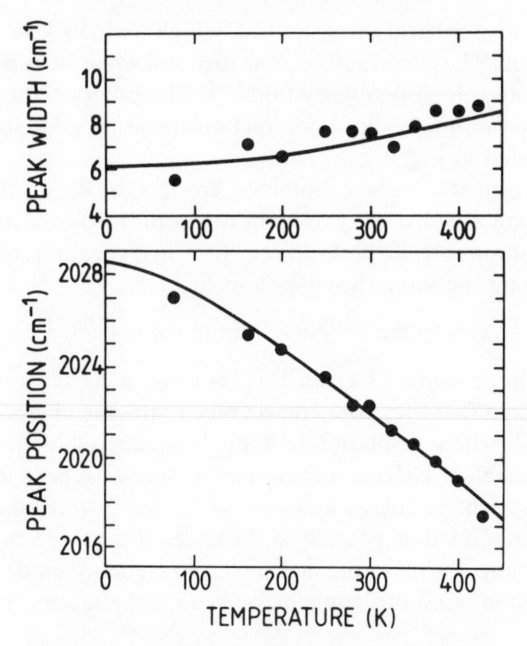

Fig. 8. The position and width of the infrared absorption peak as function of substrate temperature for the $(\sqrt{3} \times \sqrt{3})R30°$ structure of CO on Ru(001). All molecules are chemisorbed in the ontop position and the solid lines are calculated within the theory describing the anharmonic coupling to a low frequency mode. (*Reproduced by permission from Persson et al.*[41].)

the solid lines are best fits giving[42]:

$$\omega_0 = 105\,\text{cm}^{-1} \quad \delta\omega = -6\,\text{cm}^{-1} \quad \eta = 6\,\text{cm}^{-1}.$$

Comparing with the normal mode calculation and the experimentally determined value for CO/Pt(111) below, it seems likely that for the ontop bonded molecules the anharmonic coupling is to the frustrated translation. As expected, $\delta\omega$ is then negative as the C—O stretch vibration frequency decreases when going away from the ontop position.

As we are dealing with an ensemble of molecules with a large dynamical dipole moment, we have to consider the dipole–dipole interaction. The strong intensity transfer to high frequency modes that was seen for two different species in the isotopic mixtures in the previous section will also affect the peak shape for this ensemble of slightly shifted oscillators. If there were no molecule–molecule interaction the anharmonic coupling would for the ontop bonded molecules give a broadened infrared absorption peak with a low frequency tail. However, when the dipole–dipole interaction is taken into account, there is an intensity transfer to the high frequency side, making the line narrow again. The effect may become reversed when $\delta\omega$ is positive, as this gives a peak with a high frequency tail[42,52]. These important effects on the lineshape due to the dipole–dipole interaction are discussed in the next section and are illustrated in Fig. 9.

That the parameter values obtained from this fit of the theory to experimental data are physically relevant was confirmed by a study on CO in the ontop position on Pt(111). In an IRS investigation of this system Schweizer[14] using the same theory obtained:

$$\omega_0 = 50\,\mathrm{cm}^{-1} \quad \delta\omega = 3\,\mathrm{cm}^{-1} \quad \eta = 4\,\mathrm{cm}^{-1}.$$

Lahee *et al.*[46] have studied CO on Pt(111) using inelastic scattering of He atoms and showed that they could detect adsorbate–metal modes at energies below 200 cm^{-1} (with a resolution of 4 cm^{-1}), an energy range previously of course not accessible to IRS but also burried in the elastic peak in EELS. They found for this system an energy loss peak at 48 cm^{-1}, which they assign to a frustrated translation and possibly a weak feature at 133 cm^{-1} due to a frustrated rotation. Furthermore, for such a low energy mode that lies well in the one-phonon band of the metal, one can rather accurately calculate[42] $\eta = 4\,\mathrm{cm}^{-1}$.

In principle, there exists an independent way of determining ω_0, $\delta\omega$ and η. The theory predicts the existence of sidebands at $\Omega \pm 2\omega_0$[49]. The strength of these combination bands should be $(\delta\omega/\omega_0)^2/8$ of the fundamental mode with a width equal to 2η[53]. We investigated very carefully the $c(4 \times 2)$ CO/Ni(111) system without being able to detect these bands. It is not clear whether this is because the bands are too weak (1×10^{-3}) and/or too broad (40 cm^{-1}) or that it has a more fundamental reason.

To conclude, it seems that the nature of the anharmonic coupling between a high frequency intramolecular mode and a thermally excited low frequency mode is understood. It turns out that the strength of the influence on the infrared spectrum critically depends on the values of ω_0, $\delta\omega$ and η. However, we have to wait for more experimental data on these low frequency modes, probably obtained with the helium atom scattering technique, before we can make more definite conclusions.

3.4. Vibrational energy dissipation

One of the main differences between a free and an adsorbed CO molecule is that whereas the vibrational energy of the gas phase molecule can only be transferred into a photon, giving the excitation a lifetime of 30 ms, the adsorbed molecule is able to dissipate the vibrational energy into the substrate, giving the mode a lifetime in the picosecond range. This is extremely important for the energy transfer in most dynamical processes at surfaces, as in many cases the molecules in an intermediate state are vibrationally excited. For a metal substrate there exist two fundamental processes that can accommodate the energy, excitation of phonons and of electrons, the latter mechanism often called creation of electron–hole pairs.

To study a dynamical phenomenon like the vibrational damping, the most direct way is to use some kind of time resolved spectroscopy. As we are dealing with processes on the ps scale, this calls for a pulsed laser experiment. The technique is to saturate by a pump beam all molecules in their first vibrationally excited state and then with a time delayed probe beam measure the relaxation. With this approach one has been able to determine both the vibrational lifetime and the influence of anharmonic coupling to other modes in molecular solids and other systems[54]. However, the experimental problem for studies of adsorbed molecules is the same as that encountered by ordinary infrared spectroscopy some twenty years ago, namely to obtain high enough sensitivity to detect a monolayer or less on the metal surface. Cavanagh and coworkers have made great efforts in creating such a laser system in the infrared and at the time of writing they have reached the $2000\,cm^{-1}$ region with pulses of 10 ps width at a power of $15\,\mu J$[55]. Previously, in a transmission experiment they have measured the lifetime of the O—H stretch mode at $3000\,cm^{-1}$ on silica[54]. Despite these considerable achievements a lot of instrumental development has still to be done before the lifetime of a vibrational excitation of a molecule adsorbed on a metal surface can be unambiguously determined.

Another, less straightforward way to determine the vibrational lifetime is by studies of the infrared absorption peak shape. Consider a single adsorbed molecule at 0 K. The width of the peak is then determined by the lifetime broadening and in the first approximation it has a Lorentzian shape with a full width at half maximum (FWHM) $\Delta = (2\pi c\tau)^{-1}$, τ then being the lifetime. However, as usual we have to consider an ensemble of molecules at finite temperatures and then there exist other peak broadening mechanisms that must be taken into account.

First of all we must consider *inhomogeneous* broadening, which in the first approximation give a Gaussian shaped peak, but with modifications discussed below. Chemisorbed molecules tend to occupy distinct adsorption sites but if we have a large concentration of surface defects (steps, vacancies, adatoms, impurities and so on) the chemical environment will differ for the adsorbed molecules, which will give rise to a broadening of the peak. However, for the high purity materials used in surface science and with proper alignment and surface preparation the defect concentration should be less than 1%[56]. If we study a rather dense monolayer this effect should then be negligible, but it can be extremely important in studies in the low coverage regime.

Even for a high coverage monolayer on a perfect surface we can get additional inhomogeneous broadening. It should be clear from the discussion

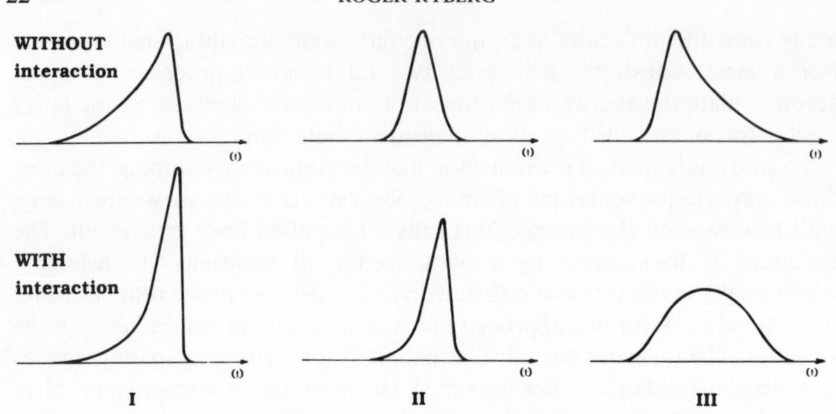

Fig. 9. The effect of the dipole–dipole interaction on inhomogeneously broadened infrared absorption peaks. Without and with the interaction turned on for peaks with I, a low frequency tail, II, a symmetric Gaussian distribution and III, a high frequency tail. (*Adapted from Persson and Hoffmann*[52].)

in section 3.2 that there can be a substantial molecule–molecule interaction. Disorder in the overlayer will then produce additional broadening, as the vibrational environment will differ among the sites. An interaction that gives rise to an upward coverage-dependent frequency shift (an increasing vibration frequency for increasing coverage) will give rise to a low frequency tail of the absorption peak (the local coverage is less than or equal to that of the full monolayer) and vice versa. However, a general effect of the dipole–dipole coupling is the intensity transfer to high frequency modes as discussed several times in the previous sections. This mechanism will push the intensity upwards, so a low frequency tail will be diminished, a symmetric absorption peak will get a low frequency tail, while a peak with a high frequency tail will be broadened[52]. These general effects are shown schematically in Fig. 9. Obviously, anyone who wants to interpret the shape of the infrared absorption peak of dense overlayers must take these effects into account and when calculating such shapes use the full theory of the dipole–dipole interaction[32].

An example of an inhomogeneously broadened peak is found for CO physisorbed on an Al(100) surface[22]. In addition to the dipole–dipole coupling there is a substantial short range interaction. The adsorbed molecules do not form any ordered structures and hence the overlayer contains a large degree of disorder. We find in Fig. 10 a spectrum with a rather Gaussian shaped peak with a high frequency tail. However, as the structure of this system is unknown it s not possible to make a more detailed interpretation of the peak shape.

From the discussion in section 3.3 it should be obvious that another peak broadening mechanism, at least at higher temperatures, is the *anharmonic coupling* to low energy modes. We discussed the origin of the broadening in that context and found in Fig. 7 that for $c(4 \times 2)$CO/Ni(111) this interaction

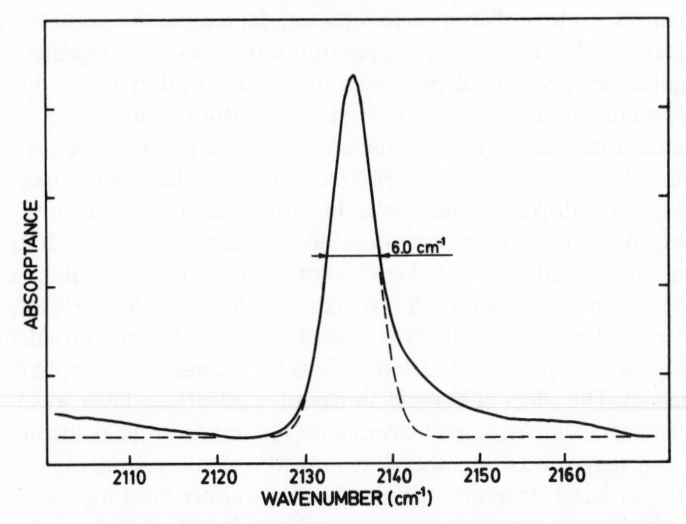

Fig. 10. The infrared absorption peak of a full monolayer of CO physisorbed on an Al(100) surface at 30 K. No ordered structures are formed and the peak is inhomogeneously broadened. The dotted line is a Gaussian distribution. (*Reproduced by permission from Ryberg*[22].)

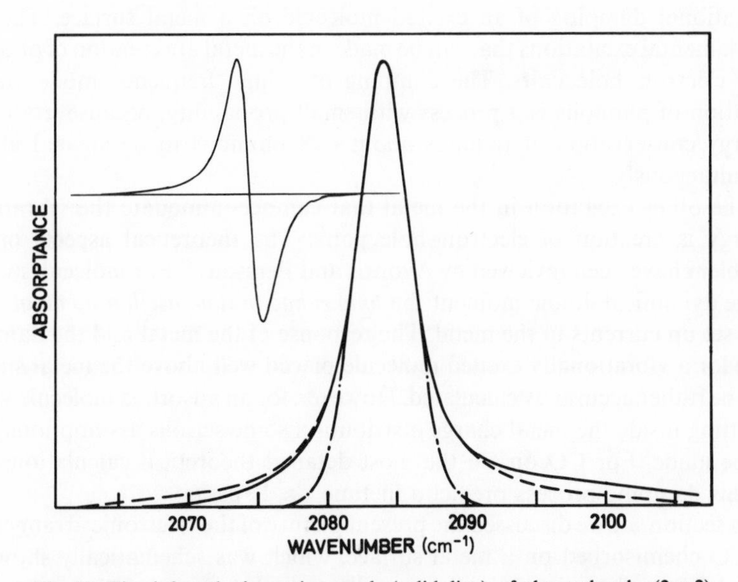

Fig. 11. The infrared absorption peak (solid line) of the ordered $c(2 \times 2)$ structure of CO on Cu(100) at 100 K. Shown also are Lorentzian (dashed) and Gaussian (dash-dotted) distributions. The recorded first derivative spectrum is shown in the inset. (*Reproduced by permission from Ryberg*[12].)

gave a considerable contribution to the width for temperatures above 70 K but also that it vanished at very low temperatures. It seems fair to say that if there is a negligible temperature dependence of the absorption peak, there is no dephasing contribution at least in that temperature region.

A system that seems free from the most obvious inhomogeneous or phase relaxation broadening is $c(2 \times 2)$CO/Cu(100)[12], where all molecules are bonded in the ontop position. Figure 11 shows the infrared absorption peak, which exhibits almost no temperature dependence in the range 20 to 130 K. The peak has a predominantly Lorentzian shape with a weak high frequency tail. The intrinsic line width $\Delta = 4.6\,\mathrm{cm}^{-1}$, which would correspond to a lifetime $\tau = 1.2\,\mathrm{ps}$ if the width was caused solely by lifetime broadening. As stated above the influence of surface defects for a dense overlayer like this can be neglected. The effect of imperfections in the adsorbate layer was tested by measurements on an incomplete overlayer. This made the peak lose most of its Lorentzian shape and it was therefore argued[12] that the weak high frequency tail of the ordered structure in Fig. 11 sets an upper limit for the inhomogeneous broadening for this system. Another observation that speaks against inhomogeneous broadening is the fact that most recent infrared investigations made with high resolution spectrometers give similar values of the peak width: $5\,\mathrm{cm}^{-1}$ on Cu(100)[12], Cu(111)[57] and Pt(111)[14] and $6\,\mathrm{cm}^{-1}$ on Ni(111)[40] and Ru(001)[41].

There exists an extensive literature on theoretical calculations of the vibrational damping of an excited molecule on a metal surface. The two fundamental excitations that can be made in the metal are creation of phonons and electron–hole pairs. The damping of a high frequency mode via the creation of phonons is a process with small probability, because from pure energy conservation, it requires about 6–8 phonons to be created almost simultaneously.

The other excitation in the metal that can accommodate the vibrational energy is creation of electron–hole pairs. The theoretical aspects of this problem have been reviewed by Avouris and Persson[58]. For molecules with a large dynamical dipole moment the *long range in time-oscillating dipole field* will set up currents in the metal. The response of the metal and the damping rate for a vibrationally excited molecule placed well above the metal surface can be rather accurately calculated. However, for an adsorbed molecule which is sitting inside the metal charge distribution some serious assumptions have to be made. For CO on Cu the most detailed theoretical calculations[59,60] of this damping process predict a lifetime $\tau \approx 10\,\mathrm{ps}$.

In section 3.1 we discussed the present picture of the electronic arrangement of CO chemisorbed on a metal surface, which was schematically shown in Fig. 3. When the molecule is vibrationally excited charge is oscillating between the 2π resonance and the metal. This gives rise to the large increase in the dynamical dipole moment, as was discussed in that section. These *local charge oscillations* also cause vibrational damping and if one uses the increase in dynamical dipole moment due to the chemisorption as a measure of the oscillating charge, in a simple model[58] for CO/Cu one can estimate $\tau \approx 2$ ps.

Langreth[61] pointed out that if the energy relaxation originates from these local charge oscillations, then the infrared absorption peak will not have a Lorentzian shape. The oscillating charge will not be in phase with the external field, which will give the absorption peak a low frequency tail. This seems to be in contradiction with the data in Fig. 11. However, the calculation was made for a single adsorbed molecule and as discussed above influences on the line shape due to the vibrational interaction can be severe. In a later work Crljen and Langreth[42] incorporated the dipole–dipole interaction of the full monolayer using the theoretical approach introduced by Persson and Ryberg[32]. The result is reproduced in Fig. 12, showing that the low frequency tail of a single molecule due to the dipole–dipole interaction becomes a high frequency tail for the full monolayer. An important consequence of the theory is that the vibrational quantities that we have assigned to the adsorbed molecule (including its image) are renormalized. Fitting the theory to the experimental spectrum in Fig. 11 gives for a *single adsorbed molecule* a line width $\Delta = 2.8\,\text{cm}^{-1}$ (instead of the apparent width of $4.6\,\text{cm}^{-1}$), which then corresponds to a lifetime $\tau = 1.9$ ps. The polarizabilities are also renormalized to $\alpha_v = 0.31\,\text{Å}^3$ and $\alpha_e = 4.3\,\text{Å}^3$, compared to $\alpha_v = 0.25\,\text{Å}^3$ and $\alpha_e = 2.5\,\text{Å}^3$ obtained from the isotopic mixtures in section 3.2.

The theory reproduces the experimental line shape of the ordered overlayer very well but is unable to explain the more complicated spectrum of an incomplete layer. More important, using the same parameter values Crljen

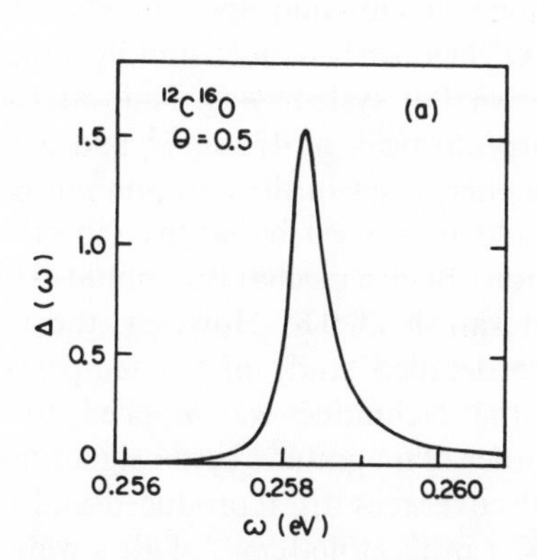

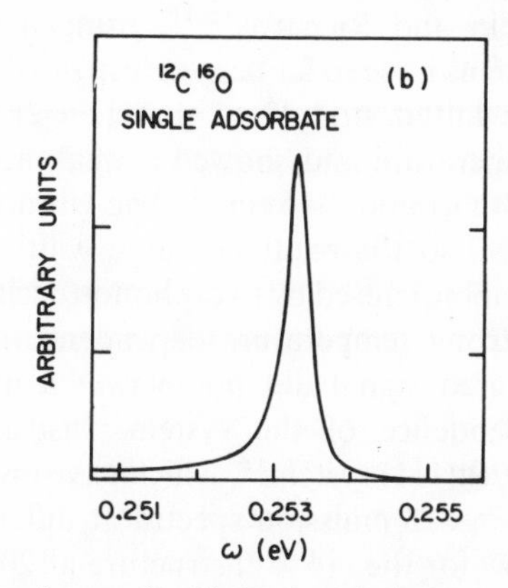

Fig. 12. The calculated infrared absorption peak, assuming lifetime broadening and vibrational damping via local charge oscillations. (a) Fitted to the experimental spectrum in Fig. 11 for an ordered overlayer incorporating the dipole–dipole interaction. (b) The same calculation for a single adsorbed molecule. (*Reproduced by permission from Crljen and Langreth*[42].)

and Langreth[42] were able to reproduce the asymmetry of the absorption peaks as function of composition of the isotopic mixtures in Fig. 4. This shows, in an independent way, both that we have a very good description of the dipole–dipole interaction and that the line width, to a significant degree, is caused by vibrational damping due to electron–hole pair excitations.

To conclude, even if there exist several processes that affect the vibrational line shape it seems probable that when most of them have been sorted out and with the good agreement between theory and experiment, the lifetime broadening for a chemisorbed CO molecule is of the order of a few cm^{-1}, corresponding to a lifetime of a few ps. The main vibrational energy relaxation mechanism is creation of electron–hole pairs caused by the local charge oscillations between the metal and the 2π molecular resonance crossing the Fermi level.

4. LOW FREQUENCY METAL–MOLECULE MODES

Studies of the metal–molecule stretch modes are very important for the understanding of adsorption systems as they are the vibrational levels in the binding energy potential. These modes will be highly excited during all kinds of surface processes like sticking, desorption and reactions. The damping rate of such a mode will definitely be important for the kinetics of the system and therefore the experimentalists have for a long time tried to extend the infrared measurements to this region below $500\,cm^{-1}$. The associated experimental problems were discussed to some extent in section 2.

The only reported infrared measurements on metal–molecule modes on single crystalline surfaces have been carried out in very nice experiments by Tobin and Richards[15,16], using a dedicated emission spectrometer. Two systems have so far been investigated, CO chemisorbed on Ni and Pt surfaces. The initial study[15] of the $c(2 \times 2)$CO/Ni(100) system was made at room temperature and showed a weak absorption peak at $475\,cm^{-1}$ and with a width of about $15\,cm^{-1}$. The vibration energy lies in the two-phonon band of Ni, so the relatively large width could be caused by strong vibrational damping caused by two-phonon excitations. Such a mechanism should exhibit a strong temperature dependence and vanish at 0 K. However, the weak infrared signal did not permit a more detailed study of the temperature dependence of the system. Instead, the technique was applied to the CO/Pt(111) system[16], which gave one order of magnitude better signal/noise. In Fig. 13 emission spectra at different coverages are reproduced and they show for the $c(4 \times 2)$ structure at 200 K a peak at $460\,cm^{-1}$ with a width of about $7\,cm^{-1}$. The highest phonon frequency is much lower for Pt than for Ni, so here the vibrational mode is above the two-phonon band and the damping rate for three-phonon excitations should be at least one order of magnetic smaller[62]. Furthermore, for the Pt system they were able to study

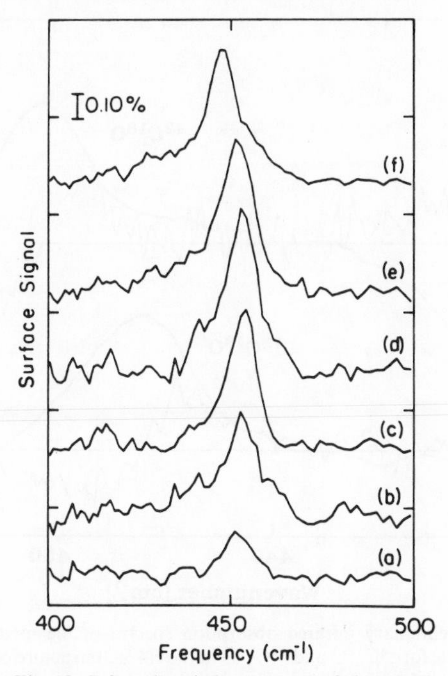

Fig. 13. Infrared emission spectra of the metal–molecule stretch mode for $^{13}C^{16}O$ on Pt(111) at 275 K. The different coverages are: (a) 0.07, (b) 0.14, (c) 0.27, (d) 0.31, (e) 0.50 (the $c(4 \times 2)$ structure) and (f) 0.55. (*Reproduced by permission from Tobin and Richards*[16].)

the temperature dependence above 200 K, which was found to be rather weak. Together with a large spread in the values of the widths this indicated that the peak width was determined by inhomogeneous broadening. It follows that for the Ni system there is a need for a temperature dependence study before anything more definite can be said about the damping mechanisms.

The inherent drawback of the emission technique is that the substrate, which acts as the light source, must be held at a relatively high temperature. This makes it difficult to make more definite studies on thermally excited modes, which often require rather low substrate temperatures. So the efforts to develop a more traditional reflection spectrometer for this region have continued, since much important information can be obtained if these thermally excited modes could be frozen out. As mentioned in section 2, during the time of writing this paper, for the first time I was able to detect the metal–molecule stretch mode in a reflection experiment, using a wavelength modulation spectrometer. Figure 14 shows preliminary spectra of $^{13}C^{18}O$ and $^{12}C^{18}O$ in the $c(4 \times 2)$ structure on Pt(111) at 100 K. The reproduced first

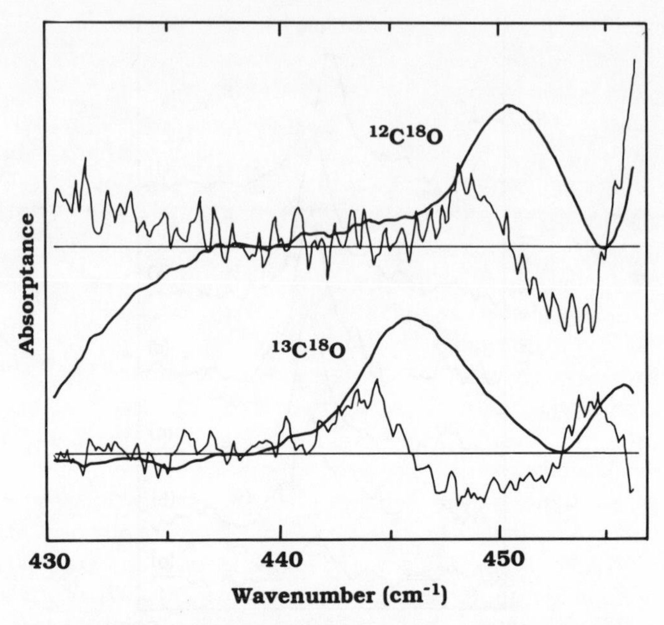

Fig. 14. Preliminary infrared absorption spectra of the metal–molecule stretch mode for $^{13}C^{18}O$ and $^{12}C^{18}O$ in the $c(4 \times 2)$ structure on Pt(111) at 100 K recorded by a reflection wavelength modulation spectrometer. The recorded first derivative spectrum has been numerically integrated. Both spectra are reproduced in the figure.

derivative spectrum is a sum of three 20-minute spectra, so the total measuring time for the spectra in Figs. 13 and 14 are comparable. Naturally, a lot of work has to be done before anything can be said about the temperature dependence but it seems very promising that new interesting information about these low frequency modes will be forthcoming in the near future.

5. ATOMIC ADSORBATES

A class of system of particular interest is atomic adsorbates as dissociated molecules play an important role in most reactions. The experimental problems for infrared spectroscopy of such atomic adsorbates are servere. First of all, by necessity they only exhibit low frequency modes, giving all the problems discussed above. Secondly, they may be sitting very deep into the surface, making the electronic screening by the conduction electrons strong and hence weakening the apparent infrared adsorption. By this substrate interaction the peaks may also be very broad due to a strong electron–hole pair damping.

Only two systems on single crystalline metal surfaces have been studied by IRS so far. Using the same approach as for the low frequency mode of CO,

Tobin and Richards[16] were able to detect the Pt–O stretching mode of atomic oxygen chemisorbed on a Pt(111) surface. Even if the vibration frequency lies in the same region (460 cm^{-1}) as the Pt–CO mode, the peak width is about 5 times larger. The integrated absorption is roughly the same but the signal/noise is, due to the large width, so poor that it prevents a more detailed study of this interesting system. If this broadening should be attributed to inhomogeneities or has a more fundamental origin, it is then still to be investigated.

A very special case is adsorbed atomic hydrogen. Due to its small mass the stretch modes can at some surfaces have a frequency around 1000 cm^{-1} and hence fall in a region accessible to ordinary reflection spectrometers. Chabal and coworkers have, using a Fourier transform infrared (FTIR) spectrometer, extensively studied atomic H and D chemisorbed on W(100) and Mo(100) surfaces[44,63,64]. These systems are rather complicated, as they exhibit H coverage-dependent surface reconstructions. We will here only consider the hydrogen saturated systems, which produce an unreconstructed metal surface. The hydrogen occupies every bridge site, giving a densely packed overlayer with two H on every metal atom. The IRS data for H and D on W(100)[44] are reproduced in Fig. 15, showing a broad peak at 1070 (767) cm^{-1} and a sharp, derivative-like peak at 1270 (915) cm^{-1}. Similar results have also been

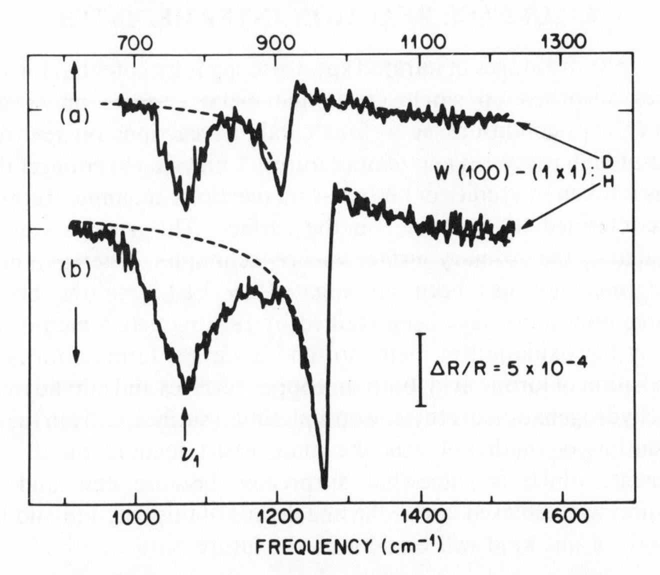

Fig. 15. Infrared absorption spectra of a full monolayer of H and D on W(100) at 100 K. The peak at 1070 (767) cm^{-1} is the symmetric stretch mode and the sharp peak at 1270 (915) cm^{-1} is assigned to the second harmonic of the wag mode. The dotted line is calculated using the theory for an electron–hole pair damped mode by Langreth[61]. (*Reproduced by permission from Chabal*[44].)

obtained on Mo(100)[63]. The peak around $1000\,cm^{-1}$ can safely be assigned to the symmetric stretch mode, even if the origin of the large width of about $100\,cm^{-1}$ is not yet understood. The assignment of the sharp peak at 1270 (915) cm^{-1}, on the other hand, is less straightforward. Chabal[44] suggests that it originates from the second harmonic of the wag mode at $645\,cm^{-1}$. Surprisingly, this sharp peak is not seen in the EELS spectra[65] but only the (by symmetry) infrared forbidden asymmetric stretch mode[66] at $1290\,cm^{-1}$. However, the asymmetric lineshape is very well described within the theory of Langreth[61] for an electron–hole pair damped mode, in this case calculated for a single adsorbed species, i.e. assuming no vibrational interaction between he H atoms. Even if the assignment of the mode still has to be solved, it seems likely that for this mode we observe the same vibrational damping mechanism as for the C—O stretch discussed in section 3.4, that is via electron–hole pair excitation caused by some local charge oscillation.

Obviously, the development of infrared spectrometers that can study adsorbate modes in the low energy regime will not only give interesting information about the low frequency modes of adsorbed molecules but will also make it possible to probe the equally important field of atomic adsorbates.

6. SURFACE REACTION INTERMEDIATES

One of the advantages of infrared spectroscopy is its potential for studies of molecules adsorbed on single crystalline metal surfaces under ultrahigh vacuum (UHV) conditions as well as catalytic reactions on real dispersed catalysts at high pressures and temperatures. A first step to connect these two extremes is taken by studies of low pressure reactions at temperatures where a reaction intermediate is stable on the surface. This species can then be investigated by the ordinary surface science techniques. Whereas a number of such intermediates has been investigated by EELS, to the best of my knowledge only three have been studied by IRS, namely a methoxy species formed by the oxidation of methanol and a surface formate formed in the decomposition of formic acid, both on copper surfaces and ethylidyne formed by the dehydrogenation of ethylene on a platinum surface. Only in the work on the oxidation of methanol was the main effort focused on the reaction mechanisms, which is somewhat surprising, because new and relevant information was achieved about the fine details of this reaction and hopefuly more work of this kind will be done in the future.

6.1. Methoxide on Cu(100)

In the decomposition or oxidation of methanol, methoxide (CH_3O) is formed as a surface intermediate on a number of metal surfaces[67]. The

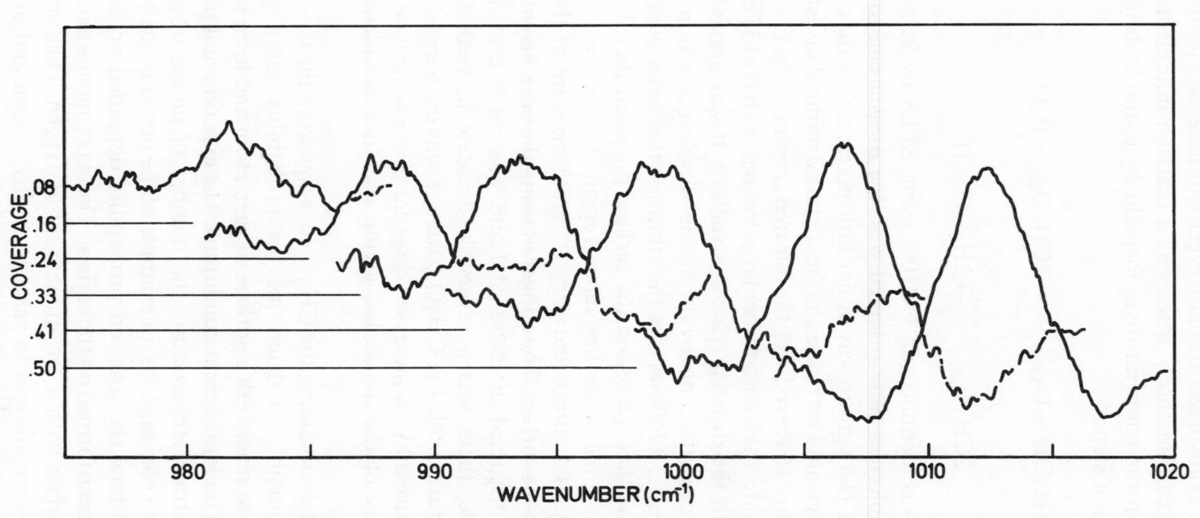

Fig. 16. Infrared spectra of the C—O stretch mode of different coverages of CH_3O on Cu(100) at 100 K. The large frequency shift is mainly caused by the dipole–dipole interaction. (*Reproduced by permission from Ryberg*[39].)

oxidation on a copper surface was initially studied by Wachs and Madix[68] using thermal desorption spectroscopy. The two most important conclusions in their work were that adsorbed atomic oxygen was necessary for the reaction and that there existed a methoxy species as a reaction intermediate, which decomposed above room temperature to formaldehyde and hydrogen, giving the two main reaction steps;

$$2CH_3OH + O_{ads} \xrightarrow{>150K} 2CH_3O_{ads} + H_2O \tag{4}$$

$$2CH_3O_{ads} \xrightarrow{>325K} 2H_2CO + H_2 \tag{5}$$

The methoxide was identified on Cu(100) using EELS by Sexton[69], who showed that the molecules were oriented with the oxygen end towards the metal and that the CH_3O group was rather little perturbed by the chemisorption. With this background we made a rather detailed infrared investigation of this reaction[39]. Figure 16 reproduces the infrared spectra of the C—O stretch mode of the methoxide, showing a rather large frequency shift of 37 cm^{-1} from low coverages up to the ordered $c(2 \times 2)$ structure. It was argued that this frequency shift was caused mainly by the dipole–dipole interaction. From the discussion in section 3.2 it follows that the vibration frequency would then be rather insensitive to other species on the surface, provided they do not have vibrational modes in the same frequency region. Hence, we can use the vibration frequency as a direct and very accurate measure of the concentration of CH_3O on the surface. The danger of using the peak height as such a measure was also discussed in section 3.2 and was really evident for this mode. For example, there was a reversible decrease in peak height for increasing temperature, while the CH_3O coverage was unchanged. This was interpreted to be caused by an increased disorder in the adsorbate layer due to a higher mobility of the molecules, giving rise to a broadening of the infrared spectrum.

With this method of obtaining the CH_3O coverage from the C—O stretch frequency it was possible to determine the probability for a gas phase methanol molecule to react with a surface oxygen atom and form methoxide according to Eq. (4) under different conditions. The experimental parameters used were the substrate temperature, the number of preadsorbed oxygen atoms and the CH_3O coverage. As an example of the power of the technique, we show in Fig. 17 how the reaction probability depended on how much CH_3O already had been formed on the surface. The experiment was conducted by predosing the surface with a certain amount of oxygen. Then small, fixed doses of methanol were allowed to react with this oxygen and an infrared spectrum was recorded after each dose. The CH_3O coverage was determined from the vibration frequency. The experiment was made at two different substrate temperatures as shown in the figure. The reaction probability at a

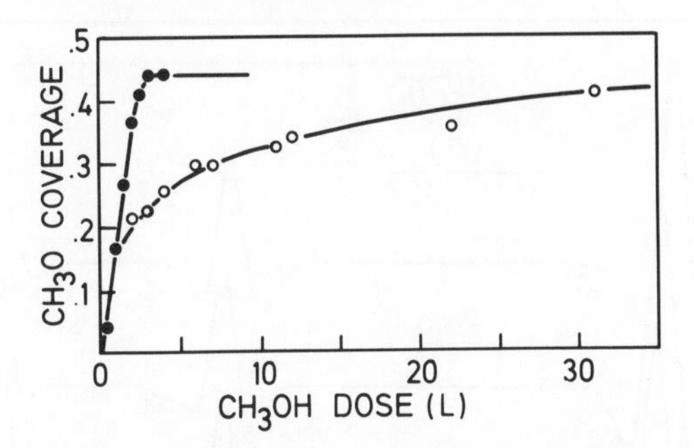

Fig. 17. The CH_3O coverage determined from the C—O vibration frequency as function of CH_3OH dose at a substrate temperature of 200 K (solid circles) and 300 K (open circles). The reaction probability in Eq. (4) at a certain CH_3O coverage is the derivative of the curve at that point. (*Reproduced by permission from Ryberg*[39].)

certain CH_3O coverage is the derivative of the curve at that point. We find at 200 K that the probability is constant until all oxygen atoms have reacted. At 300 K we start out with the same probability, which however decreases for increasing CH_3O coverage and finally very many methanol doses were required to react off all oxygen atoms. The interpretation of this behaviour was that at 200 K the mehoxide molecules are rather immobile, while at 300 K they rapidly migrate along the surface and are able to block the reaction.

Coupled with LEED investigations (both the oxygen atoms and the methoxide formed ordered structures), the following model for the reaction was suggested, which strongly emphasized structural factors. Both the oxygen atoms and the methoxide probably occupy hollow sites. The reaction can only occur if there exist empty nearest neighbour sites adjacent to the oxygen atom. The reaction probability can be decreased in two ways; by adsorbing an excess of oxygen (or any other species occupying the hollow site) or by increasing the substrate temperature, making the formed CH_3O mobile on the surface and able to block these sites.

Another example of the power of infrared spectroscopy in this context became evident in the investigations of the C—H stretch modes[70]. Previously, it had always been assumed that methoxide, like free methanol, has a symmetric CH_3 group. Around $3000\,cm^{-1}$ the vibrational spectrum would then consist of two peaks, one symmetric and one triply degenerate asymmetric stretch mode. However, the infrared spectra reproduced in Fig. 18 show three peaks. The first important conclusion was that the methoxide was tilted on the surface. The infrared surface selection rule discussed in section 2 states that it is only vibrational modes with a nonvanishing component of its dynamical dipole moment normal to the surface that can be excited by the

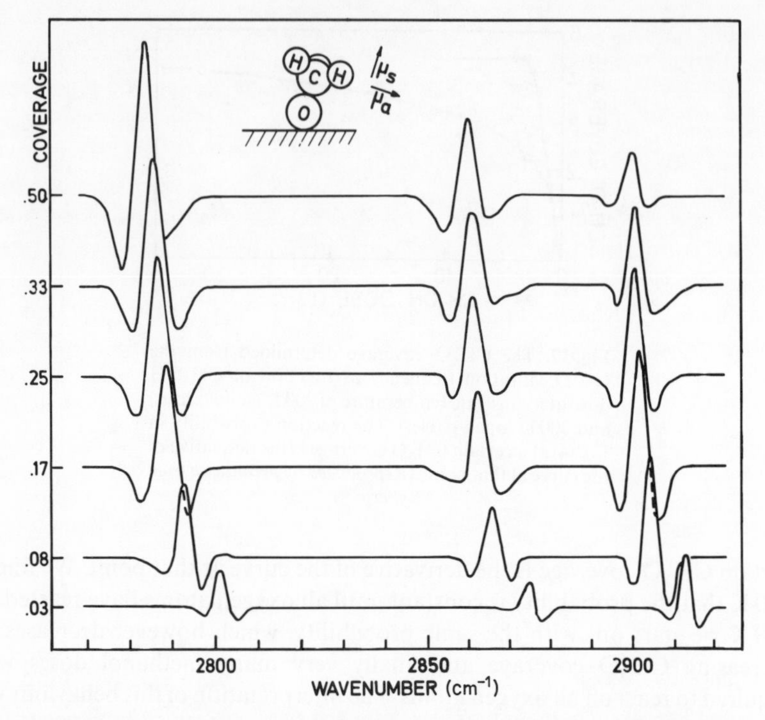

Fig. 18. Infrared spectra of C—H vibrations of different coverages of CH_3O on Cu(100) at 100 K, showing the symmetric (at 2800 cm^{-1}) and two asymmetric stretch modes. Inset shows a tilted chemisorbed methoxide molecule and the orientations of the dynamical dipole moments μ. (*Reproduced by permission from Ryberg*[70].)

infrared radiation. If CH_3O were oriented with the C—O axis normal to the surface, only the symmetric C—H stretch mode would be observed, as is seen from the orientations of the dynamical dipole moments of the modes in the insert in the figure. Such an orientation had been suggested based on UPS[71] and near edge X-ray fine-structure spectroscopy (NEXAFS)[72] work but here the infrared spectroscopy gave a contradicting answer. The NEXAFS work has later been redone[73], now indicating a molecule tilted $30 \pm 10°$ from the surface normal. A similar value of $35 \pm 5°$ has also been obtained for a Cu(110) surface[74].

Still it is surprising that three peaks are observed in Fig. 18. Methoxide chemisorbed in the hollow position on a Cu(100) surface has a lower symmetry than the CH_3O group along, so if there were a substantial interaction between the CH_3 group and the metal, the degeneracy could be broken giving rise to two asymmetric stretch mode peaks. More important, in the earlier works on these systems one had not been aware of a careful infrared investigation, combined with a valence force-field calculation of gas-phase methanol[75]. This work showed that because of the interaction with the hydroxyl end the CH_3 group is not symmetric and the infrared spectrum of free methanol does show three C—H stretch modes. It turned out[70] that this asymmetry was preserved for the chemisorbed methoxide and by comparing CH_3O and CD_3O it was found that the CH_3 group was very little affected by the chemisorption. Furthermore, by using the calculated dynamical dipole moments of methanol, we could estimate from the peak heights in Fig. 18 a tilt angle of 30°, that is about the same that was deduced from the latest X-ray data[73]. This agreement could be rather fortuitous or indicate that the CH_3O group makes a chemisorption bond on the copper surface that electronically is rather similar to the missing hydrogen bond of methanol. The only problem with the interpretation above is that the recorded asymmetric modes are much weaker for CD_3O than for CH_3O[70], indicating either a much smaller tilt angle for the deuterated species or that the chemisorption system is not yet completely understood.

6.2. Ethylidyne on Pt(111)

The adsorption of ethylene on platinum has been subject to a number of investigations. While being stable at low temperatures, upon heating the molecules decompose to hydrogen and carbon. It was early realized that a stable surface intermediate was formed at room temperature. Finally, after some controversy Steininger *et al.*[76], using thermal desorption and EELS, were able to show that on Pt(111) an ethylidyne, CCH_3, intermediate was formed between 220 K and 340 K, giving the dehydrogenation sequence;

$$2C_2H_{4ads} \xrightarrow{>220K} 2CCH_{3ads} + H_2 \tag{6}$$

$$2CCH_{3ads} \xrightarrow{>340K} 4C + 3H_2 \tag{7}$$

This ethylidyne intermediate has recently been subject to two IRS investigations, by Chesters and McCash[77] and in a more detailed study by Malik *et al.*[78] Their spectra show three absorption peaks above 800 cm^{-1}, which in light of the EELS work[76] could be assigned to the C—C stretch mode at 1120 cm^{-1}, the symmetric CH_3 bend mode at 1340 cm^{-1} and the symmetric CH_3 stretch mode at 2885 cm^{-1}. In contrast to the surface methoxy discussed in the previous section, there is no peak associated with the asymmetric CH_3 stretch mode[76] at 2950 cm^{-1}. This shows, in line with the discussion above, that the CCH_3 is oriented with the C—C axis normal to the surface (as indicated in Fig. 19) and verifies at the same time the validity of the infrared surface selection rule.

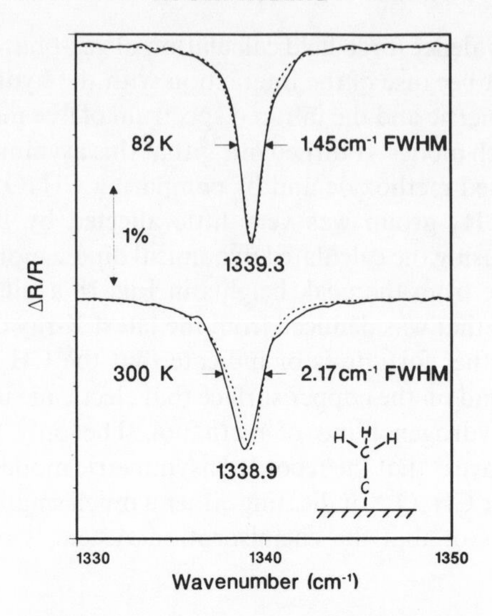

Fig. 19. Infrared spectra of the symmetric CH_3 bend mode of ethylidyne (CCH_3) on Pt(111) at 82 K and 300 K. The dotted lines are calculated assuming a vibrational–rotational coupling. Inset shows the suggested orientation. (*Reproduced by permission from Malik et al.*[78].)

The FTIR spectra of the symmetric CH_3 bend mode from Malik *et al.*[78] are reproduced in Fig. 19. They show a very narrow absorption peak, with an intrinsic width at 82 K of only 1.4 cm^{-1}, which is the narrowest peak yet observed for an adsorbed molecule. It has been argued from a NMR study[79] that the ethylidyne species exhibit free rotations around the C—C axis even at 80 K. Therefore, Malik *et al.*[78] considered the selection rules for a coupling between these free rotations and the symmetric bend mode. As this absorption peak obviously is unusually little affected by inhomogeneous, lifetime or phase relaxation broadening, it would be a good candidate for investigating such a coupling. However, they do not observe any sidebands or fine structure due to the rotational modes, but only a weak temperature dependence in peak width and position, as seen in the figure. They conclude that this behaviour does not fit in a simple way into the picture of a rotational–vibrational coupling but also that the IRS data are inconclusive to the existence of free rotations for this species.

6.3. Formate on Cu(110)

Formic acid decomposes generally on metal surfaces to carbon dioxide and hydrogen. A thermal desorption study on Cu(110)[80] showed that the reaction could only go via a surface formate in the two steps;

$$2CHOOH \xrightarrow{>270k} 2HCOO_{ads} + H_2 \tag{8}$$

$$2HCOO_{ads} \xrightarrow{>430K} 2CO_2 + H_2 \tag{9}$$

Sexton[81] made an EELS study on this reaction on the Cu(100) surface, identifying the surface formate and determining its symmetry. An IRS

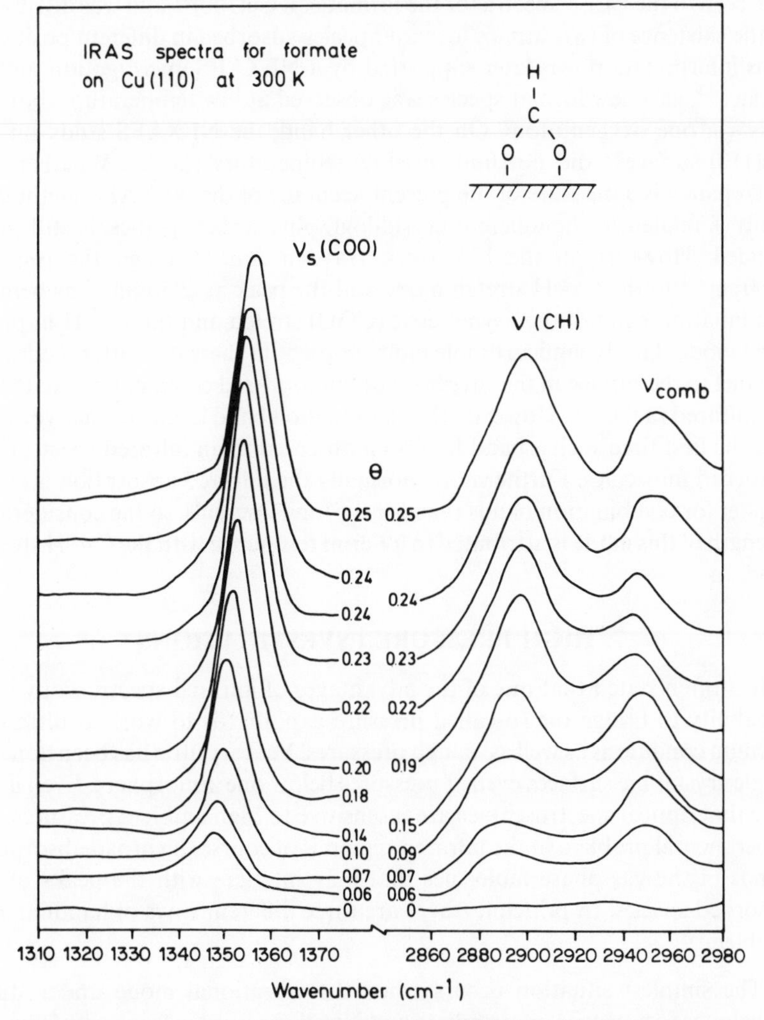

Fig. 20. Infrared spectra of the surface formate (HCOO) at different coverages on Cu(110) at 300 K. Inset shows the suggested orientation. (*Reproduced by permission from Hayden et al.*[43].)

investigation on Cu(110) was done by Hayden *et al.*[43]; the infrared spectra are reproduced in Fig. 20. The strong absorption peak around 1350 cm^{-1} is the symmetric (COO) stretch mode. The absence of the asymmetric mode around 1600 cm^{-1} shows, in line with the discussion above, that the chemisorbed surface formate is oriented with two symmetric oxygen bonds, as indicated in the figure. This orientation has also been confirmed by a NEXAFS investigation[82]. It is interesting to see the appearance of two peaks in the 3000 cm^{-1} region, as HCOO only has one C—H stretch mode. A doublet was also seen in the EELS spectra for the formate on Cu(100)[81] and was attributed to the existence of two surface formate species adsorbed in different positions. This interpretation was later supported by a NEXAFS investigation on this surface[73], as a new format species was observed at low temperatures bonded only via one oxygen atom. On the other hand, the NEXAFS study on the Cu(110) surface[82] did not find this low temperature species. Whether this discrepancy is a measure of the present accuracy of the NEXAFS method or really is related to the different crystallographic surface planes, is still to be decided. However, in the IRS work Hayden *et al.*[43] assign the peak at 2900 cm^{-1} to the C—H stretch mode and the peak at 2950 cm^{-1} as being a combination band of the asymmetric (COO) stretch and the C—H in plane bend mode. The dynamical dipole moments of both these two latter modes are parallel to the surface in the suggested orientation and could not be excited by the infrared radiation. However, the combination band is infrared active and it was the first time such a band had been observed in an infrared spectrum of adsorbed molecules. Furthermore, normally the infrared absorption is much weaker for combination bands than for the fundamentals, so the considerable strength of this mode is attributed to a Fermi resonance with the C—H stretch mode.

7. HIGH PRESSURE INVESTIGATIONS

It is often stated that one of the advantages of infrared spectroscopy is its capability to bridge the so-called pressure gap, that is to work at ultrahigh vacuum conditions as well as at high pressures. Yet very little has been done on single crystalline surfaces even at pressures below one atmosphere. Even if it is true that optical spectroscopies are insensitive to high ambient pressures, the experimental problem in the infrared region is of course the broad absorption bands of the gas phase molecules that may interfere with the peaks of the adsorbed species. In principle, there are three different ways of handling this problem:

1. The simplest situation occurs when the vibrational mode under study belongs to a surface intermediate and is well removed in frequency from the modes of the gas phase molecules. The measurements are then straightforward giving a sensitivity/resolution similar to the UHV situation.

2. Otherwise one can take advantage of the infrared surface selection rule mentioned in section 2, that it is only the p-component of the light that can excite the vibrational mode. The ordinary spectrometer can then be used just with a polarizer inserted in the beam. By subtracting the spectra for p- and s-polarized light one will hopefully be left with a spectrum characteristic of the surface molecules. The problem with this method is that the output, especially for grating spectrometers, is polarized, which will give a residual spectrum even in the absence of a surface. Further, when working at high pressures the method demands a very high dynamical range of the detection system, as the gas phase absorption may be orders of magnitude larger than the signal of interest.
3. This latter problem is overcome by the design of a true double beam instrument with one beam reflected on the surface and the reference beam just probing the surrounding gas. Using two matched detectors, the signals can be subtracted directly. The obvious problem is that there exist no absolutely identical beams or detectors, so in this case also we will be left with a background spectrum. However, at really high pressures this is probably the only fruitful approach.

Even if these problems with overlapping bands can be avoided, at high enough pressures the gas phase absorption will be so strong that the sensitivity will be detector noise limited. This problem can only be handled by keeping the optical path length in the pressure cell as small as possible.

When the development of dedicated infrared spectrometers for surface studies started some ten years ago, some of them were designed as more or less complete ellipsometers, which in principle are insensitive to the ambient gas phase molecules. Fedyk *et al.*[83] detected CO adsorbed on an evaporated Cu film at 4 torr, while Golden *et al.*[84] reported work at 100 torr. More recently, Burrows *et al.*[85] used a Fourier transform spectrometer and the polarizer approach above to study the reaction-rate oscillations in the oxidation of CO on a large Pt polycrystalline foil at pressures up to one atmosphere. With this rapid FTIR spectrometer they obtained a time resolution of 0.6 s at a sensitivity of 5% of a full CO monolayer.

As yet, the only work that has been reported on a single crystalline surface has been done by Hoffmann and Robbins[86]. They used the same approach as Burrows *et al.*[85] to study the methanation reaction ($CO + 3H_2 \rightarrow CH_4 + H_2O$) on a Ru(001) surface. Figure 21 shows the spectra of the C—O stretch mode taken at 50 torr of a mixture of CO and H_2 at temperatures up to 600 K. The decrease in peak height at 500 K indicates the buildup of a passivating carbon layer. The 'negative' absorption around 2140 cm^{-1} is due to imperfect canceling of the gas phase CO band.

Obviously, this field will become very important in the future, both for studies on its own right and to make connection between UHV data and work

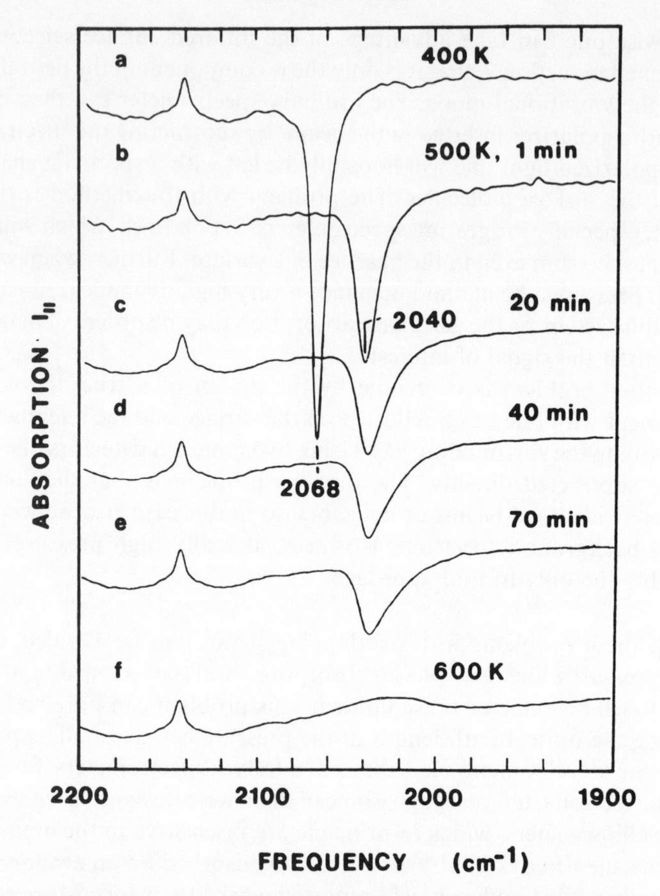

Fig. 21. Infrared vibrational spectra of a Ru(001) surface exposed to a CO/H_2 mixture (1:3) at 50 torr as function of temperature and reaction time. The shift between 400 K and 500 K is caused by the change in the steady state CO coverage. (*Reproduced by permission from Hoffmann and Robbins*[86].)

in reaction cells. It is probably here where much of the future instrumental development of infrared spectroscopy will take place.

8. CONCLUSION

During the last decade infrared spectroscopy has developed into a very important tool for studies of adsorbed molecules. Very rapid FTIR spectrometers, which are easily adapted to a reflection experiment, are commercially available. Recently a dispersive, modulation spectrometer dedicated for surface studies has also come on the market. In the same way as happened when commercial EELS spectrometers evolved some years ago, this will probably lead to a rapid increase in the number of groups that use IRS routinely.

It has been the purpose of this review to show that many of the fundamental physical processes that affect the infrared spectra of adsorbed molecules are, to a large extent, understood. The importance of the vibrational interaction is demonstrated and there exists a theory which give a good description of this mechanism. The effects of an anharmonic coupling to low frequency modes have been revealed. The spectroscopists have finally agreed on a value of the intrinsic width of the absorption peak for a chemisorbed CO molecule and as yet nothing speaks against the interpretation that it is to a large extent a measure of the vibrational lifetime.

Still there exist many interesting fundamental processes that are unexplored. The 'chemical' effects, i.e. why the vibration frequency differs among the crystallographic planes, adsorption sites, metal surface and sometimes as a function of coverage, are mainly not understood. Here we probably have to wait for improvement of the theoretical calculations. On the experimental side the most important task is to continue to increase the usable energy range of reflection spectrometers below $500\,cm^{-1}$. This will allow studies of the low frequency metal–molecule modes in more detail and also open the interesting and fundamental field of studies of atomic adsorbates. A quite different experimental problem is to develop a dedicated spectrometer able to work at ambient pressures above one atmosphere. This would make it possible to connect UHV work and studies of heterogeneous catalytic reactions at 'high' pressures. This important field of surface reactions and reaction intermediates is mainly unexplored.

Obviously, there is still much interesting work to be done on infrared spectroscopy of molecules adsorbed on metal surfaces in the future, both for those interested in instrumental development as well as those who will use the technique as one of many surface science tools.

Acknowledgements

I want to thank my friends and collaborators over the years, Bo Persson and Stig Andersson, for critically reading this manuscript. The work has been supported by the Swedish National Board for Technical Development and the Swedish National Science Research Council.

References

1. Hoffmann, F. M., *Surface Sci. Reports*, **3**, 107 (1983).
2. Hollins, P., and Pritchard, J., *Progr. Surface Sci.*, **19**, 275 (1985).
3. Ueba, H., *Progr. Surface Sci.*, **22**, 181 (1986).

4. Darville, J., in *Vibrations at Surfaces* (Eds. R. Caudano, J.-M. Gilles and A. A. Lucas), Plenum, New York, 1982.
5. Darville, J., in *Vibrations at Surfaces*, (Eds. C. R. Brundle and H. Morawitz), Elsevier, Amsterdam, 1983.
6. Darville, J., in *Vibrations at Surfaces* (Eds. D. A. King, N. V. Richardson, and S. Holloway), Elsevier, Amsterdam, 1986.
7. Persson, B. N. J., *Solid State Comm.*, **30**, 163 (1979).
8. Ryberg, R., *Surface Sci.*, **114**, 627 (1982).
9. Hoffmann, F. M., Levinos, N. J., Perry, B. N., and Rabinowitz, P., *Phys. Rev.* B, **33**, 4309 (1986).
10. Schweizer, E., Nagel, J., Braun, W., Lippert, E., and Bradshaw, A. M., *Nucl. Instr. and Meth.* A, **239**, 630 (1985).
11. Richards, P. L., and Tobin, R. G., in *Methods of Surface Characterization* 4 (Eds. J. T. Yates, Jr. and T. E. Madey) Plenum, New York, 1986.
12. Ryberg, R., *Phys. Rev.* B, **32**, 2671 (1985).
13. Andersson, S., Persson, B. N. J., Gustafsson, T. and Plummer, E. W. *Solid State Commun.*, **34**, 473 (1980).
14. Schweizer, E., Thesis (Berlin, 1986).
15. Chiang, S., Tobin, R. G., Richards, P. L., and Thiel, P. A., *Phys. Rev. Lett.*, **52**, 648 (1984).
16. Tobin, R. G., and Richards, P. L., *Surface Sci.*, **179**, 387 (1987).
17. Efrima, S., and Metiu, H., *Surface Sci.*, **108**, 329 (1981).
18. Müller, W., and Bagus, P. S., *J. Vac. Sci. Technol.*, A, **5**, 1053 (1987).
19. Rogozik, J., and Dose, V., *Surface Sci.*, **176**, L847 (1986).
20. Johnson, P. D., and Hulbert, S. L., *Phys. Rev.*, B, **35**, 9427 (1987).
21. Müller, W., and Bagus, P. S., *J. Vac. Sci. Technol.* A, **3**, 1623 (1985).
22. Ryberg, R., *Phys. Rev.* B, **37**, 2488 (1988).
23. Persson, B. N. J., *Surface Sci.*, **116**, 585 (1982).
24. Nyberg, C., and Tengstål, C. G., *Phys. Rev. Lett.*, **50**, 1680 (1983).
25. Karlsson, P.-A., Mårtensson, A.-S., Andersson, S. and Nordlander, P., *Surface Sci.*, **175**, L759 (1986).
26. Black, J. E., *Surface Sci.*, **116**, 240 (1982).
27. Rahman, T. S., Black, J. E., and Mills, D. L., *Phys. Rev.* B, **25**, 883 (1982).
28. Kuhlenbeck, H., Neumann, M., and Freund, H.-J., *Surface Sci.*, **173**, 194 (1986).
29. Bare, S. R., Griffits, K., Hofmann, P., King, D. A., Nyberg, G. L., and Richardson, N. V., *Surface Sci.*, **120**, 367 (1982).
30. Hayden, B. E., Robinson, A. W., and Tucker, P. M., *Surface Sci.*, **192**, 163 (1987).
31. Hammaker, R. M., Francis, S. A., and Eischens, R. P., *Spectrochim. Acta*, **21**, 1295 (1965).
32. Persson, B. N. J., and Ryberg, R., *Phys. Rev.*, B, **24**, 6954 (1981).
33. Mahan, G. D., and Lucas, A. A., *J. Chem. Phys.*, **63**, 1344 (1978).
34. Persson, B. N. J., and Liebsch A., *Surface Sci.*, **110**, 356 (1981).
35. Scheffler, M., *Surface Sci.*, **81**, 562 (1979).
36. Andersson, S., and Persson, B. N. J., *Phys. Rev. Lett.*, **45**, 1421 (1980).
37. Hermann, K., Bagus, P. S., and Bauschlicher, C. W., Jr., *Phys. Rev.*, B, **30**, 7313 (1984).
38. Pfnür, H., Menzel, D., Hoffmann, F. M., Ortega, A., and Bradshaw, A. M., *Surface Sci.*, **93**, 431 (1980).
39. Ryberg, R., *J. Chem. Phys.*, **82**, 567 (1985).

40. Persson, B. N. J., Ryberg, R., *Phys. Rev.*, B, **32**, 3586 (1985).
41. Persson, B. N. J., Hoffmann, F. M., and Ryberg, R., *Phys. Rev. B*, **34**, 2266 (1986).
42. Crljen, Z. and Langreth, D. C., *Phys. Rev. B*, **35**, 4224 (1987).
43. Hayden, B. E., Prince, K., Woodruff, D. P., and Bradshaw, A. M., *Surface Sci.*, **133**, 589 (1983).
44. Chabal, Y. J. *Phys. Rev. Lett.*, **55**, 845 (1985).
45. Richardson, N. V., and Bradshaw, A. M., *Surface Sci.*, **88**, 255 (1979).
46. Lahee, A. M., Toennies, J. P., and Wöll, Ch. *Surface Sci.*, **177**, 371 (1986).
47. Gadzuk, J. W., and Luntz, A. C., *Surface Sci.*, **144**, 429 (1984).
48. Trenary, M., Vram, K. H., Bozso, F., and Yates, J. T., Jr., *Surface Sci.*, **147**, 269 (1984).
49. Persson, B. N. J., and Ryberg, R., *Phys. Rev. Lett.*, **54**, 2119 (1985).
50. Marks, S., Cornelius, P. A., and Harris, C. B., *J. Chem. Phys.*, **73**, 3069 (1980).
51. Erley, W., Wagner, H., and Ibach, H., *Surface Sci.*, **80**, 612 (1979).
52. Persson, B. N. J., and Hoffmann, F. M., *Proceedings Vibrations at Surfaces*, Elsevier, Amsterdam (1987).
53. Persson, B. N. J., private communication.
54. Heilweil, E. J., Casassa, M. P., Cavanagh, R. R., and Stephenson, J. C., *J. Chem. Phys.*, **82**, 5216 (1985).
55. Cavanagh, R. R., Heilweil, E. J., and Stephenson, J. C., *Proceedings Vibrations at Surfaces*, Elsevier, Amsterdam (1987).
56. Linke, U., and Poelsema, B., *J. Phys. E: Sci. Instrum.*, **18**, 26 (1985).
57. Chesters, M. A., *J. Elec. Spec. Rel. Phenon.*, **38**, 123 (1986).
58. Avouris, P., and Persson, B. N. J., *J. Phys. Chem.*, **88**, 837 (1984).
59. Eguiluz, A. G., *Phys. Rev. Lett.*, **51**, 1907 (1983).
60. Liebsch, A., *Phys. Rev. Lett.*, **54**, 67 (1985).
61. Langreth, D. C., *Phys. Rev. Lett.*, **54**, 126 (1985).
62. Persson, B. N. J., *J. Phys.* C, **17**, 4741 (1984).
63. Prybyla, J. A., Estrup, P. J., Ying, S. C., Chabal, Y. J., and Christman, S. B., *Phys. Rev. Lett.*, **58**, 1877 (1987).
64. Arrecis, J. J., Chabal, Y. J., and Christman, S. B., *Phys. Rev.* B, **33**, 7906 (1986).
65. Woods, J. P., and Erskine, J. L., *J. Vac. Sci. Technol.*, **A5**, 435 (1987).
66. Biswas, R., and Hamann, D. R., *Phys. Rev. Lett.*, **56**, 2291 (1986).
67. Richter, L. J., and Ho, W., *J. Chem. Phys*, **83**, 2165 (1985).
68. Wachs, I. E., and Madix, R. J., *J. Catal.*, **53**, 208 (1978).
69. Sexton, B. A., *Surface Sci.*, **88**, 299 (1979).
70. Ryberg, R., *Phys. Rev.* B, **31**, 2545 (1985).
71. Hoffmann, P., Mariani, C., Horn, K., and Bradshaw, A. M., *Proceedings of 3rd ECOSS Conference*, Cannes (1980).
72. Stöhr, J., Gland, J. L., Eberhardt, W., Outka, D., Madix, R. J., Sette, F., Koestner, R. J., and Doebler, U., *Phys. Rev. Lett.*, **51**, 2414 (1983).
73. Outka, D. A., Madix, R. J., and Stöhr, J., *Surface Sci.*, **164**, 235 (1985).
74. Bader, M., Puschmann, A., and Haase, J., *Phys. Rev.* B, **33**, 7336 (1986).
75. Serrallach, A., Meyer, R., and Günthard, Hs. H., *J Mol. Spectrosc.*, **52**, 94 (1974).
76. Steininger, H., Ibach, H., and Lewald, S., *Surface Sci.*, **117**, 685 (1982).
77. Chesters, M. A., and McCash, E. M., *Surface Sci.*, **187**, L639 (1987).
78. Malik, I. J., Brubaker, M. E., Mohsin, S. B., and Trenary, M. *J. Chem. Phys.*, **87**, 5554 (1987).
79. Wang, P. K., Slichter, C. P., and Sinfelt, J. H., *J. Phys. Chem.*, **89**, 3606 (1985).

80. Ying, D. H. S., and Madix, R. J., *J. Catal.*, **61**, 48 (1980).
81. Sexton, B. A., *Surface Sci.*, **88**, 319 (1979).
82. Crapper, M. D., Riley, C. E., Woodruff, D. P., Puschmann, A., and Haase, J., *Surface Sci.*, **171**, 1 (1986).
83. Fedyk, J. D., Mahaffy, P., and Dignam, M. J., *Surface Sci.*, **89**, 404 (1979).
84. Golden, W., Dunn, D. S., and Overend, J., *J. Phys. Chem.*, **82**, 843 (1978).
85. Burrows, V. A., Sundaresan, S., Chabal, Y. J., and Christman, S. B., *Surface Sci.*, **180**, 110 (1987).
86. Hoffmann, F. M., and Robbins, J. L., *J. Vac. Sci. Technol.*, **A5**, 724 (1987).

Advances in Chemical Physics
Edited by K. P. Lawley

MOLECULAR DESORPTION FROM SOLID SURFACES: LASER DIAGNOSTICS AND CHEMICAL DYNAMICS

DAVID S. KING and RICHARD R. CAVANAGH

Center for Atomic, Molecular and Optical Physics, National Institute of Standards and Technology, Gaithersburg, MD 20899, USA

CONTENTS

Abstract

Review of recent state resolved experiments exploring thermal and laser driven desorption processes at metal surfaces. (Prepared January, 1988)

1. INTRODUCTION

The fundamental chemical and physical forces which influence the interactions of molecules with solid surfaces are an area of vigorous research. The models of chemical reactions at surfaces initially developed by Langmuir, Hinschelwood, Ely, and Rideal were based on thermodynamics and statistical mechanics where it is possible to describe chemical response by averaging over the many degrees of freedom. With the advent of beam scattering, it became possible to investigate the dependence of atom–surface encounters on the incident beam kinetic energy and/or angle. With recent advances in laser diagnostics, it has become possible to illuminate the dependencies of chemical and physical processes at surfaces on the internal degrees of excitation. These dependencies are sensitive to subtle details of the gas–surface interaction potential and allow one to test many of the basic assumptions of statistical theories. This revolution in surface dynamics has been driven by the ultilization of laser-based state-resolved diagnostics to map out the extent of rotational, vibrational, and electronic energy transfer resulting from carefully orchestrated gas–surface interactions.

Application of the chemical physics techniques of molecular beam scattering and of laser-based state-resolved diagnostics to surface science questions has been a child of the 1980s. This has been due, in part, to advances in commercial apparatus and to the formation of collaborative efforts between traditionally gas phase and surface science oriented researchers. Since the 1960s, crossed molecular beam scattering and photodissociation experiments have provided dramatic insight into the important physical interactions that govern reaction dynamics, such as the classic $F + H_2 \rightarrow HF + H$ example[1,2]. For the gas phase community, studies of molecule–surface interactions allow for the replacement of a point scatter (e.g. an atom) with a fixed solid plane. Surface scientists have developed tremendously sophisticated tools for evaluating equilibrium structure and kinetics under ultra high vacuum conditions. The wedding of lasers and molecular beam techniques provides the opportunity to extend the characterization of the molecule–surface interaction from the bottom of the chemisorption well to desorption, that region so important for most chemical processes. Over the last several years many review articles have been written on molecule–surface dynamics, including comprehensive theoretical discussions[3], and a recent review of molecule–surface dynamics emphasizing vibrational energy transfer[4]. Direct beam scattering results have been extensively discussed in numerous review articles[5]. These experiments have been of critical importance for directing our thoughts regarding molecule–surface dynamics. It is not the intention of this chapter to cover all the material in this area, but rather to concentrate on a series of related experiments involving 'thermal' activation of an adsorbate. Section 2 will briefly discuss the experimental techniques and types of information obtained in state-resolved studies. Sections 3 and 4 will discuss the results of thermal desorption and laser-induced desorption studies.

As is frequently the case in emerging disciplines, it is advantageous to fully explore a limited number of well characterized systems. These results then form the basis for hypothesis and further experimentation. There are two systems that have provided major foci for molecule–surface studies; both involve NO as the molecular constituent. Ag(111) has served as the substrate of choice for beam scattering experiments while Pt(111) has been used in desorption studies. These choices are predicted in part on pragmatic considerations. Adsorbate residence times must be very short if beam scattering studies are to interrogate interactions with the clean substrate (i.e. there must be at most a very shallow chemisorption well, such that experiments are conducted in the zero-coverage limit). Conversely, for studies of desorption one needs substantial binding interaction. There is, of course, a continuum of interactions that unite thermal desorption and direct scattering as limiting cases. One important intermediate condition allows the study of trapping/desorbing trajectories in beam scattering experiments using pulsed molecular beams and maintaining the surface at elevated temperature such that isothermal desorption cleans the surface between beam pulses.

Although we want to direct attention to thermally driven processes, the nature of laser–surface interactions allows for transient, nonthermal phenomena to occur, becoming manifest in the observed dynamics. Such transient phenomena might include adsorbate interaction with hot electrons or high energy phonons resulting from nonspecific excitations of the metal substrate, or unquenched photochemical processes resulting from direct adsorbate-localized excitations. Evidence for such interactions is included in section 4 for comparison to the 'thermal' laser-induced desorption (LID) studies.

One of the exciting developments of the past two years has been the implementation of state-resolved techniques to probe H_2 desorbing from metal surfaces. A wealth of kinetic energy and internal state distribution data is just now becoming available. The range and complexity of possible hydrogen–metal interactions (encompassing the roles of surface, subsurface, and bulk hydrogen) preclude their inclusion in this chapter. A listing of state-resolved molecule/surface experiments is presented in the Appendix. This includes references for beam scattering and H_2 desorption studies not considered in this chapter.

2. EXPERIMENTAL ASPECTS

Good reviews of the basic tenets and methodologies exist for both thermal desorption[6] and laser-induced desorption[7] where the desorbed molecules are detected using a mass spectrometer. These techniques have proven very sensitive tools for probing adsorbate interactions, surface coverages, desorption kinetics, speciation, reaction or diffusion kinetics and depth profiling. Such techniques have also been used for obtaining the angular and kinetic energy distributions of scattered and desorbed species. Although this information gives a complete description for atom–surface interactions, there is a whole new dimension of complexity involved in describing molecule–surface interaction potentials and the resulting dynamics due to the internal energy states associated with the molecular adsorbates.

To unravel this added level of complexity, laser techniques have been used to map out state-to-state reaction dynamics for molecule–surface interactions. The two general approaches are laser-induced fluorescence (LIF) and resonance-enhanced multiphoton ionization (REMPI). Both utilize the information contained in laser excitations of molecules from their ground to excited electronic states. These excited molecules then either: (1) spontaneously fluoresce and the resulting fluorescence photons are detected (LIF); or (2) undergo transitions to an autoionizing continuum by absorption of one (or more) additional photons and the resulting ions/electrons are detected (REMPI). The common characteristic that both techniques share is that the probe laser (given sufficient resolution) only interacts with molecules in a particular quantum state. For the NO molecule one may choose an individual set of (vibrational, electronic, rotational and fine structure) states. The particular internal state being probed is defined by the laser wavelength. The resulting LIF or REMPI signals are directly proportional to the number density of molecules in the probe volume. Calculations based on spectroscopic formulations can give absolute densities (within a factor of two) when the appropriate transition dipole moments are known. Calibrations using a thermalized gas sample can be used to define an appartus constant which can be used subsequently to estimate sample densities[8].

There are claimed advantages and disadvantages for both techniques. Both have sensitivities of about $10^6\,cc^{-1}$ per quantum state for moderate power laser sources. Of course increased laser power provides increased sensitivies in either case. One advantage (we feel) to the LIF technique for NO detection is that a single calibration of the apparatus constant (photon collection and detection efficiencies) is valid for pressure from 10^{-12} torr to one atmosphere.

A useful means for compiling or comparing the results of a large number of experiments is to fit the observed populations to some distribution function. A reasonable first try for thermally activated processes would seem to be a Boltzmann function, where the population in a particular rotational level (J)

of energy E_{int} is

$$P(J) \propto (2J + 1)\exp(-E_{int}/kT_R) \quad (1)$$

where $(2J + 1)$ is the (magnetic sublevel) degeneracy of the Jth rotational level. For a thermal distribution, a plot of $\ln[P(J)/(2J + 1)]$ against E_{int} produces a line with a slope $-kT_R$.

Such an analysis provides the population distribution $A_0^{\{0\}}(J)$. Higher-order moments of the multipole distribution may be obtained for each rotational level of the sample[9]. The signal dependence on the direction of polarization of linearly polarized light can provide the quadrupole moment of the alignment distribution, $A_0^{\{2\}}(J)$. Similarly, using elliptically polarized excitation the degree of orientation can be obtained. The moments of the multipole distribution (population, alignment and orientation) can be combined with translational and vibrational characterization to provide a sensitive probe of the molecule–surface interaction potential, providing a wealth of dynamical information which is not accessible from state-averaged diagnostics.

3. THERMAL DESORPTION

3.1. NO/Pt(111): characterization

A wide range of studies on the NO/Pt system have well characterized the appropriate chemistry, binding states, and desorption kinetics of NO on Pt. Temperature programmed desorption (TPD) studies on the (111) basal plane have distinguished three major desorption features with desorption peaks in the temperature range $200 < T_{peak}(K) < 400$. These are depicted in Fig. 1a which shows TPD spectra following five different initial dose conditions[10]. Although more open (100) and (110) faces of Pt have desorption features at surface temperatures $T_S < 400$ K, they also exhibit a desorption feature around 450 K that has been ascribed to recombinative desorption[11]; the requisite O_a and N_a species being supplied through NO dissociation. There is very little NO dissociation on Pt(111).

The results of a electron energy loss spectroscopy (EELS) study of NO/Pt(111) are included in Fig. 1b[12]. A loss feature at 1490 cm^{-1} became apparent at low NO coverage. The intensity of this feature appeared to increase with the filling of the $T_{peak} = 380$ K TPD feature. At higher NO coverage a second loss feature at 1710 cm^{-1} grew in intensity concurrent with the filling of the $T_{peak} = 355$ K desorption feature. Through analogy to metal-nitroso compounds these features have been assigned to NO species with N bonded to either two Pt metal atoms (bridge bonded) or a single Pt atom (atop), respectively. Similar observations have been made using reflection/absorption infrared spectroscopy[13].

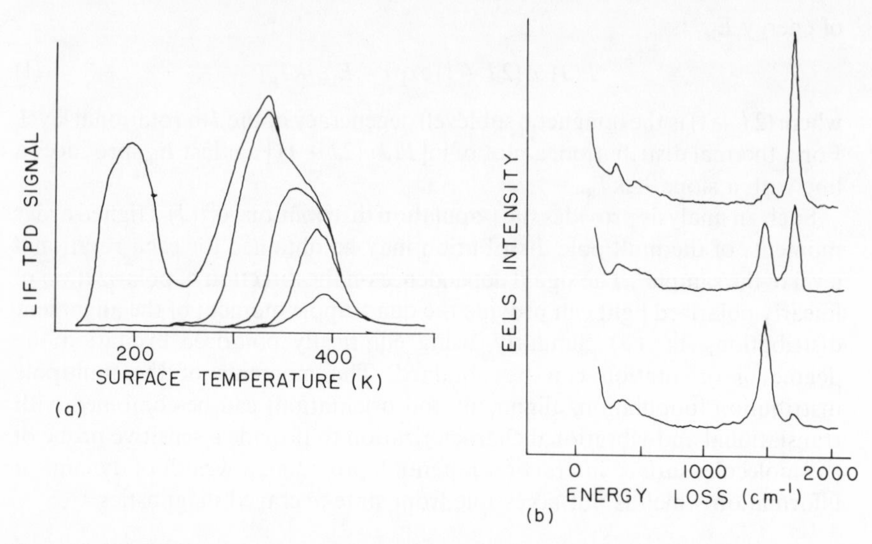

Fig. 1. (a) Thermal desorption spectra of NO/Pt(111) for initial coverages of $\Theta = 0.025$, 0.08, 0.15, 0.25 and 0.40. (b) Electron energy loss spectra for NO/Pt(111) for coverages of 0.08, 0.25 and 0.40. The two main loss features are 1490 and 1710 cm^{-1}. (*Adapted from Refs. 10 and 11.*)

Low energy electron diffraction (LEED) studies for NO/Pt(111) have shown an ordered $p(2 \times 2)$ structure developing at fairly low exposures[14]. This pattern has been shown to persist up to a point where the two high T_{peak} features have become filled. Stoichiometrically, a full $p(2 \times 2)$ overlay would correspond to a coverage of $\Theta = 0.25$. This number is in agreement with coverage estimates deduced by comparing NO and CO TPD experiments using mass spectrometric detection[11]. At low T_s, increased exposure to NO results in the third, low T_{peak} TPD feature which corresponds to an increase in coverage up to $\Theta \approx 0.4$. The nature of the binding state(s) occupied by this increased coverage is unclear. Although a broadening in LEED pattern has been noted upon this increased coverage, the concurrent changes in the overall LEED pattern and the EEL spectrum were small. This weakly bound NO might either occupy random (disordered) binding sites (atop or bridge) in the first layer, or it might actually participate in the formation of a partial second layer. These high coverage species have been characterized by adsorption energies E_a (high coverage) ≈ 9–12 kcal/mol, which are much stronger than generally attributed to physisorbed species[15].

The kinetics of NO on Pt surfaces have been explored by many authors both in TPD and beam scattering types of experiments. In a series of sophisticated modulated beam scattering experiments, Serri *et al.*[16] (referred to subsequently in the text as STC) attempted to reconcile the apparently disparate results reported previously[11,17–19]. In their experiments they noted that for

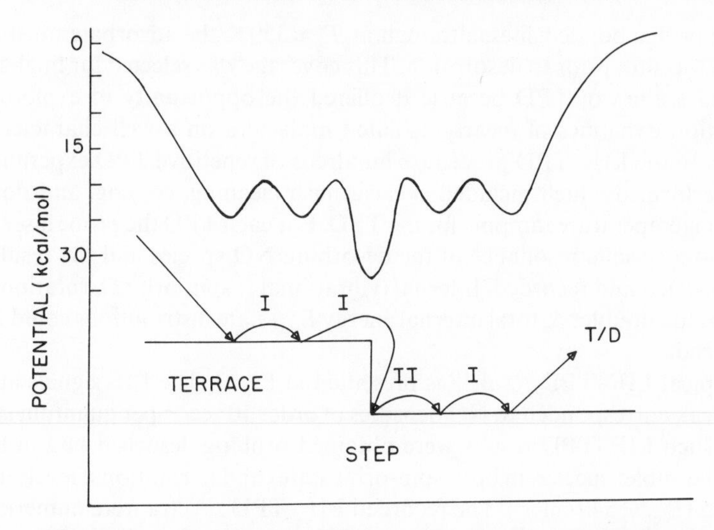

Fig. 2. Kinetics scheme for NO/Pt(111) including terrace diffusion (I), trap-to-terrace jumps (II), and thermal desorption T/D. (*Adapted from Ref. 16.*)

$\Theta < 10^{-2}$ the desorption kinetics were not first order in NO coverage and depended dramatically on low coverages of oxygen. In a kinetics model they postulated that step-defects played an important role in controlling such low coverage kinetics. Higher coverages of NO or trace amounts of oxygen block these active sites. Under these conditions the desorption kinetics become determined by the terrace kinetics. In all cases, desorption was assumed (by microscopic reversibility) to proceed from the terraces; the steps only acted as holding sites and therefore contributed indirectly to the adsorption and desorption processes. Schematically this model is represented in Fig. 2. The kinetics picture has a first-order desorption rate with a $10^{16}\,s^{-1}$ pre-exponential (reflecting the hindered rotation of the adsorbate) and a 25 kcal/mol activation energy. Surface diffusion was modeled with an attempt frequency of $5 \times 10^{13}\,s^{-1}$ and a 5 kcal/mol diffusion barrier. The step-to-terrace kinetics were modeled assuming similar entropy factors for both sites, with a pre-exponential of $10^{12}\,s^{-1}$ and an activation energy of 14 kcal/mol (in agreement with low coverage TPD results). In the 500–700 K temperature range this gives very facile diffusion. The adsorbates diffuse hundreds of angstroms in typical residence times of a few tens of nanoseconds, thereby having the possibility for encountering many steps.

3.2. Dynamics: low coverage

TPD spectra similar to the lowest exposure trace in Fig. 1 correspond to initial NO coverages of $\Theta \approx 0.025$. At low T_S these adsorbates primarily occupy bridge bonded sites, although at $T_S \approx 350$ K the adsorbate must also sample top sites prior to desorption. This coverage was selected for final-state resolved studies of TPD because it offered the opportunity to explore the desorption dynamics of (nearly) isolated molecules on a well characterized surface. In the LIF–TPD procedure hundreds of repetitive TPD experiments were performed which included cyclical flash cleaning, cooling, and dosing and then temperature ramping for the TPD. For each TPD the probe laser was tuned to a specific resonance of the desorbing NO species and the resulting LIF collected and recorded. Internal (vibrational v; spin-orbit Ω, rotational J, and lambda doublet Λ; total internal energy E_{int}) state distributions could then be probed.

A typical LIF–TPD result[10] is presented in Fig. 3. The LIF signals at the TPD peak corresponded to NO densities of order 10^6 cc^{-1} per quantum state. Many such LIF–TPD results were obtained probing desorbed NO in both lambda doublet species, in both spin-orbit states and in rotational levels up to $J = 17.5$ ($E_{int} = 540$ cm^{-1}). The recorded LIF–TPD spectra were numerically integrated over the temperature range $350 \leqslant T_S(\mathrm{K}) \leqslant 410$. These integrated signals (normalized to the probe laser power and divided by appropriate spectroscopic line strengths) gave the relative internal populations of NO.

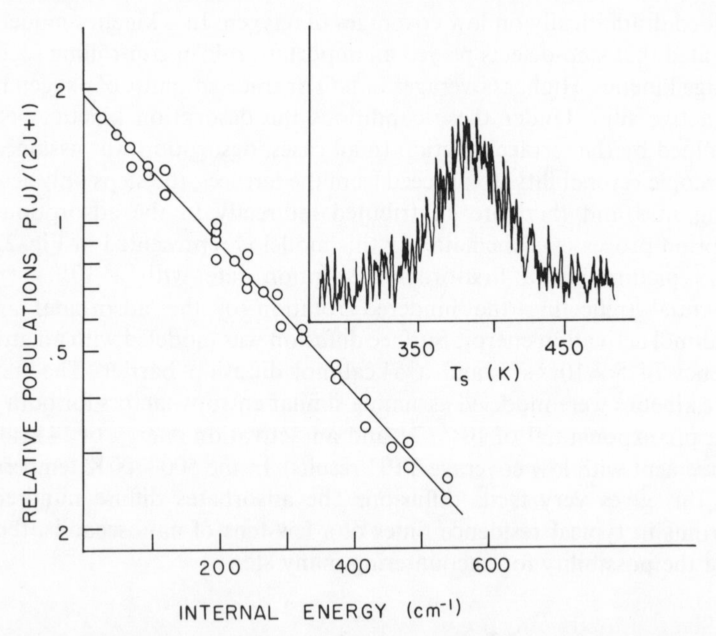

Fig. 3. Boltzmann plot of the internal state distribution from TPD for NO/Pt(111) for $\Theta_{init} = 0.025$. The fitted line corresponds to a value $T_R = 360$ K. Inset: LIF-TPD result, probing NO($v = 0$; F_1, $J = 14.5$, $\Pi(A'')$; $E_{int} = 375$ cm^{-1}).

These results are included in Fig. 3 in the familiar Boltzmann representation giving a value $T_R = 360$ K (solid line), essentially equal to the average value of T_S over the integration range.

The desorption flux is so low under these conditions that no gas phase collisions occurred between molecular desorption and LIF probing. Phase space treatments[20] of final-state distributions for dissociation processes where exit channel barriers do not complicate the ensuing dynamics often result in nominally 'thermal' distributions. In the phase space treatment a loose transition state is assumed (e.g. one resembling the products) and the conserved quantities are total energy and angular momentum; the probability of forming a particular final state of (E, J) is obtained by analyzing the number of ways to statistically distribute the available (E, J).

More informative are the stochastic trajectory simulations run by Muhlhausen *et al.* (MWT), on empirical interaction potential surfaces for scattering and desorption[21]. Although the major thrust was to understand the direct beam scattering results of NO/Ag(111), extension of these calculations allows for comparison to the desorption of NO from Pt(111)[21]. Important insights derived from the NO/Ag(111) calculations were:

1. Rotationally mediated trapping is important. The coupling between rotational and kinetic energy arises from the orientational dependence of the attractive portion of the potential.
2. Despite the orientational dependence of the sticking probability, at low incident kinetic energies all molecules are reoriented into favorable binding orientations during the collision. However, they might not actually stick.
3. The apparently statistical distribution observed in the calculations for low J arose from several kinds of trajectories that combined to produce the observed distribution. The existence, position and magnitude of the rotational rainbow is sensitive to the orientational anisotropy of the attractive interaction potential (washed out by averaging), and does not result from anisotropies in the repulsive interaction region.

The MWT model can be applied to the thermal desorption process. As long as the surface-bound molecule can exchange energy of average magnitude $\Delta E \approx kT_S$ with the substrate heat bath, one should expect desorbed molecules to exhibit complete translational and rotational accommodation. As ΔE falls below kT_S (at increasing T_S), incomplete accommodation should become manifest. This effect factors into the MWT model as a dynamical correction through translational and rotation energy dependencies in the sticking probability.

Explicit results are only presented by MWT for NO/Ag(111). To the extent that the relationship $\Delta E/kT_S$ scales with E_a, deviations from full accommodation might be expected for NO/Pt(111) at $T_S > 750$ K. An additional mechanism for reducing the rotational energy of the desorbed species arises from strong coupling, induced by the surface, between (frustrated) rotational and translational motion. This provides the requisite energy for overcoming the binding potential. Since the strong orientational forces of the attractive interaction potential most strongly couple translational motion along the surface normal, $\hat{\mathbf{n}}$, to rotational motion with angular momentum $\mathbf{J} \perp \hat{\mathbf{n}}$, there should be a net alignment of desorbing molecules. This result would be independent of equilibrium binding geometry. There was no evidence for either rotational cooling nor alignment in the low coverage TPD experiment for NO/Pt(111) at $T_S = 380\,K$.

3.3 Adsorbate–adsorbate interactions

Experiments similar to those described for low coverages of NO were performed for initial NO coverages ranging from $\Theta = 0.025$ to $\Theta \leqslant 0.40$. The trend to decreasing T_{peak} with increasing coverage (see Fig. 1) is consistent with a decrease in binding energy $E_a \sim 34$ kcal/mole (low coverage) to $E_a \sim 25$ kcal/mole to $E_a \sim 10$ kcal/mole (high coverage). In every instance the populations in states (Ω, J, Λ) always scaled as a temperature with $T_R \approx T_S$ over the range of interation. A typical LIF–TPD result is shown in Fig. 4 for an

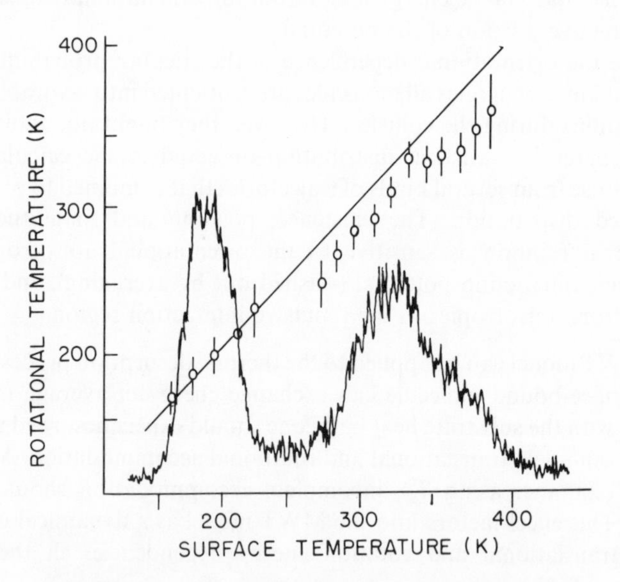

Fig. 4. Laser-induced fluorescence detected TPD for NO/Pt(111) for $\Theta_{init} = 0.40$ (saturation coverage). NO($v = 0$; F_1, $J = 4.5$, $\Pi(A')$; $E_{int} = 40\,cm^{-1}$) was probed. The instantaneous rotational temperatures obtained for 20 K intervals in T_S are plotted against T_S. The solid line corresponds to full rotational accommodation, $T_R = T_S$.

initial coverage $\Theta = 0.4$. Integration of this data over the range $190 < T_S(\mathrm{K}) < 210$ and $280 < T_S(\mathrm{K}) < 400$ gave Boltzmann rotational and spin-orbit population distributions described by the parameters $T_R = 193 \pm 8\,\mathrm{K}$ and $T_R = 318 \pm 13\,\mathrm{K}$, respectively.

In a simple TPD kinetics picture each resolved desorption feature is given a set of macroscopic kinetics parameters (ν, E_a). An important question is whether the rotational state distribution of the desorbed NO is characteristic of a particular binding state. This question was addressed by analyzing the $\Theta = 0.4$ LIF–TPD data for successive intervals of widths $\Delta T_S = 20\,\mathrm{K}$. The best fit rotational temperatures to the data were obtained by nonlinear least squares procedures and are plotted in Fig. 4 against flux weighted average T_S. Clearly, T_R follows T_S across this entire temperature range. Interestingly, across each desorption feature T_R reflects the instantaneous temperature of the surface.

The influence of coadsorbed CO on the thermal desorption of NO/Pt(111) is complex, depending on relative NO/CO coverages and dose sequence[10,22]. Under most conditions NO desorbs prior to significant CO desorption. At low NO coverage on an otherwise CO-saturated surface, NO desorption occurs at $T_{peak} = 300\,\mathrm{K}$, substantially shifted from the value of $T_{peak} = 380\,\mathrm{K}$ for neat, low coverage NO. For higher NO coverage, NO desorption occurs over the range $150 \leqslant T_S(\mathrm{K}) \leqslant 400$. In all the LIF–TPD experiments with NO/CO coadsorbed, the rotational populations distributions were Boltzmann-like with values of $\alpha_{ROT} \equiv T_R/T_S$ varying between $1.0 \pm .1$ at $T_S = 215\,\mathrm{K}$ to 0.9 ± 0.07 at $T_S = 320\,\mathrm{K}$.

TPD and EELS experiments have shown coadsorbed NO/NH_3 to form a surface stabilized complex on Pt(111)[23]. This assertion was based in part on the very large vibrational shifts observed for the neat (NO or NH_3) systems compared to the mixed system, and in part on the simultaneous desorption of NO and NH_3 from the mixed system at values of T_S higher than typical of desorption from the neat systems at comparable initial coverages. Kinetically, desorption from this system has been characterized as reaction limited, i.e. decomposition of the $NO–NH_3$ complex is the rate-limiting step in the desorption process. It was hoped that state-resolved diagnostics would distinguish between a concerted process (simultaneous dissociation of the complex and desorption) versus a step-wise process through a rotational state distribution which was not accommodated with the surface.

A TPD investigation of mixed systems with various ratios of NO/NH_3 indicated a one-to-one stoichiometry in the complex[14]. Excesses of NO and/or NH_3 are not complexed. NO state-resolved TPD spectra were obtained for coadsorbed NO and NH_3 where there was an initial excess NO coverage. This coverage was prepared by predosing the cold Pt(111) ($T_S = 132\,\mathrm{K}$) with a saturation coverage of NH_3. This sample was flash annealed to 250 K, during which some of the NH_3 was desorbed. The remaining NH_3 coverage was approximately $\Theta_{NH_3} \approx 0.1$. The sample was then saturated with NO at 132 K. This procedure gave reproducible NO and NH_3 TPD spectra. The resulting NO TPD spectra indicated nearly equal amounts of NH_3-complexed NO ($T_{peak} = 355$ K) and uncomplexed NO ($T_{peak} = 325$ K).

Population distributions obtained either by integrating state-resolved LIF–TPD data across each of the two major NO desorption features or within $\Delta T_S = 20$ K increments were always Boltzmann. The latter results were characterized by rotational temperatures $90 \pm 5\%$ of the surface temperature.

The high degree of rotational accommodation in the NO–NH_3/Pt(111) system ($\alpha_{ROT} = .90 \pm .05$) appears slightly lower than that obtained for neat NO and for NO–CO mixtures ($\alpha_{ROT} = .95 \pm .05$). However, this difference in rotational accommodation is within the expected accuracy of these experiments. The important implication of these results (i.e. the insensitivity of rotational accommodation of NO to complexation) is that desorption of initially NH_3-complexed NO proceeds sequentially. The complex dissociates and the newly formed NO_a remains adsorbed for some period of time, during which it can exchange energy with the Pt(111) heat bath. Using the kinetics of STC, one calculates a residence time for NO/Pt(111) at 360 K of nominally 1 s, long enough to fully equilibrate with the surface heat bath prior to desorption. Thus, any barriers to formation/dissociation of this complex must be below the zero potential of the desorbed products.

3.4. Signature of TPD dynamics

There are several features common to the desorption of NO discussed above:

1. Thermal distributions of NO(v; Ω, J, Λ; E_{int}) states were observed, wherein the population in any level was determined by the internal energy and the parameter T_R, and independent of spin-orbit state or lambda doublet species. This is in contrast to the rotational rainbows, the propensities for preferential population in the $\Pi(A')$ lambda doublet species and the F_1 spin orbit state which were observed in direct inelastic scattering of NO/Ag(111).
2. No significant alignment was observed; i.e. the LIF–TPD signal-to-noise levels for $J \leqslant 12.5$ placed limiting values of < 0.1 on the alignment parameter $|A_0^{\{2\}}(J)|$. ($A_0^{\{2\}}$ can vary from 2 for $\mathbf{J} \parallel \hat{\mathbf{n}}$, to -1 for $\mathbf{J} \perp \hat{\mathbf{n}}$; a value of 'zero' describes an isotropic distribution.)

These results are uniformally consitent with expectations based on transition state theory. In comparison to the dynamical corrections introduced by MWT for modeling scattering/desorption phenomena one must conclude for the above systems (with $9 \leqslant E_{ads}$ (kcal/mol) $\leqslant 34$) that there is neither evidence for a strong dependence of sticking probability on rotational energy nor any (frustrated) rotational–translational coupling mechanism that might preferentially align the desorbing NO.

3.5. Higher T_S

It is impractical to attempt to explore conventional TPD dynamics for NO/Pt(111) at values of $T_S > 400$ K since this would require extreme heating rates. Two techniques have been utilized to extend state-resolved dynamics to higher T_S. These include trapping/desorbing beam scattering and surface reactions (i.e. as observed for the NO–NH_3 coadsorption system).

Beam scattering of NO/Pt(111) has been reported over the range $220 \leqslant T_S(K) \leqslant 1200$[24,25]. At the lower surface temperatures the scattered flux of NO showed two components: a specular component and a trapping/desorbing component. The specular component retained some degree of memory of initial beam conditions in its scattering angle and kinetic and internal energies. The fraction characterized as trapping/desorbing exhibited a cosine angular distribution about the surface normal, with kinetic and internal energy distributions dependent on T_S and independent of initial beam conditions. The contribution of the specular peak to total scattered flux increased as T_S increased. Above $T_S \approx 400$ K, there can be little steady state NO coverage and scattering must be on low coverage Pt(111). As shown in the STC model, the incident trapping/desorbing molecules must initially trap at terrace sites, and may migrate substantial distances across the surface (possibly visiting defect sites), before finally desorbing from terrace sites. At the highest $T_S \approx 1200$ K explored, one calculates site hopping times ≈ 0.2 ps for comparison to estimated residence times of ≈ 20 ps.

Rotational population distributions were measured for trapping/desorbing and for the specular scattering flux over this range of surface temperatures. These results are summarized in Fig. 5. Also included in Fig. 5 are results for the desorption of NO produced through the oxidation of NH_3 on Pt(111). In both the experiments by Asscher *et al.*[26], at $T_S \approx 800$ K and by Hsu *et al.* (HSL)[27], for $800 \leqslant T_S(K) \leqslant 1300$ the observed lack of rotational accommodation was comparable to that of the beam scattering experiments. Not only do these oxidation results extend the range of T_S investigated and confirm the leveling-off in T_R to a limiting valve $T_R^{\lim} \approx 400$ K; they also confirm that the surface oxidation of NH_3 proceeds via a sequential process. Noticeable in Fig. 5 is the appreciable rotational cooling for trapping/desorbing species at $T_S > 500$ K. Only slight translational cooling was observed in related, mass spectrometer detected time-of-flight experiments[25a]. Although the reported rotational and translational cooling appears to corroborate the trajectory simulation of MWT, these simulations (admittedly for NO/Ag(111)) predict a comparable degree of cooling for both degrees of freedom due to the strong coupling of translational and rotational degrees of freedom. The MWT model would also strongly indicate, in the presence of such pronounced 'rotational cooling', that these desorbing molecules should be preferentially aligned with $\mathbf{J} \parallel \hat{\mathbf{n}}$. This was not observed, even for the HSL high temperature oxidation results where $\alpha_{ROT} \approx 0.3$.

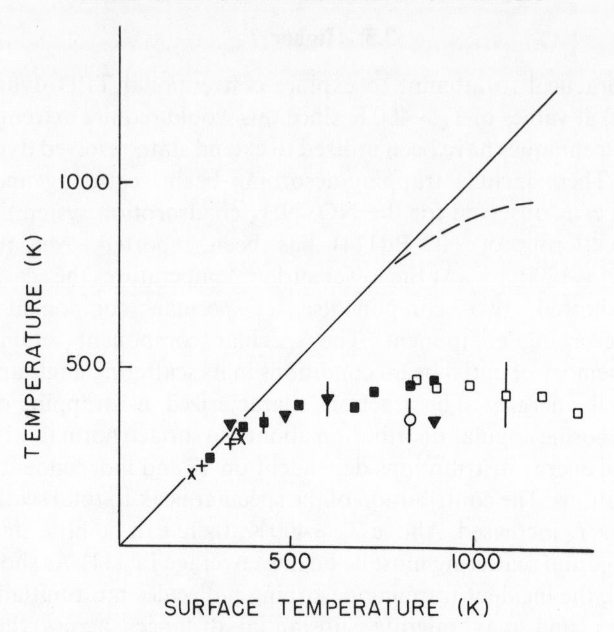

Fig. 5. Rotational temperatures of NO desorbing from Pt(111). The data are representative of data published for (×) neat thermal desorption[10a], (+) thermal desorption in the presence of coadsorbed CO[10b], (solid squares)[24] and (solid triangles)[25] trapping/desorption in molecular beam scattering, (open triangle) reaction limited desorption from NO–NH_3 complexes[14], (open circle)[26] and (open square)[27] NH_3 oxidation reactions. The solid line is for full accommodation. The dashed curve represents results for translational energy measurements in direct inelastic scattering[18b].

The concept of rotational cooling may be considered in two fashions. Simplistically, it might occur because the average energy exchange per collision between the surface and adsorbates at levels of energy near the desorption threshold is less than kT_S. Thermal fluctuations, therefore, cannot maintain the requisite Boltzmann population in those high-energy states which are most efficiently depopulated by desorption. Alternatively, as a result of microscopic reversibility, dynamical constraints in the rotational energy dependence of the sticking probability produce desorbing species that appear rotationally 'cold'. From either viewpoint, the experimental observables would depend critically on the detailed nature of the molecule–surface interaction potential – especially the anisotropy of the attractive well. Thus the presence and degree of 'rotational cooling' should be highly system dependent.

Beam studies of the trapping/desorbing type discussed above have been performed for NO interacting with a wide range of surfaces, including oxided Pt, graphite, Ge and Ge–O. In all cases (see Appendix) the results were qualitatively similar. Deviations in $T_R < T_S$ were apparent at $T_S \geqslant 350$ K and T_R approached a limiting value $T_R^{lim} \approx 400$ K for Pt–O, Ge and Ge–O, and $T_R^{lim} \approx 250$ K for graphite. If the MWT model were operative in all these instances such similarities would not be expected.

There is one example of a substantial lack of rotational accommodation in a normal thermal desorption experiment[28]. For NO/Ru(001) the major TPD feature occurs at $T_{peak} = 450$ K. State-resolved LIF–TPD studies from saturation coverages of NO on Ru(001) obtained a Boltzmann rotational population distribution described by a temperature $T_R = 235$ K, corresponding to a rotational accommodation coefficient $\alpha_{ROT} \approx 0.5$. NO decomposes on Ru(001) at temperatures $T_S > 300$ K. The source for this depressed α_{ROT} might be the dynamical competition between molecular desorption and dissociation for those molecules near or above the desorption threshold. In a simplistic model, if large-amplitude, hindered rotational motion is required for dissociation (which yields adsorbed $O_a + N_a$) then the thermally activated pool of molecular adsorbates with high angular momenta will be depleted through decomposition, resulting in an apparently rotationally cold group of desorbing molecules.

One can, on a plot such as given in Fig. 5, draw a smooth curve through all Pt(111) data for thermal desorption, trapping/desorbing and oxidation-reaction produced NO. For the low T_S thermal desorption studies one can characterize the initial chemisorption state reasonably well. For the low coverage high T_S conditions germane to the beam and oxidation reactions the NO adsorbates are highly mobile on the surface. As an alternative to the simple 'rotational cooling' picture developed by MWT and favored by HSL, one should consider the adsorption state responsible for these high T_S desorptions. STC argued in their model that all adsorption and all desorption proceeded from terrace sites. This reproduced the kinetics obtained for moderate T_S. At higher T_S the molecules spend little time at any given terrace site. Using the STC kinetics at $T_S = 1000$ K one calculates a terrace-to-terrace jump rate of $4 \times 10^{12}\,s^{-1}$ (this also equals the terrace-to-defect jump rate), a defect-to-terrace jump rate of $9 \times 10^8\,s^{-1}$, and desorption rates from terraces and defects of 3.7×10^{10} and $4 \times 10^8\,s^{-1}$, respectively. Two factors become obvious: (1) residence times at terrace sites are only 2×10^{-4} that at defects; (2) from any terrace site diffusion is 100-fold more probable than desorption, whereas desorption and diffusion out of defects have nearly equal probability. Within this kinetic model, a substantial amount of the desorption must be occurring from defect sites at high surface temperatures (e.g. 1000 to 1200 K). This complicates comparison of such high T_S data to lower T_S thermal desorption results which are insensitive to defect sites (i.e. the defects are essentially always filled).

3.6. Alignment

Very exciting experiments by Jacobs *et al.* have discerned a substantial degree of alignment in NO molecules trapping/desorbing from a Pt(111) surface[29]. These experiments were performed with the surface at 545 K. The surface was dosed via a pulsed molecular beam and the isothermally desorbed NO leaving along the surface normal was probed via REMPI. The degree of alignment was measured by monitoring REMPI signals while rotating the plane of polarization of the (linearly polarized) probe laser.

The polarization dependencies in the signal were reduced to an ensemble average in the quadrupole alignment term $A_0^{\{2\}}(J)$, which was dependent on final rotational state. For final states of low angular momentum (e.g., $J < 12.5$) no alignment was observed, consistent with the lower-T_S thermal desorption results. For final states $J > 12.5$ a slight positive alignment ($\mathbf{J} \parallel \hat{\mathbf{n}}$) was observed, as shown in Fig. 6. The magnitude of this alignment increases with increasing J, up to a value $A_0^{\{2\}}$ $(J = 35.5) = +0.14 \pm 0.04$, for the highest J measured. Assuming the spatial distribution of J to be described by an ellipsoid, this value of $A_0^{\{2\}}$ would correspond to 20% more $J = 35.5$ molecules desorbing like 'helicopters' than 'cartwheels'. Preferential alignment $\mathbf{J} \perp \hat{\mathbf{n}}$ was observed in the direct, rotationally inelastic scattering of NO/Ag(111) for $\Pi(A')$ species.

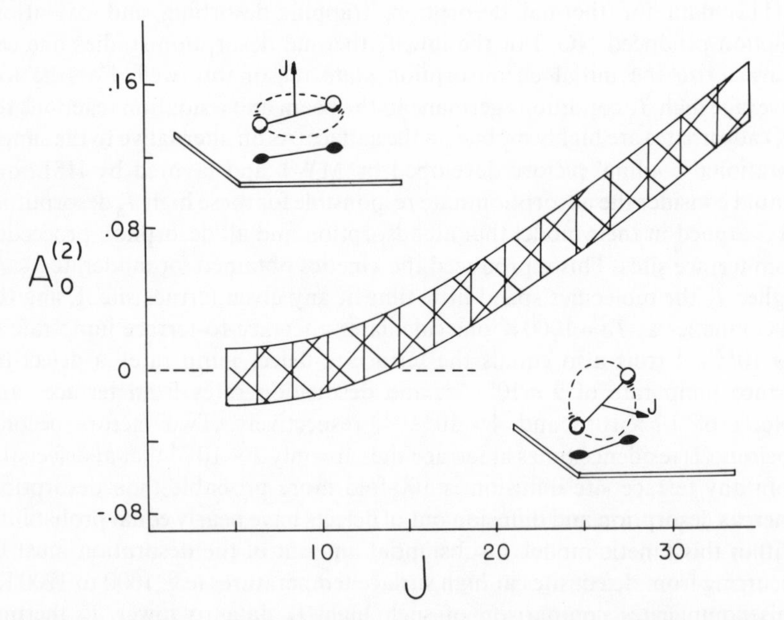

Fig. 6. Alignment results reported for NO trapping/desorption from Pt(111). At high J, $A_0^{\{2\}}$ was positive implying a propensity for helicopter-type desorbing molecules. (*Adapted from Ref. 29.*)

Calculations by Corey *et al.* (CSAL)[30] implicated electron orbital effects as the source of the molecular alignment. In the experiment by Jacobs, both lambda doublets exhibited the same degree of alignment; therefore, this same mechanism cannot be dominant.

3.7. Orientation

In a beam-surface scattering experiment, if the scattered molecules show a preference for rotationeither clockwise or counter-clockwise about the normal to the scattering plane, they are defined to be oriented. Although one might fruitfully conjecture how an anisotropic interaction potential might produce aligned desorbing diatomics, it is not quite so easy to picture why a diatomic would become both aligned and oriented when scattered from an idealized 'flat' surface. Yet such an effect has been observed by Sitz *et al.* (SKZ)[31] for the direct-inelastic scattering of N_2 from Ag(111). The molecular beam was incident on the crystal along the [2, 1, 1] azimuth and the scattered N_2 was probed via a two-photon resonance + two-photon REMPI ionization process. The final state population distribution, $A_0^{\{0\}}(J)$, was obtained by scanning the laser wavelength. The alignment [$A_0^{\{2\}}(J)$ and $A_0^{\{4\}}(J)$] was determined by sitting on a particular J-transition and rotating the linear polarization of the probe laser using a half-wave retardation plate. The orientation [$A_{-1}^{\{1\}}(J)$ and $A_{-1}^{\{3\}}(J)$] was obtained by continuously varying the ellipticity of the probe laser polarization.

The scattering was peaked at the specular angle and all the population, alignment, and orientation results were characterized in terms of their dependence on surface temperature, incident beam kinetic energy (E_i), and angles of beam incidence Θ_i and scattering Θ_f. Although rotational rainbows were always observed, they were more pronounced for higher values of E_i or for super-specular scattering $|\Theta_f| > |\Theta_i|$. At $T_s = 540\,K$, and $\Theta_f = \Theta_i$, the magnitude of the observed alignment increased with J, approaching the $\mathbf{J} \perp \hat{\mathbf{n}}$ limiting value of $+1$ by $J = 10$. SKZ found that this alignment and the angular flux and rotation rainbow results were relatively well reproduced by scattering models for rigid rotors on a flat surface.

Such a model did not account for the observation of orientation. For specular scattering, the magnitude of the observed orientation was nominally independent of J at low J (although never approaching its limiting value of ± 1.0) and rapidly fell to zero at higher values of J. SKZ rationalize the presence of this orientation (in direct scattering) as arising from in-plane forces such as friction. A simple, hand-waving picture for producing orientation has the surface corrugation distort the N_2 into a quasi-heteronuclear diatomic. Depending on the details of the interaction potential and Θ_i this can generate excess 'top-spin' or 'bottom-spin'. Although SKZ have not yet unambiguously established the direction of orientation, they have shown the degree of orientation to depend critically on J, Θ_i, Θ_f, and T_S. In this model both the shallow corrugation and anisotropy in the interaction potential are responsible for producing the orientation. It is surprising that no orientation has been observed in scattering experiments with polar molecules such as NO or from more corrugated surfaces such as Pt, Ni or Ge.

3.8. Polyatomic molecules

Most molecule–surface experiments performed with final state resolution have focussed on diatomics, for reasons of technical simplicity. Kay *et al.* courageously attacked the scattering of NH_3 from gold[32] and tungsten[33] in a series of experiments providing great insight into details of molecule–surface dynamics extending beyond diatomics. NH_3 has two rotational quantum numbers: $J \equiv$ total rotational angular momentum and $K \equiv$ projection of J onto the symmetry axis. For a given value of J, molecules with $K = 0$ can be characterized as 'tumbling' and molecules with $K = J$, as 'spinning'. Due to the substantial difference in moments of interia associated with spinning versus tumbling motion, tumbling molecules contain more rotational energy than spinning ones.

Beam scattering experiments performed for NH_3 on W(100) were best described in terms of trapping/desorbing trajectories due to the strong attractive interaction. Under conditions of $T_S = 300\,K$, the scattered NH_3 (probed by REMPI techniques) exhibited full rotational accommodation with a single rotational temperature describing the population distribution over spinning and tumbling type of levels, similar to desorption and trapping/desorption results for diatomics.

The interaction between NH_3 and Au(111) is much weaker. For $T_S = 300\,K$ with $E_i = 5.49$ kcal/mol the favored scattering trajectories were of the direct-inelastic type. In the specular scattering channel, there was an excess of population in the states $J_k = 0_0$, 1_0, and 1_1, (which were populated in the incident beam). Populations in the higher lying $J_{K=0,1}$ levels followed a Boltzmann distribution with $T_R \approx 300\,K$. These are primarily tumbling species. Surprisingly, population in the purely spinning level 3_3 was severely depressed with respect to expectations based on isoenergetic $J_{K=0,1}$ tumbling levels. Although for this example it appeared that T_R of the tumbling species equalled T_S, the same value of T_R was obtained over the range $300 \leqslant T_S(K) \leqslant 800$ and T_R was observed to scale with the incident beam normal kinetic energy, as expected for direct-inelastic scattering.

Classical trajectory calculations were run for NH_3 interacting with a smooth, rigid surface in an attempt to identify the gross features of the interaction potential which might lead to a preference for molecular tumbling in a direct scattering event. From these simulations it appeared that the long-range attractive dipole–image dipole interaction served to orient the highly polar NH_3 with its symmetry axis parallel to the surface normal. Collisions of this type are expected to be (1) rotationally inelastic, and (2) ineffective in producing spinning-type angular momentum. Given sufficient incident kinetic energy, collisions of this type, with the N atom up, might be expected to selectively result in vibrational excitation of the NH_3 umbrella mode, as observed.

3.9. Interaction potentials

There has been substantial experimental and theoretical work on the direct scattering of diatomics from surfaces. A few points are emphasized here for comparison to the results from the laser-induced desorption studies described next. In direct scattering one is concerned with a prompt interaction wherein there can be one, or at most a very few classical turning points in the molecular trajectories, and in which the scattered molecules retain some memory of incident beam conditions[34]. Much of the scattering work has been performed with NO. The nitric oxide ground state is a $^2\Pi$, in addition to rotational angular momentum, the NO molecule may have electron-spin and orbital angular momentum. CSAL have explicitly shown that two interaction potentials, (V_+ and V_-) should be considered. In scattering experiments the degeneracy between these two potentials is lifted when the internuclear axis is parallel to the surface and quantum mechanical interference effects can result in propensities for scattering into specific lambda doublet and spin-orbit states. Such propensities have been observed experimentally for NO/Ag(111). Kinematics suggest possible mechanisms for rotational rainbows and alignment, which have also been observed in direct, rotationally inelastic scattering experiments.

4. LASER-INDUCED DESORPTION

Laser–surface interactions have been utilized recently to explore a wide range of physical and chemical processes. Lasers have been used to ablate material from metallic or insulating substrates for a variety of purposes including depth profiling and etching. Several aspects of ultraviolet lasers (wavelength, pulse length, beam quality, and pulse energy) have made them particularly attractive vaporization sources for mass spectroscopy, speciation characterization, surface cleaning or thin film characterization. Many of the proven and potential uses of laser–surface interactions are particularly attractive for materials processing and analysis in the semiconductor industry.

Laser-induced desorption can be utilized quantitatively to remove adsorbates in a localized area of a surface, without substantially altering the temperature of the surrounding sample. This provides a viable technique for measuring surface diffusion kinetics[35]. Alternatively, sampling different regions of a homogeneous surface for each laser pulse provides a sensitive mechanism for following thermal reaction kinetics without the need for the nascent reaction products to thermally desorb of their own accord[36]. One aspiration in this line of research is to identify reaction intermediates (e.g. radicals) in situ and to measure their kinetics.

4.1. A simple laser-induced desorption model

Absorption of visible laser pulses by a metal is generally localized near the metal surface (optical penetration depths of a few hundred Å) and results in the formation of an electron–hole pair. This ensemble of low energy electrons (few eV, dependent on wavelength) relaxes initially by electron–electron and electron–surface scattering; subsequent electron–phonon scattering processes eventually result in the creation of a transient temperature rise in the local phonon bath. Dissipation of the initial low energy plasma has been shown to occur for Cu on a ~ 1 ps time scale[37]. Equilibration of the phonon bath might require a few tens of picoseconds, resulting in a rapid temperature jump as the photon energy is converted into thermal lattice energy.

This conversion of photon energy into thermal energy is the underlying basis of the T-jump formulations forwarded by Ready and others[38]. The transient temperature jump $\Delta T(t)$ induced by a laser pulse is

$$\Delta T(t) = \left(\frac{I_0}{K}\right)\left(\frac{\kappa}{\pi}\right)^{1/2} \int_0^t A(t-\tau)\tau^{-1/2}\,\mathrm{d}\tau \tag{2}$$

where I_0 is the maximum laser power density absorbed (Joules/cm^{2}s), $A(t)$ is the normalized laser temporal profile, K is the thermal conductivity and κ the thermal diffusivity. I_0 depends on laser pulse energy, pulse duration and sample absorption. For small values of ΔT, K and κ will be temperature independent. For a triangular pulse of rise time t_r and FWHM t_0, the maximum temperature achieved during pulsed laser irradiation is

$$\Delta T_{\max} = \frac{8}{3\mathrm{K}}\left(\frac{\kappa}{\pi}\right) I_0 \left(\frac{t_0}{4 - t_r/t_0}\right)^{1/2} \tag{3}$$

From this expression one should note that $\Delta T_{\max}$ scales linearly with I_0 (the energy absorbed) and as $(t_0)^{-1/2}$ for a given pulse shape.

Two points concerning this heating model should be emphasized. Since experimentally the laser beam diameter is much larger and the penetration depth much smaller than the thermal diffusion length, this heat transport problem may generally be regarded as one-dimensional. This development assumes that all the energy carried by the photon has been effectively coupled into lattice phonon modes[37,38]. This is a process that requires a small but finite amount of time and which is substrate dependent. Although the foregoing

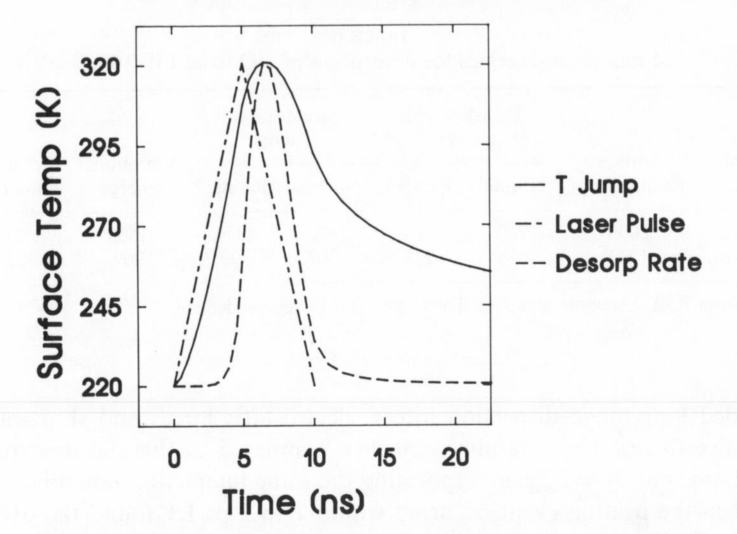

Fig. 7. Laser-induced heating model. The solid line represents the temperature transient calculated from Eq. (3) for a 5 ns FWHM laser pulse (dotted trace). The instantaneous desorption rate calculated from Eq. (4) is represented by the dashed curve.

should be valid for laser pulse durations exceeding 10 ns, it must fail for very short duration pulses.

For a thermally activated process, the desorption rate is

$$-\mathrm{d}\Theta/\mathrm{d}t = v\exp(-E_a/kT_S)\Theta^n \tag{4}$$

where Θ is the surface coverage, E_a and v are the activation energy and pre-exponential for desorption and n is the order of the desorption process. For small coverage changes, the desorption rate will maximize when the surface is at its maximum temperature. Thus, in the absence of any dynamical constraints, laser-induced desorption should result in species exhibiting internal and kinetic energies comparable to T_{max}. The results of solving Eqs. (2) and (4) for a triangular laser pulse are presented in Fig. 7.

4.2. Dynamical expectations

A simple model for the dynamics of nonresonant laser-induced desorption of adsorbates from surfaces has been formulated by Lucchese and Tully (LT)[39]. LT present the result of stochastic, classical trajectory calculations for thermal and laser-induced desorption of NO from LiF(100). For the LID simulations the initial temperature was set at 0 K and temperature jumps of several thousand degrees were driven in a few picoseconds through nonspecific heating of the substrate. The interaction potential for these calculations

TABLE 1
Final mean energies for desorption of NO from LiF(100).

Heating regime	Surface temp. (mean)	Translational energy		Rotational energy		Vibrational energy	Residence time (ps)
		Normal	Parallel	Normal	Parallel		
Thermal	2000	710	1990	1670	1870	1560	2.6
1–2 ps	1900	638	275	302	799	209	2.0

Data from R. R. Lucchese and J. C. Tully[39]; entries in degrees Kelvin.

included long-range dispersion forces, electrostatic forces and short-range repulsive forces. These results were then compared to thermal desorption simulations at $T_S = T_{max}$, incorporating the same interaction potential.

When the heating event occurred within 1 to 2 ps LT found the overall desorption kinetics were not well represented by a single linear rate expression. This they argued was due to molecules becoming highly energized, but not to levels above the desorption threshold. In such a weakly bound, activated state these molecules would not be strongly influenced by the surface atomic motion. This reduced coupling for energy transfer is similar to that postulated for translational and rotational cooling in desorption.

Results from the calculations for isothermal and laser-induced desorption are presented in Table 1. In the thermal desorption experiments the mean values of kinetic and rotational energy are lower than T_S, presumably due to the inefficiency of energy transfer from the substrate to the adsorbate which is necessary to replenish that energy lost in surmounting the binding potential. In the LID calculations, the desorbing flux distribution was more forward directed than that calculated for thermal desorption, due in part to the low mean kinetic energy parallel to the surface plane. Although the value of the normal component of kinetic energy was comparable to that obtained from isothermal desorption, the energy in all other modes was substantially less. When the LID heating rate was lowered, there was a substantial increase in parallel translational energy and broadening of the angular flux distribution, approaching the calculated isothermal desorption results.

Deviations between such matched isothermal vs laser-induced desorptions (whether in theory or experiment) must be due to incomplete equilibration of energy between the substrate and adsorbate. If the T-jump process proceeds faster than the characteristic times for energy flow, then dramatic cooling in adsorbate degrees of freedom must result[40]. The normal component of kinetic energy should have the fastest response, followed by parallel kinetic energy and rotational energy, with vibrational energy flow being slowest and strongly dependent on the stiffness of the vibrational modes involved.

Extrapolation of the picosecond simulations would imply that isothermal and laser-induced desorption results should be in qualitative agreement for nonspecific heating pulses of 1 to 10 ns duration. Of course this may not hold if all desorption occurs during the leading edge of the laser pulse or if the desorption process is driven by a nonthermal mechanism. Nevertheless, incomplete equilibration is not expected to play a major role for translational or rotational accommodations if the residence times are longer than ~ 10 ns.

There have been a large number of nanosecond measurements of normal kinetic energy distributions for molecules desorbed from molecular ices by infrared laser pulses. As shown by Schäfer and Hess (SH), the infrared laser can resonantly interact with the adsorbate vibrational spectrum[41]. The vibrationally excited molecules relax, releasing the photon energy into local lattice excitations. If the laser–molecule interaction is sufficiently intense, many absorption/relaxation steps can occur within the laser pulse resulting in substantial local heating, and subsequent desorption. The time-of-flight (TOF) spectra measured in the SH experiments always were Maxwellian in form, reflecting temperatures consistent with anticipated temperature jumps.

4.3. Examples of interactions with 'nonthermal' energy baths

The picture becomes more complicated when the desorption mechanism becomes dependent on the presence of nonthermal transients, such as high energy phonons, as implicated in the study by Ferm *et al.*[42] of the single phonon-induced desorption of HD from LiF. An HD covered LiF(100) surface was exposed to low intensity infrared radiation and the TOF of the desorbed HD molecules detected. The TOF data were fit by a thermal velocity distribution characterized by $T = 21$ K which was independent of and dramatically greater than the surface temperature ($T_s = 1.5$ to 4.2 K). This and related results were inconsistent with expectations based on desorption process induced by interactions with a thermal phonon bath. Ferm *et al.* proposed a driving mechanism wherein the incident photon was absorbed within the LiF bulk (at depths $< 6\,\mu$m, dependent on wavelength). The initial high energy optical phonons decay through a cascade-type process eventually achieving an equilibrium distribution. The lowest-branch near-zone-edge transverse acoustic phonons created in such a cascade process may exist for 10^{-4} to 10^{-9} s; long enough, the authors argue, to propagate to the surface where annihilation of these individual high energy phonons results in desorption.

The role of surface or near-surface electrons was demonstrated in the work of Ying and Ho[43] for the band-gap radiation induced desorption of NO from Si(111). Filtered Xe arc lamp light was used to tune the photodesorption wavelength across the first direct band gap of Si(111), around 400 nm. The loss of adsorbate was monitored both by mass spectroscopic detection of the desorbed species and temperature dependent, high resolution electron energy loss spectroscopy (HREELS) investigations of the remaining adsorbate coverage. In this wavelength dependent study there was clearly a dramatic increase in desorption yield that followed the broad, long wavelength Si adsorption edge. Ying and Ho argued the desorption to be driven by interaction with 'hot carriers' since the sample temperature rise was minimal in these cw experiments and the desorbing species had to surmount the adsorption energy barrier. Such a hot carrier–adsorbate interaction might result in desorption provided the initial photo-generated carriers were created within one mean free path of the surface (≈ 400 nm).

The ultraviolet laser photolysis of CH_3Br or CH_2I_2 on ionic crystals, oxides and metals presents a clear departure from 'thermal' expectations. The chosen laser excitation wavelengths in these experiments were resonant with transitions associated with highly repulsive dissociative states in the isolated molecule, resulting in fragmentation on the time scale of one-half a vibrational period (≈ 15 fs). Ultraviolet excitation of CH_3Br in the gas phase results in dissociation with near unit quantum yield, producing CH_3 fragments with most probable velocities of 6×10^5 cm/s. For 193 nm excitation, the photolysis of CH_3Br on Ni(111) was reported to occur with high quantum yield for ejection of the CH_3 fragment[44]. The amount of Br ejected was substantially less. Most of the Br photoproduct apparently remained on the surface. No photodesorbed CH_3Br was reported. The observed CH_3 velocity distribution was broader and less energetic than observed for gas phase experiments, indicative of energy transfer to the Ni substrate. In contrast, although Domen and Chuang[45] were able to demonstrate the 308 nm photodesorption of CH_2I from CH_2I_2 on Al_2O_3, they were unsuccessful in similar attempts on metallic Ag films.

4.4. Laser-induced thermal desorption: NO/Pt(foil)

Experiments were conducted in our laboratory to evaluate many of the dynamical expectations for rapid laser heating of metals. One of the aims of this work was to identify those population distributions which were characteristic of thermally activated desorption processes as opposed to desorption processes which were driven by nonthermal energy sources. Visible and near-infrared laser pulses of nominally 10 ns duration were used to heat the substrate in a nonspecific fashion. Initial experiments were performed by Burgess *et al.*[46] for the laser-induced desorption of NO from Pt(foil). Operating with a chamber base pressure 2×10^{-10} torr and with the sample at 200 K, initial irradiation of a freshly cleaned and dosed sample resulted in a short time transient (i.e. heightened desorption yield) followed by nearly steady state LID signals. The desorption yields slowly decreased with time due to depletion of the adsorbate layer at the rate of ca. 10^{-4} monolayer

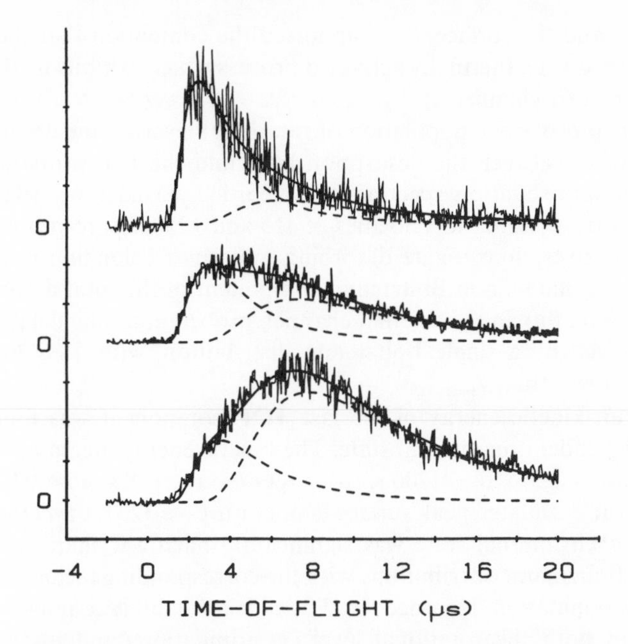

Fig. 8. Laser-induced fluorescence detected times-of-flight for NO desorbed from Pt(foil) using 532 nm laser heating. The three traces were obtained while probing (top) NO ($F_1, J = 19.5; E_{int} = 662\,cm^{-1}$), (middle) NO ($F_1$, $J = 12.5$; $E_{int} = 281\,cm^{-1}$) and (bottom) NO (F_1, $J = 3.5$, $E_{int} = 25\,cm^{-1}$). (*Adapted from Ref. 46.*)

per shot. For long term experiments this posed a problem, and therefore many of the experimental investigations were performed with a low pressure (10^{-8} torr) background of NO flowing through the chamber to replenish the NO coverage. This background gas contributed a small, time independent LIF signal that was useful for estimation of the desorption yield, corresponding to peak NO densities of approximately $10^{10}\,cm^{-3}$ (5×10^{-7} torr).

Representative TOF spectra following 532 nm heating pulses for three different rotational states of NO are presented in Fig. 8. These spectra appeared to consist of contributions from two distinct desorption channels: a *slow* 'thermal' channel and a *fast* non-Boltzmann channel. The distinctly different rotational distributions observed for these two components (see below) provided additional support for such an interpretation.

For rotational levels of $E_{int} \leqslant 400\,cm^{-1}$, the slow component of the TOF data was fitted by a Maxwellian speed distribution with a mean kinetic energy $\langle KE \rangle/2k \approx 350 \pm 50$ K (dividing by 2 to achieve Maxwell–Boltzmann equivalent temperatures)[5]. The peak surface temperature induced by the laser pulse in these experiments was calculated to be 300–320 K. The good agreement between the mean kinetic energy determined for this desorption component and the surface T_{max} supported the contention that these molecules desorbed via a thermally activated process which exhibits no dynamical biases in the exit channel.

In order to probe the population of rotational levels of the desorbed NO, the time delay between the desorption laser and the LIF probe was fixed, and rotational excitation spectra were recorded. Fixed time delays of 9.0 or 3.0 μs (corresponding to velocities of 415 and 1250 m/s, respectively) were used to selectively interrogate desorbing molecules belonging primarily to either the thermal or non-Boltzmann component of the total desorbed flux. The desorption flux in the thermal channel, probed at a time delay of 9.0 μs, was fitted well by a single Boltzmann distribution, with $T_R = 200 \pm 20$ K, somewhat lower than T_{max}.

The mean kinetic energy of the fast TOF component was found to be strongly dependent on internal state. The kinetic energy increased smoothly from $\langle KE(J = 3.5) \rangle / 2k = 1200$ K to $\langle KE(J = 19.5) \rangle / 2k = 2800$ K, greatly exceeding the calculated peak surface temperature (≈ 320 K), and the reduced width for each internal state was significantly narrower than expected for Maxwell–Boltzmann distributions with the corresponding mean energy. The rotational population recorded at 3.0 μs delay exhibited non-Boltzmann behavior in both the rotational level (J) populations and in the lack of equilibration between the spin-orbit components. There was an enhanced probability for population in low rotational levels of the F_1 spin-orbit states (e.g. F_1, $J \leqslant 6.5$; $E_{int} \leqslant 80\ \text{cm}^{-1}$). This enhanced population might represent a low energy rainbow resulting from dynamical effects in the laser-driven desorption process. For example, direct desorption of NO which is bound in standing configurations might lead to an enhanced population of low J levels. Above 80 cm^{-1} internal energy, populations in rotational levels in both spin-orbit states decreased approximately exponentially with rotational energy (although the J state distributions were clearly non-Boltzmann, this exponential fall-off can be parameterized by a 'rotational temperature', T_R ($E_{int} > 100\ \text{cm}^{-1}$) $= 425 \pm 50$ K). Total populations in the two spin-orbit states were inverted with approximately 1.6 times more population in the higher energy F_2 states than in F_1. Approximately 3% of the fast component total flux (532 nm excitation) was in $v = 1$. The TOF spectra for molecules desorbed in the $v = 1$ state were peaked near 3.0 μs. Rotational excitation spectra for NO($v = 1$) gave rotational population distributions very similar to that of the NO($v = 0$) fast component desorbed under the same conditions.

Figure 9 depicts the mean kinetic energy of desorbed NO ($v = 0$; $F_1, J = 9.5$, $\Pi(A')$; $E_{int} = 165.5\ \text{cm}^{-1}$) as a function of the temperature jump induced by the laser. It is apparent that the kinetic energy of the slow component tracked changes in the peak surface temperature effected by varying the heating laser pulse fluence for all (calculated) ΔT. The nearly quantitative correlation obtained between the peak surface temperature and the kinetic energy of the

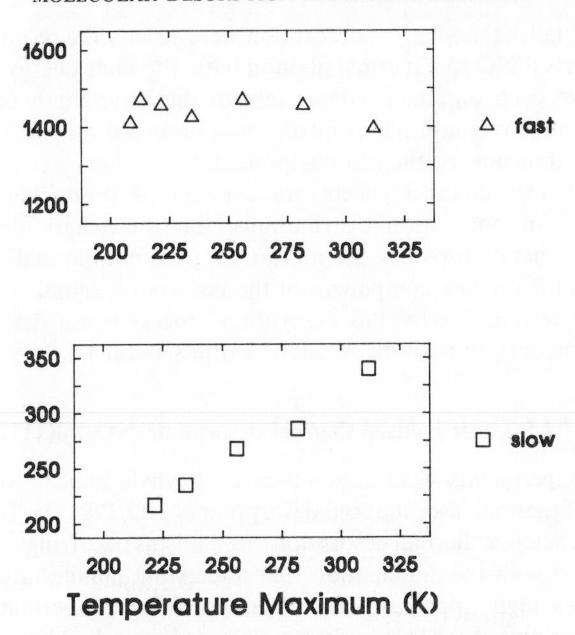

Fig. 9. Average values of the kinetic energy for LIF–LID studies of NO/Pt(foil) as a function of temperature jump ($T_{int} = 200$ K). The lower set of data correspond to the 'thermal' channel; the upper to the non-Boltzmann channel.

slow NO component is consistent with a thermally driven desorption process, with nearly complete translational energy accommodation. In contrast, the mean kinetic energy of the fast NO component was essentially independent of the magnitude of the temperature jump and did not appear to correlate at all with the thermal effects associated with the heating pulse.

In experiments designed to produce similar temperature jumps, the relative yields of NO($v = 0$) from the two desorption channels were not changed by laser heating at either 532 or 1064 nm. Therefore, it is not reasonable to associate one channel with a resonant interaction with the adsorbed species while attributing the other to a nonresonant process. Changing the heating laser wavelength from 532 to 1064 nm did not produce any significant changes in the internal state (Ω, J, Λ) distributions for either channel. The results of the thermal channel are $\alpha_{TRANS} \approx 1.0$ and $\alpha_{ROT} \approx 0.6$. These results are reasonable in terms of LT's model, a low T-jump, and 3.5 ns duration of the desorption event. The observation that the mean kinetic energy of the thermal channel remained invariant indicated that ΔT was indeed the same. Notably, the mean kinetic energy of the fast component was about 30% lower with 1064 nm excitation than with 532 nm excitation; still being dependent on J

and much higher than T_{max}. If desorption were strictly the result of couplings of the adsorbed NO to a thermal phonon bath, the same energy distributions should have been obtained independent of the wavelength of the heating laser. Such a wavelength independence was observed for the 'thermal' LID component, but not for the fast component.

Although exit channel effects are capable of producing a range of non-Boltzmann population distributions, the wavelength dependence of the kinetic energy provides an indication that nonthermal activation is responsible for the fast component of the desorption signal. The activation mechanism responsible for this desorption process is not determined from these experiments, but will be re-addressed in section 4.6.

4.5 Laser-induced thermal desorption: NO/Pd(111)

Similar experiments were undertaken by Prybyla *et al.*[47] to explore the dynamics of thermal laser-induced desorption of NO/Pd(111). Their goal was to fully characterize thermal desorption mechanisms occurring at high surface temperatures with the anticipation that interesting manifestations of molecular physics might appear for large T-jumps. In these experiments, a $T_{init} = 300$ K NO-saturated Pd(111) surface, maintained by pulsed beam dosing, was irradiated. The heating laser pulses were ≈ 8 ns FWHM and of near top-hat spatial profile. This resulted in a transient T-jump to $T_{max} = 1150 \pm 150$ K, determined by transient reflectivity measurements, and the desorption of nominally $\leqslant 10^{-2}$ of the local adsorbate coverage, determined by calibrations made by back-filling the UHV scattering chamber with a known NO pressure.

The kinetic energy and angular distributions for desorbing molecules in various NO(v; Ω, J, Λ) levels were measured using REMPI detection. The experimental TOF spectra appeared to be Maxwellian in shape, corresponding to average kinetic energies $\langle KE \rangle/2 = 570 \pm 50$ K. The angular distribution appeared approximately, but somewhat broader than, cosine. Both the kinetic energy and angular distribution were observed to be independent of (v; J, Λ; E_{int}).

The rotational population distributions were Boltzmann in nature, characterized by $T_R = 640 \pm 35$ K. This seems substantially lower than T_{max}, yet somewhat larger than the temperature associated with the translational degree of freedom. The lambda doublet species were statistically populated. The population ratio of $v = 1/v = 0$ was roughly 0.09, consistent with a vibrational temperature $T_V = 1120 \pm 35$ K. The same rotational and spin-orbit distributions were obtained for molecules desorbed in $v = 1$ as for $v = 0$ levels. Finally, there was no dependence in the J-state distributions on desorption angle.

Prybyla *et al.* believe these results to give valuable insight into details of the

NO–Pd(111) interaction potential. Asserting the validity of microscopic reversibility, these results illuminate the detailed energy dependence of the sticking probability for gas phase NO on a 1150 K Pd(111) surface with the following interesting aspects. The observation of $T_V \approx T_{max}$ implies both that vibrational energy flow between the adsorbate internal mode and the surface phonon bath is facile, and that vibrational excitation plays no role in the adsorption/desorption event. The flatter-than-cosine angular distribution is indicative of a one-dimensional barrier which is overcome at the expense of *z*-direction kinetic energy. In the LT modeling, surmounting such a barrier would be accompanied by translational and rotational 'cooling', as observed. This cooling should be most severe for translational motion along the surface normal and for the strongly coupled rotations with $\mathbf{J} \perp \hat{\mathbf{n}}$. Preliminary experiments by Prybyla *et al.* imply that, in fact, the kinetic energy increases with increasing angle way from $\hat{\mathbf{n}}$; however, rotational alignment results are not yet available. The independence of results on $(v; J, \Lambda)$ implies the sticking probability to be proportional to the product of a set of uncorrelated distribution functions.

4.6. Desorption through nonthermal channels: NO/Pt(111)

The results of the NO/Pd(111) work and the 'thermal' channel on the Pt foil support one's naive expectation for nanosecond duration laser-induced thermal desorption. The non-Boltzmann results on Pt(foil) do not. In the foil experiments, the surface cleanliness and binding states were not well characterized. Results of more recent experiments conducted on a well characterized Pt(111) single crystal[48] provide substantial insight into the excitation process and the desorption mechanism associated with the non-Boltzmann signals seen in the NO/Pt(foil) LID experiments. The LID experiments were performed in a UHV chamber with a 10^{-10} torr base pressure. A Q-switched YAG laser provided excitation wavelengths of the laser fundamental (1064 nm, 1.17 eV, 15 ns FWHM), second harmonic (532 nm, 2.34 eV, 11 ns FWHM) and third harmonic (355 nm, 3.51 eV, 9 ns). Pulse energies were adjusted to produce equivalent 100 K temperature jumps[38]. Under these irradiation conditions about 10^{-5} of the local adsorbate coverage was desorbed on each shot (the sample was redosed regularly to replenish the desorbed molecules).

The observed TOF spectrum depended both on surface preparation (i.e., coverage) and on the specific final state probed. Figure 10 (top) presents a TOF obtained for a sample saturated with NO at $T_{init} = 120$ K, using 532 nm for surface excitation and probing NO($v = 0$; F_1, $J = 6.5$, $\Pi(A')$; $E_{int} = 80\,\text{cm}^{-1}$). The bottom trace was obtained under similar dosing conditions but with the crystal subsequently thermostated at 220 K. The obvious loss of intensity for slowly moving molecules (long flight times)

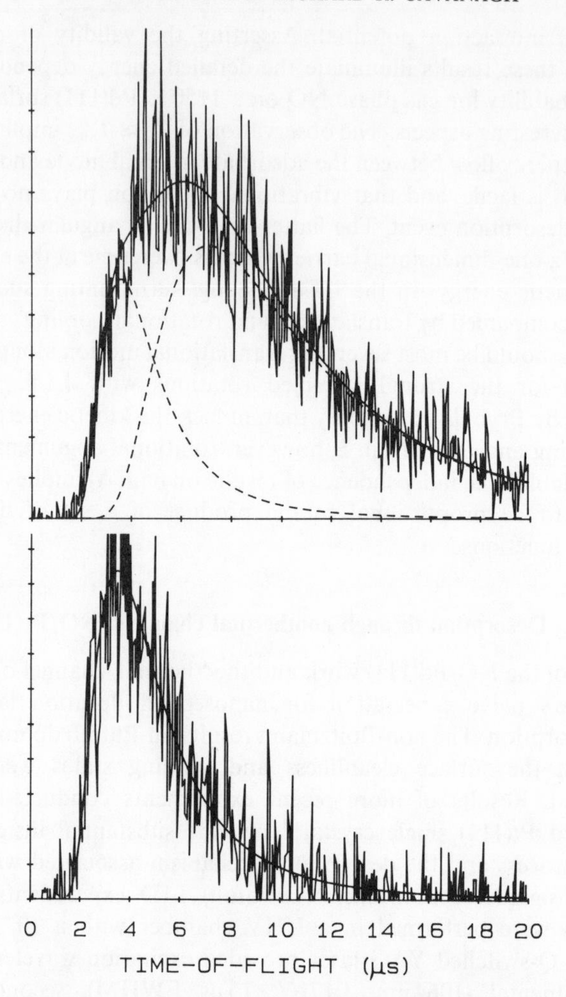

Fig. 10. LIF–LID traces for NO/Pt(111) detecting NO($v=0$; F_1, $J=6.5$, $\Pi(A')$; $E_{int}=80\,cm^{-1}$) at a flight distance of $\Delta z=3.81$ mm. The top curve was obtained for a surface saturated with NO at 120 K, corresponding to $\Theta=0.4$. The dashed curves are fits to the two desorption components; the solid curve is their sum. The lower curve was obtained following annealing at 220 K, corresponding to $\Theta=0.25$. The heating laser wavelength was 532 nm.

resulted from the isothermal desorption of molecules in the corresponding low energy binding state (i.e. the $T_{peak}=200$ K TDP feature; see Fig. 1).

The slow component in Fig. 10 was characterized by a mean kinetic energy of 190 ± 20 K, essentially equal to T_{max}. The dynamics exhibited by this slow

LID component were very similar to that observed in our work on Pt(foil) and by Prybyla *et al.* on Pd(111); that is, moderate to high degrees of rotational and translational accommodation. These results were independent of excitation laser wavelength and confirmed the validity of the T_{max} calculations. The remainder of this section will deal with the dynamics of the non-Boltzmann channel.

Fitting of the 220 K data (lower trace) gave an average kinetic energy $\langle KE \rangle / 2k = 1100$ K substantially in excess of the calculated $T_{max} = 320$ K. This non-Boltzmann desorption flux exhibited an angular distribution sharply peaked along the surface normal. Measurements were made for F_1 and F_2 rotational levels in $v = 0$ of total internal energy up to 738 cm^{-1}, and for the F_2 levels in $v = 1$. Trends in this data imply that NO($v = 0$; F_2) had higher kinetic energies than NO($v = 0$; F_1) species of either the same J or E_{int}; for NO($v = 0$; F_2), $\langle KE \rangle / 2k$ increased slightly with J. The kinetic energies of the NO($v = 1$; F_2) were indistinguishable from those of the NO($v = 0$; F_1), within the $\pm 10\%$ precision of these measurements.

Excitation spectra recorded for $\lambda_{ex} = 532$ nm at a time delay of 3.7 μs (corresponding to desorbed molecules with speeds of 10^5 cm/s) were analyzed to give the NO($v = 0$; Ω, J, Λ; E_{int}) population distribution shown in Fig. 11. The symbols distinguish populations in the two spin-orbit states. The population in the higher energy F_2($\Delta E_{so} = 124$ cm^{-1}) state was ca. four times larger than F_1. If the relative populations were thermal, determined by $T_{max} = 320$ K,

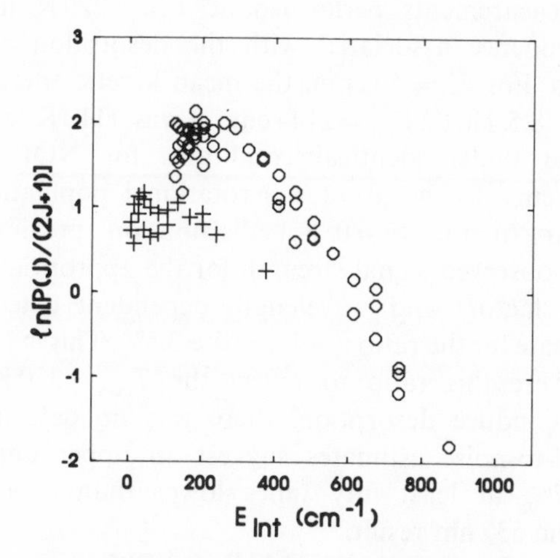

Fig. 11. Population distributions observed in the non-Boltzmann laser-induced desorption channel for NO/Pt(111). Populations in the higher (open circles) F_2 and lower energy (+) F_1 spin-orbit states are distinguished. The heating laser wavelength was 532 nm.

this ratio should have been $F_2/F_1 = 0.55$. In addition to this spin propensity, the rotational population distributions were clearly non-Boltzmann. It almost appears as though there was equal probability of forming desorbed species in all rotational levels, up to some critical energy, $E_{int} \approx 300\,cm^{-1}$, above which $P(J)$ decreases rapidly with increasing J. There was no preference for population of either lambda doublet species for levels $E_{int} \leqslant 455\,cm^{-1}$, and no alignment for levels $J \leqslant 16.5$.

Repeating these experiments using the YAG laser fundamental (1064 nm, 1.17 eV), adjusting the energy to achieve the same calculated temperature jump, gave essentially identical (Ω, J, Λ)-state distributions. The inversion of population in the two spin-orbit levels, the population plateau for internal energies below $300\,cm^{-1}$, and the rapid fall-off of rotational population for $E_{int} > 300\,cm^{-1}$, were all comparable to the observations for 532 nm excitation experiments. The measured kinetic energies for several F_2 rotational levels, ranging in E_{int} from $125\,cm^{-1}$ ($J = 1.5$) to $737\,cm^{-1}$ ($J = 18.5$), were again only moderately dependent on J. However, the kinetic energies were quantitatively different for excitations at 1064 vs 532 nm. Excitation at 532 nm gave $\langle KE \rangle/2$ centered about 1050 K, while excitation at 1064 nm gave $\langle KE \rangle/2$ centered about 725 K. That is, regardless of internal energy, excitation at 1064 nm resulted in desorbed NO which was approximately 30% less energetic than observed for the 532 nm experiments. Such a desorption channel, wherein there is a 'memory' of the photon energy used to initiate the process, must be driven by a nonthermal mechanism.

A series of measurements performed at $T_{init} = 220\,K$ investigated the wavelength dependence associated with the desorption of vibrationally excited molecules. For $\lambda_{ex} = 532\,nm$, the mean kinetic energy obtained for $NO(v = 1; F_2, J = 8.5, \Pi(A'); E_{int} = 2144\,cm^{-1})$ was 1110 K, essentially equal to that obtained under identical conditions for $NO(v = 0; F_2, J = 8.5, \Pi(A'); E_{int} = 257\,cm^{-1})$. The $(v = 1, F_2)$ rotational population distribution was also similar to the $(v = 0, F_2)$ distribution previously described. Normalizing the observed signal strength for the appropriate spectroscopic Franck–Condon factors and wavelength dependent detector efficiencies provides an estimate for the ratio $v = 1/v = 0 \approx 3.5\%$. This is 130 times greater than expected were this ratio to reflect the $T_{max} = 320\,K$. When using $\lambda_{ex} = 1064\,nm$ to induce desorption, there was no detectable $NO(v = 1)$ desorbed. Signal-to-noise estimates suggest an upper limit to the ratio $v = 1/v = 0 \leqslant 0.07\%$, at least fifty times lower than expected based on comparison to the 532 nm result.

The NO LID results on Pt(111), Pd(111) and Pt(foil) are strongly marked both by their dramatic similarities and their subtle differences. All exhibit a thermal LID component with high degrees of translational and rotational accommodation. For the Pt substrates the non-Boltzmann component exhibits: (1) anomolously high, state-dependent kinetic energies; (2) spin-orbit population inversions; and (3) non-Boltzmann rotational population distributions, although both the rotational distributions and the state dependence of the kinetic energies are substantially different for Pt(111) vs Pt(foil). Unlike the direct inelastic scattering results for NO/Ag(111), there was no observed preference for either lambda doublet species, nor was there any measurable alignment ($J \leqslant 16.5$). LID results obtained from experiments on Pt(111) using various initial T_S conditions clearly show the thermal component to arise from adsorbates in the high coverage state that desorb around 200 K ($E_a \approx 9$–11 kcal/mol). Calculating the kinetic desorption rate of these species ($T_{max} = 220$ K), $\Delta t = 3.5 \times 10^{-9}$ s) one would estimate a desorption yield of 10^{-5}, consistent with that observed. The origin of the nonthermal LID component appears to be the binding site which STC characterized by $E_a = 25$ kcal/mol. The measured desorption yield of this channel, however, are much larger than the 4×10^{-10} calculated for a kinetically controlled desorption mechanism.

A potential explanation for the Pt(111) nonthermal LID results involves the formation of a transient charge exchange species. In the gas phase, NO has a 0.09 eV electron affinity and exhibits strong, sharp resonances in electron capture cross-section for low energy electrons (0–2 eV)[49]. It is possible that some of the nascent photogenerated carriers scatter at the adsorbate covered surface creating $NO_a^{-\delta}$ species. These species would propagate along an ionic interaction potential for some period of time prior to subsequent charge neutralization. Such a process has been discussed in terms of harpooning reactions by various authors[50]. This process might lead to desorbed neutrals which exhibit: (1) high, wavelength-dependent kinetic energies arising from details of the ionic interaction potential; (2) rotational population distributions influenced by the sudden charge neutralization and concerted desorption; (3) spin-orbit distributions determined by symmetry considerations[30] for the non-adiabatic transition $^3\Sigma^-(NO^-) \rightarrow {}^2\Pi(NO)$; and (4) vibrational distributions strongly dependent on heating wavelength due to wavefunction overlap of $NO^{-\delta}(v')$ with $NO(v)$. The absence of $v = 1$ population for 1064 nm heating (compared to 532 nm) is a special case since the photon energy (27 kcal/mol) is insufficient to both elevate the adsorbed NO out of its 25 kcal/mole chemisorption well and provide it with 5.4 kcal/mol of vibrational energy. In this particular case, desorption must be driven by direct adsorbate interaction with hot electrons prior to any relaxation process.

4.7. Nonthermal comparisons

The foregoing experiments using visible or near-infrared lasers hoped to explore desorption dynamics within a framework characterized by a thermal desorption picture. There are two complementary studies that undertook to explore nonthermal desorptions.

Budde *et al.*[51] have recently observed the ultraviolet laser-induced desorption of NO from oxidized Ni(100). The 193 nm excitation wavelength used was resonant with gas phase NO transitions to a predissociative upper state. Desorption yields of NO from clean Ni(100) were essentially zero. Comparison of TPD results from clean and oxidized nickel surfaces indicated that an oxidized nickel surface could support a weakly bound NO state not found on clean Ni(100).

Two components were observed in the LID TOF spectra. The slower component was found to have a cosine angular flux distribution and an average kinetic energy ≈ 250 K, approximately the value of T_{max}. This component was also characterized by a rotational population distribution described by a rotational temperature $T_R = 170 \pm 20$ K. These results are comparable to the 'thermal' LID results for Pt(111), Pd(111) and Pt(foil) discussed in the previous sections.

A fast, nonthermal component was also reported. This desorption component had an angular flux distribution peaked along the surface normal. The kinetic energy distributions obtained from the TOF were strongly dependent on internal state; going from $\langle KE \rangle / 2k = 1010$ K for NO($v = 0$; F_1, low J) to values greater than 3000 K for NO($v = 0$; F_2, high J). The average kinetic energies for F_2 levels were consistently 15% higher than for F_1 levels of the same J or E_{int}. The reported internal state population distributions were essentially Boltzmann in both spin-orbit manifolds, decaying with a rotational temperature $T_R \approx 300$ K, which was independent of T_S, as shown in Fig. 12.

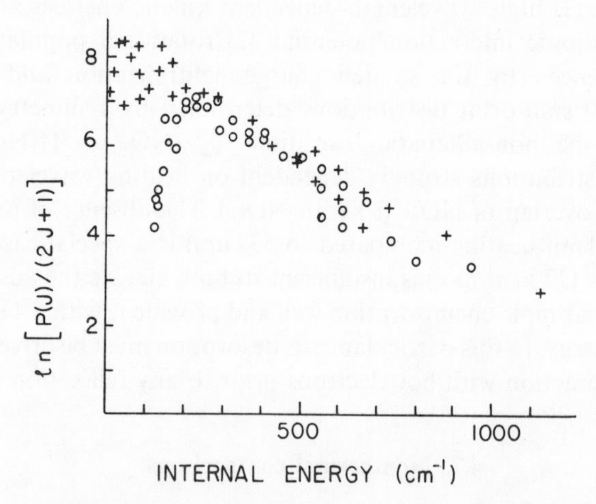

Fig. 12. Population distribution observed in the non-Boltzmann channel for the 193 nm laser-induced desorption of NO/Ni(100)–O. Populations in the higher (open circles) F_2 and lower energy (+) F_1 spin-orbit states are distinguished. (*Adapted from Ref. 51.*)

The relative population ratio F_1/F_2 was slightly higher than expected from a 300 K thermal distribution (e.g. 2.1 vs 1.8). Of particular note, in comparison to a simple Boltzmann distribution, there was a substantial absence of population in the $F_2(J < 5.5)$ levels from that expected based on a thermal (300 K) distribution. Approximately 1% of the desorbed molecules were vibrationally excited.

Budde *et al.* argued these results to be evidence of desorption from a repulsive potential based on the state-dependent kinetic energy and the dynamical depletion of population in F_2(low J) levels. This is a system that certainly has the potential for manifesting dynamics for competing adsorbate photochemical processes. Experiments on initially clean Ni(100) produced no LID signals until an oxide layer had been produced, presumably through the photodissociation of adsorbed NO. LID then proceeded from these pretreated surfaces. While photodissociation could generate a surface oxide, it might also lead to molecular desorption through caged recombination or electronic quenching processes. The available data does not identify the nature of the desorption process that result in the prompt LID component, and cannot distinguish between direct (adsorbate localized photoabsorption) versus indirect (substrate mediated) desorption mechanisms.

The visible and near-infrared LID results for NO/Pt were discussed in terms of hot electrons combined with a charge transfer mechanism. For the 193 nm LID result considered here, the photon energy is above the substrate work function, thereby providing a direct source of electrons to bathe the adsorbed NO species. Comparison of translational energy and vibrational state distributions for NO/Pt(111), NO/Pt(foil), and NO/Ni(100)–O suggests that the mechanisms driving the desorption processes in these systems might be related. However, the details of the specific interaction potentials must be substantially different to account for the disparate spin-orbit and rotational population distributions.

The results of the experiment of Burns *et al.* (BSJ)[52] on the electron stimulated desorption (ESD) of NO/Pt(111) provides a second point of comparison. A low energy ESD channel was observed for NO/Pt(111) with a threshold around 6 eV. The resulting, desorbed NO was characterized by MPI–TOF techniques to have a peak kinetic energy for NO($v = 0$) of 0.05 eV (580 K). There was a substantial amount of vibrationally excited NO desorbed, with relative populations $v = 0 : v = 1 : v = 2 : v = 3$ of 1.0:0.58:0.26:0.43. The rotational population distribution was Boltzmann in character with $T_R \approx 580$ K (somewhat dependent on vibrational level). At electron energies above 30 to 40 eV a second high energy channel was observed with the desorbed NO having average kinetic energies $\langle KE \rangle / 2k \approx .35$ eV (≈ 4000 K).

BSJ identified the 6 eV threshold for NO ESD with an excitation involving the molecular 5σ level. This 5σ orbital is not involved in the NO–metal bonding. The assignment was based on reference to ultraviolet photoelectron spectroscopy (UPS) for NO/Pt(111), UPS and inverse photoemission results for CO/Pt, and on excited state lifetime calculations (10 fs for the 5σ, much shorter for all other energetically possible excitations). Such a single particle excitation should make NO^- isoelectronic with O_2, which is a weakly bound, presumably unhindered rotor with the O—O bond parallel to the Pt(111) surface. Desorption might then progress in a two-step fashion. The NO adsorbate first goes from a strongly constrained binding geometry to a free rotor ($NO^{-\delta}$), with the second step involving a bound-continuum transition.

Burns' ESD results[53] for NO/Ni(poly) perhaps more closely resembled the LID results discussed above. Although the major emphasis of this work was for electron energies of 300–600 eV, large neutral yields were obtained from a low energy resonance around 9 eV, in agreement with ultraviolet photoelectron peaks observed for NO/Ni(111) at an 8.8 eV binding energy. In these experiments a substantially lower fraction of the desorbed NO was vibrationally excited (2–3%) than observed on Pt(111). The desorbing NO also had a higher average kinetic energy $\langle KE \rangle / 2k \approx 1200\,K$ (.1 eV), both results being very similar to the LID results on Pt(111) and Ni(100)–O. (The resolution of these ESD experiments didn't allow full exploration of the details of the rotational and spin-orbit state distributions.)

The observations of complex dynamics associated with electron-stimulated desorption or desorption driven by resonant excitation to repulsive electronic states are not unexpected. Their similarity to the dynamics observed in the visible and near-infrared LID illustrate the need for a closer investigation of the physical relaxation mechanisms of low energy electron/hole pairs in metals. When the time frame for reaction has been compressed to that of the 10^{-9} s laser pulse, many thermal processes will not effectively compete with the effects of transient low energy electrons or nonthermal phonons. It is these relaxation channels which might both play an important role in the physical or chemical processes driven by laser irradiation of surfaces, and provide dramatic insight into subtle details of molecule–surface dynamics.

5. SUMMARY

Our understanding of the dynamics of molecular interactions with surfaces has been substantially enriched through the application of state-specific diagnostics over the past several years. A number of issues related to the manner in which a molecule accommodates energy with the thermal bath of a solid have been explored, and differences between direct scattering and thermal equilibration are now better understood. Laser-based diagnostics have illuminated the complex response of adsorbate-covered surfaces to irradiation, clearly implicating nonthermal dynamics. This latter area is far from being understood, due to the diversity of energy transfer mechanisms which may be optically activated. There is still a broad range of desorption processes which remain to be investigated. Further studies of simple systems, and the extension to more complex desorbing species will certainly provide testing grounds for established ideas and new theoretical models of dynamical processes at surfaces.

6. APPENDIX

THERMAL DESORPTION

NO/Ru(001)	$T_R \ll T_S$, dynamical	R. R. Cavanagh and D. S. King, *Phys. Rev. Lett.*, **47**, 1829 (1981).
	KE dependent on J	D. S. King and R. R. Cavanagh, *J. Chem. Phys.*, **76**, 5634 (1982).
NO/Pt(111)	$T_R \approx T_S$	D. S. King, D. A. Mantell and R. R. Cavanagh, *J. Chem. Phys.*, **82**, 1046 (1985).
NO/CO/Pt(111)	Co-adsorbates $T_R \approx T_S$	D. A. Mantell, R. R. Cavanagh and D. S. King, *J. Chem. Phys.*, **84**, 5131 (1986).
$NO/NH_3/Pt(111)$	Co-adsorption Reaction rate limited	D. R. Burgess, Jr., R. R. Cavanagh and D. S. King, Surf. Sci. (in press)
		D. Burgess, Jr., D. S. King and R. R. Cavanagh, *J. Vac. Sci. Tech A*, **5**, 2959 (1987).

OXIDATION REACTION

OH/Pt(poly)	$O_2 + H_2$; high T_S $T_R \approx T_S$	M. A. Hoffbauer, D. S. Y. Hsu and M. C. Lin, *J. Chem. Phys.*, **84**, 532 (1986).
		D. S. Y. Hsu, M. A. Hoffbauer and M. C. Lin, *Surf. Sci.*, **184**, 25 (1987).

NO/Pt(111)	$O_2 + D_2$; high T_S $T_R \approx T_S$	D. S. Y. Hsu and M. C. Lin, *J. Chem. Phys.*, **88**, 432 (1988).
	$NH_3 + O_2$; $T_S = 800\,K$ $T_R < T_S$	M. Asscher, W. L. Guthrie, T.-H. Lin and G. A. Somorjai, *J. Chem. Phys.*, **80**, 3233 (1984).
	$NH_3 + O_2$ $800 < T_S(K) < 1300$ $T_R^{lim} \approx 400\,K$	D. S. Y. Hsu, D. W. Squire and M. C. Lin, *J. Chem. Phys.* **89**, 2861 (1988)

INTERNAL-STATE RESOLVED STUDIES OF RECOMBINATION DYNAMICS

D_2/Pd(poly)		L. Schröter, H. Zacharias and R. David, *App. Phys.* A, **41**, 95 (1986).
H_2/Pd(poly)		H. Zacharias and R. David, *Chem. Phys. Lett.*, **115**, 205 (1985).
D_2/Pd(100)		L. Schröter, G. Ahlers, H. Zacharias and R. David, *J. Electron Spectros. Rel. Phenom.*, **39**, 403 (1987).
H_2, D_2/Cu(110), Cu(111) + S covered		G. D. Kubiak, G. O. Sitz and R. N. Zare, *J. Vac. Sci. Tech.* A, **3**, 1649 (1985); *J. Chem. Phys.*, **81**, 6397 (1984); *J. Chem. Phys.*, **83**, 2538 (1985).
N_2/Fe		R. P. Thorman and S. L. Bernasek, *J. Chem. Phys.*, **74**, 6498 (1981).

TRAPPING DESORPTION

NO/Pt(111)		D. C. Jacobs, K. W. Kolasinski, R. J.

	Madix and R. N. Zare, *J. Chem. Phys.*, **87**, 5038 (1987).
NH_3/W(110)	B. D. Kay and T. D. Raymond, *J. Chem. Phys.*, **85**, 4140 (1986).
NO/Ir(111)	R. J. Hamers, P. L. Houston and R. P. Merrill, *J. Chem. Phys.*, **83**, 6045 (1985).

DIRECT INELASTIC SCATTERING

NO/Ag(111)	G. M. McClelland, G. D. Kubiak, H. G. Rennagal and R. N. Zare, *Phys. Rev. Lett.*, **46**, 831 (1981).
	G. D. Kubiak, J. E. Hurst, Jr., H. G. Rennagal, G. M. McClelland and R. N. Zare, *J. Chem. Phys.*, **79**, 5163 (1983).
	A. W. Kleyn, A. C. Luntz and D. J. Auerbach, *Phys. Rev. Lett.*, **47**, 1169 (1981); *J. Chem. Phys.*, **76**, 737 (1982); *Phys. Rev.* B, **25**, 4273 (1982); *Surf. Sci.*, **117**, 3 (1982); *Surf. Sci.*, **152/153**, 99 (1985).
	E. W. Kuipers, M. G. Tenner, M. E. M. Spruit and A. W. Kleyn, *Surf. Sci.*, **189/190**, 669 (1987).
N_2/Ag(111)	G. O. Sitz, A. C. Kummel and R. N. Zare, *J. Chem. Phys.* **87**, 3247 (1987); *J.*

	Vac. Sci. Tech A, **5**, 513 (1987).
NH_3/Au(110)	B. D. Kay, T. D. Raymond and M. E. Coltrin, *Phys. Rev.* B, **36**, 6695 (1987).
	B. D. Kay and T. D. Raymond, *J. Chem. Phys.*, **85**, 4140 (1986).
NO/Pt(111)	F. Frenkel, J. Häger, W. Krieger, H. Walther, C. T. Campbell, G. Ertl, H. Kuipers and J. Segner, *Phys. Rev. Lett.*, **46**, 152 (1981).
	F. Frenkel, J. Häger, W. Krieger, H. Walther, G. Ertl, J. Segner and W. Veilhaber, *Chem. Phys. Lett.*, **90**, 225 (1982).
	J. Segner, W. Veilhaber and G. Ertl, *Isr. J. Chem.*, **22**, 375 (1982).
	M. Asscher, W. L. Guthrie, T.-H. Lin and G. A. Somorjai, *Phys. Rev. Lett.*, **47**, 76 (1982); *J. Chem. Phys.*, **78**, 6992 (1983).
NO/Ge(111)	A. Mödl, H. Robota, J. Segner, W. Veilhaber, M. C. Lin and G. Ertl, *J. Chem. Phys.*, **83**, 4800 (1985); *Surf. Sci.*, **169**, L341 (1986).
NO/Cu–O	J. S. Hayden and G. J. Diebold, *J. Chem. Phys.*, **77**, 4767 (1982).
NO/C	J. Segner, H. Robota, W. Vielhaber, G. Ertl, F. Frenkel, J. Häger, W.

	Kreiger and H. Walther, *Surf. Sci.*, **131**, 273 (1983); *Chem. Phys. Lett.*, **90**, 225 (1982).

VIBRATIONAL EXCITATION/DEACTIVATION

NO/Ag(111)	C. T. Rettner, F. Fabre, J. Kimman and D. J. Auerbach, *Phys. Rev. Lett.*, **55**, 1904 (1985).
	C. T. Rettner, F. Fabre, D. J. Auerbach and H. Morawitz, *Surf. Sci.*, **192**, 107 (1987).
	J. Misewich, P. A. Roland and M. M. T. Loy, *Surf. Sci.*, **171**, 483 (1986).
NO($v = 1$)/Ag(111) and NO($v = 1$)/Ag(110)	J. Misewich and M. M. T. Loy, *J. Chem. Phys.*, **84**, 1939 (1986).
NO($v = 1$)/LiF(100)	J. Misewich and M. M. T. Loy, *J. Chem. Phys.*, **84**, 1939 (1986); *Phys. Rev. Lett.*, **55**, 1919 (1985).
	H. Zacharias, M. M. T. Loy and P. A. Roland, *Phys. Rev. Lett.*, **49**, 1750 (1982).
	J. Misewich, H. Zacharias and M. M. T. Loy, *J. Vac. Sci. Tech.* B, **3**, 1474 (1985).
NO($v = 1$)/graphite	H. Vach, J. Häger and H. Walther, *Chem. Phys. Lett.*, **133**, 279 (1987).

	STIMULATED DESORPTION	
NO/Pt(111)	Visible and near infrared excitation. Two channels. Strong spin-orbit effect.	S. A. Buntin, L. J. Richter, R. R. Cavanagh and D. S. King, *Phys. Rev. Lett.* **61**, 1321 (1988) *Chem. Phys.* (in press).
NO/Pt(poly)	Visible and near infrared excitation. Two channels.	D. R. Burgess, Jr., R. R. Cavanagh, D. S. King, *J. Chem. Phys.* **88**, 6556 (1988).
		D. R. Burgess, Jr., D. A. Mantell, R. R. Cavanagh and D. S. King, *J. Chem. Phys.*, **85**, 3123 (1986).
NO/Ni(100)–O	Ultraviolet excitation. Two channels.	D. Weide, P. Andresen and H.-J. Freund, *Chem. Phys. Lett.*, **136**, 106 (1987).
	Dynamical under-population in F_2 (low J) states	F. Budde, A. V. Hamza, P. M. Ferm, G. Ertl, D. Weide, P. Andresen and H.-J. Freund, *Phys. Rev. Lett.* **60**, 1518 (1988).
NO/Ni(poly)	Electron (300–600 eV) stimulated desorption	A. R. Burns, *Phys. Rev. Lett.*, **55**, 525 (1985).
NO/Pt(111)	Low energy electrons, 6 eV threshold	A. R. Burns, E. B. Stechel and D. R. Jennison, *Phys. Rev. Lett.*, **58**, 250 (1987).
NO/Pd(111)		J. Prybyla, T. F. Hines, J. A. Misewich and M. M. T. Loy (to be published).

Acknowledgements

We wish to acknowledge the contributions of our collaborators in this work: S. A. Buntin, D. R. Burgess, Jr., M. P. Casassa, J. W. Gadzuk, E. J. Heilweil, D. A. Mantell, L. J. Richter, and J. C. Stephenson. This work was supported in part by the US Department of Energy, Chemical Sciences Division (D.E.-AIO5-84ER13150).

References

1. Levine, R. D., and Bernstein, R. B., *Molecular Reaction Dynamics and Chemical Reactivity*, Oxford Univ. Press, 1987.
2. Ashfold, M. N. R., and Baggott, J. E., (Eds.), *Molecular Photodissociation Dynamics*, Royal Society of Chemistry, London, 1987.
3. (a) Barker, J. A., and Auerbach, D. J., *Surf. Sci. Rep.*, **4**, 1 (1985).
   (b) Gerber, R., *Chem. Rev.*, **87**, 29 (1987).
   (c) Kasai, H., Brenig, W., and Müller, H., *Z. Phys.*, **B60**, 489 (1985).
4. Houston, P. L., and Merrill, R. P., *Chem. Rev.*, **88**, 657 (1988).
5. (a) Luntz, A. C., *Physica Scripta*, **35**, 193 (1987).
   (b) Lin, M. C., and Ertl, G., *Ann. Rev. Phys. Chem.*, **37**, 587 (1986).
   (c) Häger, J., Shen, Y. R., and Walther, H., *Phys. Rev.*, **A31**, 1962 (1985).
   (d) Comsa, G., and David, R., *Surf. Sci. Reports*, **5**, 145 (1985).
6. (a) Menzel, D., in *Chemistry and Physics of Solid Surfaces IV* (Ed. R. Vaneslow and R. Howe), Springer, New York, 1982, p. 101.
   (b) Redhead, P. A., *Vacuum*, **12**, 203 (1962).
   (c) Brenig, W., *Physica Scripta*, **35**, 329 (1987).
7. (a) Becker, C. A., Cowin, J. P., Wharton, L., and Auerbach, D. J., *J. Chem. Phys.*, **67**, 3394 (1977).
   (b) Wedler, G., and Ruhmann, H., *Surf. Sci.*, **121**, 464 (1982).
   (c) Burgess, D. R., Jr., Viswanathan, R., Hussla, I., Stair, P. C., and Weitz, E., *J. Chem. Phys.*, **79**, 5200 (1983).
   (d)Hall, R. B., *J. Phys. Chem.*, **91** 1007 (1987).
8. King, D. S., and Cavanagh, R. R., in *Chemistry and Structures at Interfaces; New Laser and Optical Techniques* (Eds. R. B. Hall and A. B. Ellis), VCH Publishers, 1986 p. 25.
9. (a) Greene, C. H., and Zare, R. N., *Annual Rev. Phys. Chem.*, **33**, 119 (1982).
   (b) Greene, C. H., and Zare, R. N., *J. Chem. Phys.*, **78**, 6741 (1983).
10. (a) King, D. S., Mantell, D. A., and Cavanagh, R. R., *J. Chem. Phys.*, **82**, 1046 (1985).
    (b) Mantell, D. A., Cavanagh, R. R., and King, D. S., *J. Chem. Phys.*, **84**, 5131 (1986).
    (c) Burgess, D., Jr., Mantell, D. A., Cavanagh, R. R., and King, D. S., *J. Chem. Phys.*, **85**, 3123 (1986).
11. (a) Gorte, R. J., and Schmidt, L. D., *Surf. Sci.*, **111**, 260 (1981).
    (b) Gorte, R. J., Schmidt, L. D., and Gland, J. L., *Surf. Sci.*, **109**, 367 (1981).
    (c) Banholzer, W. F., and Masel, R. I., *Surf. Sci.*, **137**, 339 (1984).
12. Ibach, H., and Lehwald, S., *Surf. Sci.*, **76**, 1 (1978).
13. Hayden, B. E., *Surf. Sci.*, **131**, 419 (1983).
14. (a) Burgess, D. R., Jr., King, D. S., and Cavanagh, R. R., *J. Vac. Sci. Tech.* **A 5**, 2959 (1987).
    (b) Burgess, D. R., Jr., Cavanagh, R. R., and King, D. S., *J. Chem. Phys.*, **88**, 6556 (1988).
15. Morrison Roy, S., *The Chemical Physics of Surfaces*, Plenum Press, 1977.
16. Serri, J. A., Tully, J. C., and Cardillo, M. J., *J. Chem. Phys.*, **79**, 1530 (1983).
17. Campbell, C. T., Ertl, G., and Segner, J., *Surf. Sci.*, **115**, 309 (1982).
18. Lin, T.-H., and Somorjai, G. A., *Surf. Sci.*, **107**, 573 (1981).
    Guthrie, W. L., Lin, T.-H. Ceyer, S. T., and Somorjai, G. A., *J. Chem. Phys.*, **76**, 6398 (1982).
19. Serri, J. A., Cardillo, M. J., and Becker, G. E., *J. Chem. Phys.*, **77**, 2175 (1982).
20. (a) Pechukas, P., and Light, J. C., *J. Chem. Phys.*, **42**, 3281 (1965).
    (b) Pechukas, P., Light, J. C., and Rankin, C., *J. Chem. Phys.*, **44**, 794 (1966).

21. (a) Muhlhausen, C. W., Williams, L. R., and Tully, J. C., *J. Chem. Phys.*, **83**, 2594 (1985).
(b) Tully, J. C., in *Kinetics of Interface Reactions* (Eds. M. Grunze and H. J. Kreuzer), Springer, Berlin, 1987, p. 37.
22. (a) Gorte, R. J., and Gland, J. L., *Surf. Sci.*, **102**, 348 (1981).
(b) Park, Y. O., Masel, R. I., and Stolt, K., *Surf. Sci.*, **131**, L385 (1983).
23. (a) Gland, J. L., and Sexton, B. A., *J. Catal.*, **68**, 286 (1981).
(b) Gland, J. L., and Sexton, B. A., *Surface Sci.*, **94**, 355 (1980).
24. Segner, J., Robota, H., Vielhaber, W., Ertl, G., Frenkel, F., Häger, J., Krieger, W., and Walther, H., *Surf. Sci.*, **131**, 272 (1983).
25. (a) Asscher, M., Guthrie, W. L., Lin, T.-H., and Somarjai, G. A., *Phys. Rev. Lett.*, **49**, 76 (1982).
(b) Asscher, M., Guthrie, W. L., Lin, T.-H., and Somarjai, G. A., *J. Chem. Phys.*, **78**, 6992 (1983).
26. Asscher, M., Guthrie, W. L., Lin, T.-H., and Somorjai, G. A., *J. Phys. Chem.*, **88**, 3233 (1984).
27. Hsu, D. S. Y., Squire, D. W., and Lin, M. C., *J. Chem. Phys.*, **89**, 2861 (1988).
28. Cavanagh, R. R., and King, D. S., *Phys. Rev. Lett.*, **47**, 1829 (1981).
29. Jacobs, D. C., Kolasinski, K. W., Madix, R. J., and Zare, R. N., *J. Chem. Phys.*, **87**, 5038 (1987).
30. (a) Corey, G. C., Smedley, J., Alexander, M. H., and Liu, W.-K., *Surf. Sci.*, **191**, 203 (1987).
(b) Smedley, J., Corey, G. C., and Alexander, M. H., *J. Chem. Phys.*, **87**, 3218 (1987).
(c) Alexander, M. H., *J. Chem. Phys.*, **80**, 3485 (1984).
31. Sitz, G. O., Kummel, A. C., and Zare, R. N., *J. Chem. Phys.*, **87**, 3247 (1987).
32. Kay, B. D., Raymond, T. D., and Coltrin, M. E., *Phys. Rev.* B, **36**, 6695 (1987).
33. Kay, B. D., and Raymond, T. D., *J. Chem. Phys.*, **85**, 4140 (1986).
34. Polyani, J. C., and Wolf, R. J., *Ber. Bunsenges. Phys. Chem.*, **72**, 356 (1982).
35. (a) George, S. M., DeSantolo, A. M., and Hall, R. B., *Surf. Sci.*, **159**, L425 (1985).
(b) Hall, R. B., and DeSantolo, A. M., *Surf. Sci.*, **137**, 421 (1984).
36. (a) Deckert, A. A., and George, S. M., *Surf. Sci.*, **182**, L215 (1987).
(b) Hall, R. B., *J. Phys. Chem.*, **91**, 1007 (1987).
37. (a) Schoenlein, R. W., Lin, W. Z., Fujimoto, J. G., and Eesley, G. L., *Phys. Rev. Lett.*, **58**, 1680 (1987).
(b) Eesley, G. E., *Phys. Rev. Lett.*, **51**, 2140 (1983).
(c) Elsayed-Ali, H., Pessot, M., Norris, T., and Mourou, G., in *Ultrafast Phenomena, V5* (Eds. G. R. Fleming and A. E. Siegman), Springer-Verlag, New York, 1986, p. 264.
38. (a) Ready, J. F., *Effects of High Power Laser Radiation*, Academic, New York, 1971.
(b) Bechtel, J. H., *J. Appl. Phys.*, **46**, 1585 (1975).
(c) Burgess, D., Jr., Stair, P. C., and Weitz, E., *J. Vac. Sci. Tech.*, **A4**, 1362 (1986).
39. Lucchese, R. R., and Tully, J. C., *J. Chem. Phys.*, **81**, 6313 (1984).
40. Zare, R. N., and Levine, R. D., *Chem. Phys. Lett.*, **136**, 593 (1987).
41. (a) Schäfer, B., and Hess, P., *Chem. Phys. Lett.*, **105**, 563 (1984).
(b) Schäfer, B., and Hess, P., *Appl. Phys.*, B, **38**, 3058 (1985).
42. Ferm, P. M., Kurtz, S. R., Pearlstein, K. A., and McClelland, G. M., *Phys. Rev. Lett.*, **58**, 2602 (1987).
43. Ying, Z., and Ho, W., *Phys. Rev. Lett.*, **60**, 57 (1988).
44. Marsh, E. P., Tabares, F. L., Schneider, M. R., and Cowin, J. P., *J. Vac. Sci. Tech.* A, **5**, 519 (1987).
45. Domen, K., and Chuang, T. J., *Phys. Rev. Lett.*, **59**, 1484 (1987).

46. Burgess, D. R., Jr., Cavanagh, R. R., and King, D. S., *Surf. Sci.*, (in press).
47. Prybyla, J., Hines, T. F., Misewich, J. A., and Loy M. M. T., (to be published).
48. Buntin, S. A., Richter, L. J., Cavanagh, R. R., and King, D. S. *Phys. Rev. Lett.*, **61**, 1321 (1988); Richter, L. J., Buntin, S. A., King, D. A. and Cavanagh, R. R., *J. Chem. Phys.*, **89**, 5344 (1988).
49. Schultz, G. J., *Rev. Mod. Phys.*, **45**, 423 (1973).
50. (a) Gadzuk, J. W., *Comments At. Mol. Phys.*, **16**, 219 (1985).
(b) Gadzuk, J. W., and Nørskov, J. K., *J. Chem. Phys.*, **81**, 2828 (1984).
(c) Gadzuk, J. W:, and Holloway, S., *Phys. Rev.*, B, **33**, 4298 (1986).
(d) Newns, D., *Surf. Sci.*, **171**, 600 (1985).
(e) Antoniewicz, P. R., (to be published).
51. Budde, F., Hamza, A. V., Ferm, P. M., Ertl, G., Weide, D., Andresen P., and Freund, H.-J., *Phys. Rev. Lett.*, **60**, 1518 (1988).
52. Burns, A. R., Stechel, E. B., and Jennison, D. R., *Phys. Rev. Lett.*, **58**, 250 (1987).
53. Burns, A. R., *Phys. Rev. Lett.*, **55**, 325 (1985).

Advances in Chemical Physics
Edited by K. P. Lawley

MONTE CARLO CALCULATIONS ON PHASE TRANSITIONS IN ADSORBED LAYERS

K. BINDER

Institut für Physik, Johannes-Gutenberg-Universität-Mainz, D-6500 Mainz, Postfach 3980, Federal Republic of Germany

and

D. P. LANDAU

Center for Simulational Physics, The University of Georgia, Athens, GA 30602, USA

CONTENTS

Abstract

We review Monte Carlo calculations of phase transitions and ordering behavior in lattice gas models of adsorbed layers on surfaces. The technical aspects of Monte Carlo methods are briefly summarized and results for a wide variety of models are described. Included are calculations of internal energies and order parameters for these models as a function of temperature and coverage along with adsorption isotherms and dynamic quantities such as self–diffusion constants. We also show results which are applicable to the interpretation of experimental data on physical systems such as H on Pd(100) and H on Fe(110). Other studies which are presented address fundamental theoretical questions about the nature of phase transitions in a two-dimensional geometry such as the existence of Kosterlitz–Thouless transitions or the nature of dynamic critical exponents. Lastly, we briefly mention multilayer adsorption and wetting phenomena and touch on the kinetics of domain growth at surfaces.

1. INTRODUCTION: WHAT QUESTIONS CAN BE ANSWERED BY MONTE CARLO SIMULATIONS?

Since the first study of the equation of state of a two-dimensional hard disk model fluid[1], the Monte Carlo simulation of interacting many-body systems has become a standard tool of statistical physics and physical chemistry[2–4]. Starting from some model Hamiltonian, many configurations of the system under study are generated. In a simulation at thermal equilibrium, these configurations are generated with a probability according to their Boltzmann statistical weight; hence, estimates for the thermal averages of any observables are readily obtained from arithmetic averages over the values of these quantities in the generated configurations. Given enough statistical effort (i.e. enough computer time), these estimates can be made as accurate as desired, at least in principle.

Why does one often resort to a Monte Carlo simulation, rather than performing an analytic calculation for the considered model? The reason is that for most models of statistical mechanics thermal averages cannot be calculated exactly, but rather drastic approximations are required, which are often known to be fairly inaccurate. This is a particular cumbersome problem in surface physics: in the two-dimensional geometry of adsorbed layers, statistical fluctuations turn out to be much more important than for bulk three-dimensional systems, and thus simple closed-form approximations such as the mean field theory or its variants fail rather badly. Not only are these theories very inaccurate, as far as critical exponents for second-order phase transitions are concerned, but also their predictions on the phase diagrams of the studied models often are incorrect (e.g. Binder and Landau[5]). This inadequacy of some approximate theory is not always apparent when one compares it to experiment: the model contains adjustable parameters (e.g. the strengths of various interaction constants), and fitting an inaccurate theory to experimental data its inadequacy may be obscured by systematically wrong values for the fitted constants. Thus, often theories can be much more stringently tested by a simulation, since the latter can use precisely the same model assumptions as the analytic work with no adjustable parameters whatsoever entering the comparison. On the other hand, one also may compare directly simulation results to experimental data, and hence check whether a particular model faithfully describes the desired properties of the real material. And last but not least, the simulation gives insight, in arbitrary detail, into the system at a truly microscopic level: e.g. looking at typical instantaneous configurations of the simulated system may elucidate conjectures made by an analytic theory.

In this chapter, we shall give examples illustrating this interplay between theory, simulation, and experiment: both static (section 3) and dynamic (section 4.1) critical phenomena in adsorbed monolayers will be treated, and other system properties of interest, such as adsorption isotherms, superstructure scattering intensities, and phase diagrams (section 3). Nonequilibrium problems, such as surface diffusion (section 4.2) and growth of ordered domains, will also be addressed (section 4.3). While some earlier reviews exist[6–11], the field has been growing rapidly, and an exhaustive description would fill an entire book. Rather than aiming at completeness, we shall instead give a few examples showing how such simulations are done and evaluated. For simplicity only, these examples are taken mainly from the authors' own research – we apologize to many colleagues whose contributions are equally relevant but are only mentioned occasionally in this review.

In the next section, we give some details on typical models that are studied, or problems that arise (correlations between subsequently generated configurations, finite size effects, self-averaging of observables, etc.) and how they are overcome (more details can be found in Refs. 2–4).

2. TECHNICAL BACKGROUND

2.1. Models: lattice gas models, Potts models, clock models, etc.

One might expect that a natural model for an adsorbed monolayer at a surface would be to let the adsorbate atoms interact via suitable pairwise potentials (e.g. Lennard Jones forces if adsorption of rare gas atoms on grafoil (exfoliated graphite) is considered[12–15]), the effect of the substrate being described by a corrugation potential (which has the symmetry and periodicity of the surface onto which the adatoms are adsorbed). Although such an approach is already a serious simplification (the substrate dynamics, e.g. phonons, are neglected; adatoms may diffuse into the substrate or evaporate from it into the 'vacuum'; in many cases – e.g. chemisorbed layers[16] – the treatment should be quantum-mechanical rather than classical, etc.), we shall simplify matters further by assuming that the wells of the corrugation potential are rather steep and high. The locations of the minima of the corrugation potential thus form a well-defined lattice, at which the occupation probability density of adatoms is sharply peaked. Then we may neglect deviations of the adatom positions from the sites of this 'preferred lattice' altogether, introducing the *lattice gas model* which has as single degree of freedom an occupation variable c_i for lattice site i, with $c_i = 1$ if at site i there is an adatom, $c_i = 0$ if site i is empty (multiple occupancy of the lattice sites being forbidden). The coverage θ of the monolayer then is given by a thermal average $\langle \cdots \rangle_T$ summed over all N available lattice sites

$$\theta = (1/N) \sum_{i=1}^{N} \langle c_i \rangle_T \tag{1}$$

Each adatom is bound to the substrate with the binding energy ε (assumed here not to depend on temperature T and coverage). In addition, we assume energies φ_{nn}, φ_{nnn} and φ_3 are won if adatoms occupy nearest, next nearest and third nearest neighbor sites (see Fig. 1 for examples). As an example for non-pairwise interactions, we assume that the energy of any triangle of occupied sites is given by $-2\varphi_{nn} - \varphi_{nnn} - \varphi_t$, φ_t being a three-body interaction term. Again these interaction parameters are assumed to be independent of temperature and coverage. The total (configurational) energy of the system is

$$\mathscr{H} = -\varepsilon \sum_{i=1}^{N} c_i - \sum_{i \neq j} \varphi_{ij} c_i c_j - \sum_{i \neq j \neq k} \varphi_t c_i c_j c_k. \tag{2}$$

Here, $\varphi_{ij} = \varphi_{nn}, \varphi_{nnn}, \varphi_3$, when i, j are nearest, next nearest and third nearest neighbors, respectively; the second sum in Eq. (2) runs over all these pairs once, while the third sum runs over all nearest neighbor triangles once. Of course, one can consider lattice gases with other choices of the interactions or other lattices (e.g. Refs. 9, 17,18). We treat the cases of Fig. 1 just to have explicit examples, and furthermore these models have been suggested to describe

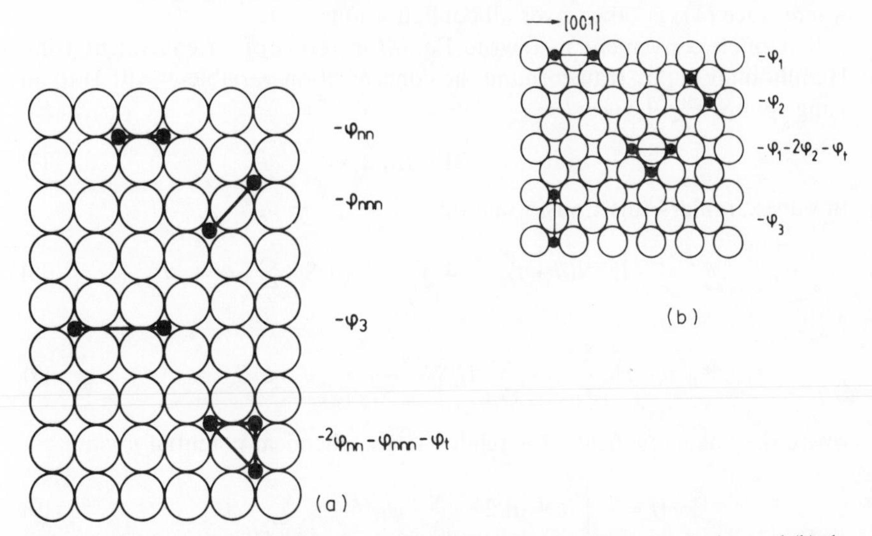

Fig. 1. Interaction energies φ between adsorbate atoms on (a) the square lattice, and (b) the centered rectangular lattice. These figures are for the purpose of qualitative illustration only and no assumptions are made about the actual geometry of adsorption sites on the substrate surface. (*From Binder et al.*[8].)

chemisorption systems such as H on Pd(100)[19,20] and H on Fe(110)[21,22]. The average adsorption energy per lattice site U_{ads} for Eq. (2) is

$$U_{ads} = \langle \mathscr{H} \rangle_T / N = -\varepsilon\theta - \frac{1}{N}\sum_{i \neq j} \varphi_{ij} \langle c_i c_j \rangle_T - \frac{1}{N}\sum_{i \neq j \neq k} \varphi_t \langle c_i c_j c_k \rangle_T \tag{3}$$

Eqs. (2), (3) are appropriate for the canonical ensemble, with T and θ the fixed independent variables. However, if the adsorbed layer is in thermal equilibrium with an external adsorbate gas reservoir, it is the pressure of this gas which is controlled experimentally, rather than θ. Since the chemical potential of the gas (μ_{gas}) equals the chemical potential μ of the layer, μ rather than θ is the proper independent variable. It then is preferable to use the grand-canonical ensemble, subtracting a term $\mu N_a (N_a = \sum_i c_i =$ number of adatoms in the layer) from Eq. (2), yielding a Hamiltonian

$$\mathscr{H}' = \mathscr{H} - \mu N_a = -(\varepsilon + \mu)\sum_{i=1}^{N} c_i - \sum_{i \neq j} \varphi_{ij} c_i c_j - \sum_{i \neq j \neq k} \varphi_t c_i c_j c_k \tag{4}$$

The relation between θ and μ at constant T is the 'adsorption isotherm',

$$\mu / k_B T = [1/(N k_B T)](\partial F / \partial \theta)_T \tag{5}$$

F being the free energy of the system,

$$F = -k_B T \ln \operatorname{Tr} \exp[-\mathscr{H}\{c_i\}/k_B T] \quad (c_i = 0, 1) \tag{6}$$

where trace (Tr) is taken over all configurations.

It is often convenient to rewrite Eq. (4) in terms of an equivalent Ising Hamiltonian $\mathscr{H}_{\text{I}}$ by transforming the concentration variable $c_i = (0, 1)$ to an Ising spin $S_i = \pm 1$ via

$$c_i = (1 - S_i)/2 \tag{7}$$

In our example[20] this transformation yields

$$\mathscr{H}' = -1/2N(\mu + \varepsilon) - 1/4 \sum_{i \neq j} \varphi_{ij} - 1/8 \sum_{i \neq j \neq k} \varphi_{\text{t}} + \mathscr{H}_{\text{I}} \tag{8}$$

with

$$\mathscr{H}_{\text{I}} = -H \sum_{i=1}^{N} S_i - \sum_{i \neq j} J_{ij} S_i S_j - \sum_{i \neq j \neq k} J_{\text{t}} S_i S_j S_k \tag{9}$$

where the 'magnetic field' H is related to the chemical potential μ via

$$H = -\left[(\varepsilon + \mu)/2 + \sum_{j(\neq i)} \varphi_{ij}/4 + \sum_{j \neq k(\neq i)} \varphi_{\text{t}} \right] \tag{10}$$

and the effective two- and three-spin exchange constants J_{ij}, J_{t} are given by

$$J_{ij} = \varphi_{ij}/4 + \sum_{k(\neq i, j)} \varphi_{\text{t}}, \qquad J_{\text{t}} = -\varphi_{\text{t}}/8. \tag{11}$$

The coverage θ then is simply related to the magnetization m of the Ising magnet,

$$\theta = (1 - \langle m \rangle_{\text{T}})/2, \qquad \langle m \rangle_{\text{T}} = (1/N) \sum_{i=1}^{N} \langle S_i \rangle_{\text{T}} \tag{12}$$

The transformation to the (generalized) Ising model is useful since it clearly brings out the symmetries of the problem: Eq. (9) is invariant under the transformation

$$H, J_{\text{t}}, \{S_i\} \rightarrow -H, -J_{\text{t}}, \{-S_i\} \tag{13}$$

which transforms (via Eq. (12)) θ into $1 - \theta$. Thus the phase diagrams for positive and negative values of J_{t} are related: the phase diagram for $-J_{\text{t}}$ is obtained from that for $+J_{\text{t}}$ by taking its mirror image around the axis $\theta = 1/2$. If $J_{\text{t}} = 0$ (pairwise interaction model), the phase diagram is symmetric in the (T, θ) phase around the line $\theta = 1/2$, and the adsorption isotherm, Eq. (5), is antisymmetric around the point $\theta = 1/2$, $\mu = \mu_c$, where μ_c is the chemical potential corresponding to $H = 0$ (Eq. (10)).

In the noninteracting case (or 'infinite' temperature) the model is analyzed very simply: $\langle c_i c_j \rangle = \theta^2$, $\langle c_i c_j c_j \rangle = \theta^3$, and thus

$$U_{\text{ads}}(T \rightarrow \infty) = -\varepsilon\theta - 1/2\,\theta^2 \sum_{i(\neq j)} \varphi_{ij} - 1/3\,\theta^3 \sum_{i \neq j(\neq k)} \varphi_{\text{t}} \tag{14}$$

Since in this limit the 'magnetization process' of the Ising model is just given by the Brillouin function ($h \equiv H/k_BT$ remaining finite)

$$m = \tanh h, \qquad h = 1/2 \ln[(1+m)/(1-m)] = 1/2 \ln[(1-\theta)/\theta] \tag{15}$$

one obtains Langmuir's isotherm[15] from Eqs. (5), (15):

$$(\mu + \varepsilon)/k_BT = \ln[\theta/(1-\theta)] \tag{16}$$

Note that these results are general and hold for arbitrary lattice structure, range of interactions, etc., and thus are not restricted to the cases shown in Fig. 1.

In order to give an example at this stage, Fig. 2 shows adsorption isotherms calculated for the lattice gas with nearest neighbor attractive interaction (a), with nearest neighbor repulsion but next nearest neighbor attraction of the same strength (b), and the nearest neighbor model with non-additive interactions (c)[23]. In the latter model the strength of a bond φ_{ij} between an atom and any of its neighbors depends linearly on the number m of its neighbors, $\varphi_{ij} = \varphi[Pz - 1 - m(P-1)]/(z-1)$, where z is the coordination number and P a parameter describing the nonadditivity of the interactions (this model is equivalent to Eq. (2) if one allows for several kinds of three-body interactions). Figure 3 shows a few examples for the coverage dependence of the ordering energies. In these examples, the ground state behavior at $T = 0$ is found directly from the Hamiltonian, by working out the energy for regular structures such as the completely empty ($\theta = 0$) or completely filled ($\theta = 1$) lattice gas, or regular orderings such as the $c(2 \times 2)$ structure ($\theta = 1/2$). The straight lines connecting these particular states in Fig. 3 represent two-phase coexistence regions. Adsorption isotherms (Fig. 2) and adsorption energies (Fig. 3) at finite nonzero temperatures have been calculated by Monte Carlo simulations.

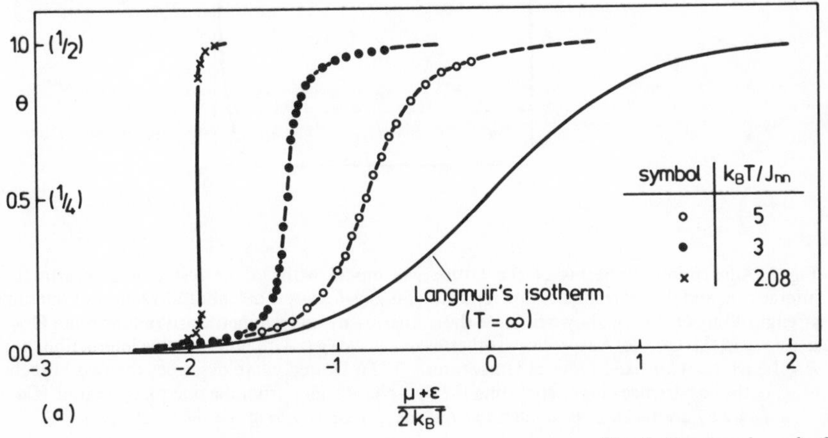

Fig. 2. *Continued overleaf*

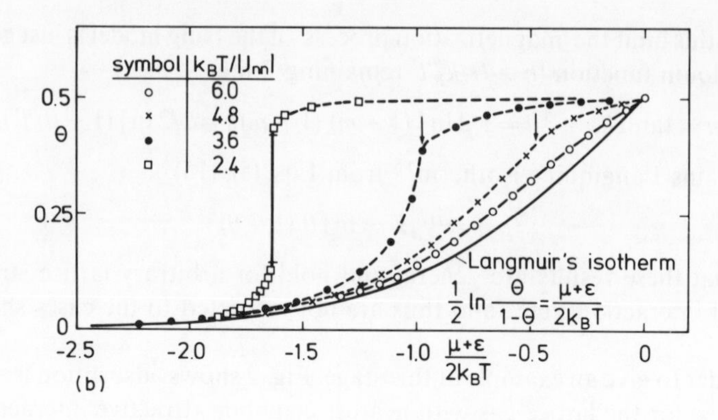

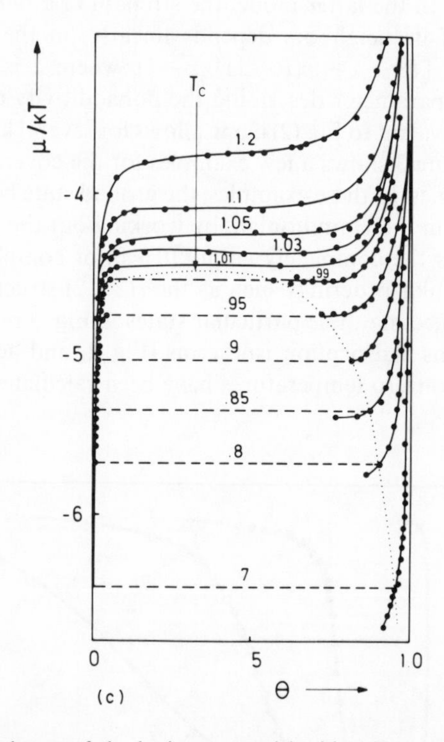

Fig. 2. Adsorption isotherms of the lattice gas model with: (a) nearest neighbor attractive interaction, and (b) nearest neighbor repulsion and next nearest neighbor attraction of the same strength. Only $\Theta < 1/2$ is shown because the isotherms are antisymmetric around the point $\Theta = 0$, $\mu + \varepsilon = 0$ in this case. (c) Adsorption isotherms of a lattice gas with non-additive interactions with $P = 1.5$ (cf. text) for fixed reduced temperature, T/T_c. Dotted curve describes the two branches Θ_{coex} of the coexistence curve separating the two-phase region from the one-phase region. (*Cases (a), (b) are taken from Binder and Landau*[20], *case (c) from Milchev and Binder*[23].)

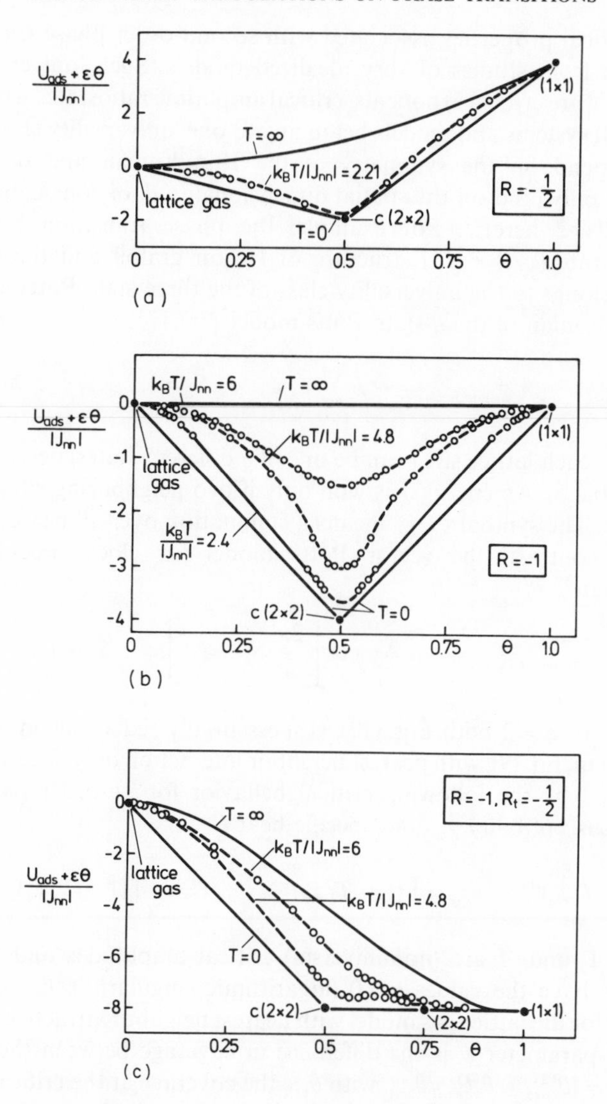

Fig. 3. Adsorption energy plotted versus coverage for $R \equiv J_{nnn}/J_{nn} = -1/2$, $R_t \equiv J_t/J_{nn} = 0$ (a), $R = -1$, $R_t = 0$ (b), and $R = -1$, $R_t = -1/2$ (c). (*From Binder and Landau*[20].)

For many systems, in particular physisorbed monolayers such as noble gases (He, Ar, Kr, etc.) adsorbed on grafoil[12–15], a lattice gas model description is not microscopically realistic, since the corrugation potential is rather weak, forming only rather shallow minima at the preferred lattice sites. However, rather than studying a more realistic model, it is still possible to infer certain critical properties associated with second-order phase transitions of the system from studies of very idealized models. Such 'universal' critical properties[24] are critical exponents, critical amplitude ratios, etc., which are the same for all systems and models belonging to one 'universality class'[24]. These classes depend on the symmetry of the Hamiltonian and of the order parameter only (and on the spatial dimensionality d, of course, but we only consider $d=2$ here)[24]. For example, the phase transition between the commensurate ($\sqrt{3}\times\sqrt{3}$) structure of He on grafoil and the disordered phase[25] belongs to the universality class of the three-state Potts model[26,27]. The Hamiltonian of the q-state Potts model[28,29] is

$$\mathscr{H}_{q\text{-state Potts}} = -J\sum_{\langle i,j\rangle}\delta_{S_iS_j}, \qquad S_i = 1,2,\ldots,q \tag{17}$$

In Eq. (17), each lattice site i can be one of q different states, described by the spin variable S_i. An energy J is won only if two neighboring sites are in the same state. The symbol $\langle i,j\rangle$ means a summation over all nearest neighbor pairs. In contrast, the vector Potts model (or clock model) has the Hamiltonian[29,30]

$$\mathscr{H}_{q\text{-state vector Potts}} = -J\sum_{\langle i,j\rangle}\cos\left[\frac{2\pi}{q}(S_i - S_j)\right], \qquad S_i = 1,2,\ldots,q \tag{18}$$

Note that for $q=2$ both Eqs. (17), (18) essentially reduce again to the Ising Hamiltonian, Eq. (9), with nearest neighbor interaction only. The latter model is described by the following critical behavior for its order parameter ψ, ordering susceptibility χ_ψ, and specific heat C:

$$\psi = \hat{B}(1-T/T_c)^\beta, \qquad \chi_\psi = \hat{\Gamma}(1-T/T_c)^{-\gamma}, \qquad C = \frac{\hat{A}}{\alpha}[(1-T/T_c)^{-\alpha}-1] \tag{19}$$

where $\hat{B}$, $\hat{\Gamma}$ and $\hat{A}$ are (nonuniversal) critical amplitudes and the critical exponents have the values $\alpha=0$ (logarithmic singularity), $\beta=1/8$, $\gamma=7/4$. Note that for the lattice gas model with nearest neighbor attractive interaction the order parameter ψ is the difference in coverage between the coexisting phases, $\psi = (\theta^{(2)}_{\text{coex}} - \theta^{(1)}_{\text{coex}})/\theta_{\text{crit}}$, with θ_{crit} the coverage at the critical point (see Fig. 2c) and $\chi_\psi = (\partial\theta/\partial\mu)T|_{\theta=\theta_{\text{crit}}}$. Also C has to be looked at $\theta=\theta_{\text{crit}}$. Note that while T_c, θ_{crit} depend on parameters contained in the Hamiltonian (e.g. $\theta_{\text{crit}}=1/2$ for a pairwise interaction model but differs from 1/2 for a nonpairwise model, see e.g. Fig. 2c), the critical exponents are independent of such 'irrelevant' details – it is only required that we have a scalar (one-component) order parameter. In contrast, the three-state clock models have a two-component order parameter[26,27], and have different exponents[31], $\alpha=1/3$, $\beta=1/9$, $\gamma=13/9$, and the three-component four-state Potts and clock models have the exponents[31] $\alpha=2/3$, $\beta=1/12$, $\gamma=7/6$.

There are many variants of these models introduced so far; e.g., one may simulate Ising models with an anisotropic range of the interaction: the ANNNI (anisotropic next nearest neighbor Ising) model[32-35] has in one lattice direction next nearest neighbor interaction of different sign than the nearest neighbor interaction, and this competition leads to the occurrence of an incommensurate phase. Such phases are very interesting in two dimensions since they cannot maintain true long-range order: rather their order-parameter correlation function decays to zero according to a power law for all temperatures in the range $0 < T \leqslant T_c$. At T_c, they disorder with a transition of the Kosterlitz–Thouless type[36] (see also sections 3.2, 3.3 below). Also the Potts models have been considered with different choices of the interaction, e.g. of antiferromagnetic type[29]. If one adds a phase shift ζ to the argument of the cosine in Eq. (18), the chiral clock model results[37] which has been taken as a candidate for a new type of critical behavior[38]. Next we draw attention to simulations of lattice models with continuous rather than discrete degrees of freedom, such as the ϕ^4 model[39-42]

$$\mathscr{H}_{\phi^4} \sum_i [r/2\ \phi^2(i) + u/4\ \phi^4(i)] + \sum_{\langle i,j \rangle} c/2\ [\phi(i) - \phi(j)]^2, \qquad -\infty < \phi(i) < +\infty \quad (20)$$

where $r < 0, u > 0, c > 0$ are constants, or the anisotropic Heisenberg antiferromagnet[43]

$$\mathscr{H}_{\text{Heisenberg}} = J \sum_{\langle i,j \rangle} [(1-\Delta)(S_{ix}S_{jx} + S_{iy}S_{jy}) + S_{iz}S_{jz}] - H_{\parallel} \sum_i S_{iz} \quad (21)$$

where each lattice site carries a three-component unit vector (S_{ix}, S_{iy}, S_{iz}) in the direction of the magnetic moment, Δ measures the uniaxial anisotropy, and $H_{\parallel}$ is a field applied in the direction of the easy axis. Equation (21) might be relevant for monolayers of a magnetic material epitaxially grown on a nonmagnetic substrate; in Eq. (20) the degree of freedom ϕ(i) experiencing a double-well potential (first sum on the right-hand side of Eq. (20)) may represent a lattice-dynamical displacement of an adsorbed atom from a preferred lattice site position, and thus Eq. (20) may model structural phase transitions[39,44] at surfaces. It is also interesting to consider lattice gas models for the coadsorption of two different kinds of atoms. This case occurs, for instance, if CO molecules are adsorbed on body-centered cubic (bcc) transition metal (001) surfaces (e.g., Mo, W, Fe)[45-51]. The resulting adsorbed layer forms a $c(2 \times 2)$ structure on the underlying substrate (see Fig. 4)[52]. A lattice gas description in analogy to Eq. (2) now needs two occupation variables C_i^A, C_i^B for the two different atomic species. Again, it is convenient to transform to pseudo-spin variables S_i but these now have three possible values: $S_i = \pm 1, 0$. Choosing $C_i^A = (S_i^2 + S_i)/2$, $C_i^B = (S_i^2 - S_i)/2$, the case $S_i = +1$ corresponds to an A-atom, the case $S_i = -1$ to a B-atom, and $S_i = 0$ to an empty site. A simple special case of the resulting generalized Blume–Emery–Griffiths model[53] is

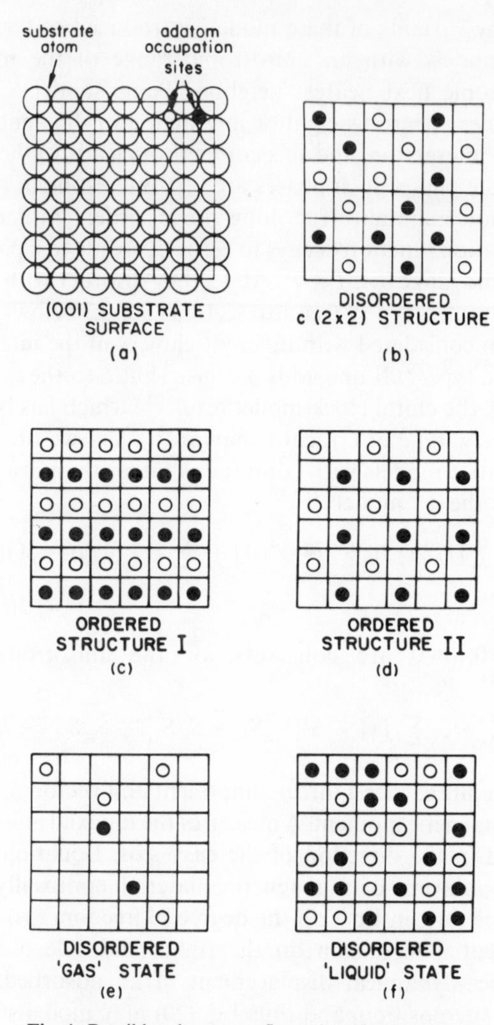

Fig. 4. Possible adatom configurations for the coadsorption of two atomic species (e.g. C,O) on the square lattices of preferred adsorption sites on (100) surfaces of b.c.c. transition metals. The two atomic species are denoted by small open or filled circles, respectively. (a) shows the top layer of the substrate and possible adsorption sites; the solid lines connect centers of the substrate atoms in this layer. (b) shows the $c(2 \times 2)$ structure with random occupation of the sites by the two species; (c) ordered structure I (the (2×1) structure); (d) ordered structure II [ordered $c(2 \times 2)$ structure]; (e) and (f) show the disordered 'lattice gas' and 'lattice liquid' states, respectively. (*From Lee and Landau*[52].)

the Hamiltonian[52]

$$\mathcal{H} = K_{nn} \sum_{\langle i,j \rangle_{nn}} S_i^2 S_j^2 + J_{nnn} \sum_{\langle i,j \rangle_{nnn}} S_i S_j + \Delta \sum_i S_i^2 \tag{22}$$

which refers to the case where $\langle C_i^A \rangle = \langle C_i^B \rangle$. The biquadratic exchange K_{nn} in Eq. (22) couples only nearest neighbors, while the standard exchange J_{nnn} is restricted to next nearest neighbors. The total concentration is controlled by the 'single-ion' term Δ. We shall return to this model in section 3.6. Finally, we consider lattice gas models for multilayer adsorption[54–59] and wetting phenomena[60–64]: here the Hamiltonian is again of the type of Eq. (2), but the system now is a semi-infinite three-dimensional system, bounded by the substrate. Performing again the transcription to the Ising spins, the Hamiltonian becomes (assuming that the binding potential ε to the substrate is felt only in the layer immediately adjacent to it), for the case of nearest-neighbor interaction which may differ in the layer adjacent to the substrate (J_S) from its value in the bulk (J):

$$\mathcal{H} = -J \sum_{\substack{\langle i,j \rangle \\ i \text{ or } j \in \text{bulk}}} S_i S_j - H \sum_i S_i - J_S \sum_{\substack{\langle i,j \rangle \\ i \text{ and } j \in \text{surface layer}}} S_i S_j - H_1 \sum_{i \in \text{surface layer}} S_i \tag{23}$$

2.2. A brief description of the Monte Carlo method in general

Extensive reviews of the Monte Carlo method can be found in Refs.[2–4]; here we restrict ourselves to brief comments. Suppose we wish to study the Ising Hamiltonian, Eq. (9). Then a Monte Carlo simulation amounts to:

1. Choice of lattice linear dimensions L_X, L_y (usually $L_X = L_y = L$, apart from physically anisotropic situations, such as the ANNNI model[32,33]). Only finite lattices can be simulated. Usually boundary effects are diminished by the choice of periodic boundary conditions, but occasionally studies with free boundaries are made[65]. Note that L_X, L_y must be chosen such that there is no distortion of the expected orderings in the system; e.g. for the model of[66] where due to a third nearest neighbor interaction superstructures with unit cells as large as 4×4 did occur, L must be a multiple of 4.
2. Choice of initial condition. To avoid the slow growth of ordered domains out of disordered initial states[11,41] and to avoid that the system gets trapped in metastable states[7,67], it is preferable to start the simulation in the appropriate perfectly ordered configuration. The various perfectly ordered states that a model can have are usually found from an analysis of the ground state which may be tedious but in most cases is rather straightforward (for examples, see Refs. 20, 22, 66). If for the chosen system parameters (temperature T, chemical potential μ) the system is in a disordered phase, the system will relax towards this state smoothly, even if it started out fully ordered, if the order–disorder transition is of second order. In the case of

first-order transitions, however, the ordered phase may still be metastable in the disordered region, and similar metastability is encountered when first-order transitions between various ordered phases occur. Hence, one should watch out for such hysteresis, performing runs at the same (T, μ) with different initial conditions (which usually include the perfectly empty or perfectly full lattice gas, respectively). The actual location of a phase boundary in cases where hysteresis occurs will be discussed in section 2.5.

3. After the linear dimensions, initial conditions, and the independent thermodynamic variables (T, μ) have been specified, the Monte Carlo simulation repeats again and again the following steps: (a) a lattice site is selected to be considered for a spin flip (one may select these sites in arbitrary regular orders or at random – the latter choice being preferable if a simulation of the dynamics of relaxation processes is intended); (b) the energy change $\delta\mathscr{H}$ associated with the spin flip is computed from the Hamiltonian and from this, one obtains the transition probability $W = \exp(-\delta\mathscr{H}/k_B T)$; (c) a random number z uniformly distributed between zero and unity is drawn and compared to W. If $W > z$, the spin flip is accepted and thus one has generated a new system configuration. If $W < z$, this trial spin flip is rejected, and the old configuration is counted once more for the averaging.
4. Suitable observables of the configuration (e.g. its energy per site E, magnetization per site m, appropriate order parameters ψ, etc.) are stored for later analysis. Of course, since subsequent configurations differ at most by a single spin flip, they are highly correlated, and hence it often is preferable to calculate these observables not after every 'Monte Carlo step' (MCS) (as the procedure 1 to 3 is called) but much less often, in order to save computing time. For example, one may do this analysis once every MCS/site or once every 10 MCS/site, etc., depending on the 'correlation times' in the problem.

It can be shown[1-5] that asymptotically (i.e. in the limit where the number M of generated configurations tends to infinity) this procedure generates system configurations $\vec{x}_\nu$ with a probability proportional to the Boltzmann weight, $P_{\text{eq}}(\vec{x}) = \exp[-\mathscr{H}(\vec{x})/k_B T]/Z$. Thus thermal averages are just calculated as simple arithmetic averages:

$$\bar{m} = \frac{1}{M - M_0} \sum_{\nu = M_0 + 1}^{M} m(\vec{x}_\nu), \quad \bar{E} = \frac{1}{M - M_0} \sum_{\nu = M_0 + 1}^{M} E(\vec{x}_\nu), \ldots \tag{24}$$

where we have anticipated that it is useful to exclude the first M_0 configurations from the averaging, since usually they are not yet characteristic for the desired thermal equilibrium. In practice, the appropriate value of M_0 is chosen from observing the actual 'time' evolution of the observables $m(\vec{x}_\nu)$, $E(\vec{x}_\nu)$, etc., with the 'observation time' $t = \nu/N$ in the simulation ('time' unit: one MCS per site). Apart from statistical errors, $\bar{m}$ and $\bar{E}$ then agree with the corresponding exact averages $\langle m \rangle_T$, $\langle E \rangle_T \equiv \langle \mathscr{H}_{\text{Ising}} \rangle_T / N$, etc., in the canonical ensemble.

Sampling the fluctuations of $m(\vec{x}_\nu)$ around $\bar{m}$, of $E(\vec{x}_\nu)$ around $\bar{E}$, of $\psi(\vec{x}_\nu)$ around $\bar{\psi}$, etc., one can obtain response functions such as the 'susceptibility' $\chi \equiv \partial \langle m \rangle_T / \partial H$, specific heat $C \equiv \partial \langle E \rangle_T / \partial T$ at constant H, ordering susceptibility $\chi_\psi \equiv \partial \langle \psi \rangle_T / \partial H_\psi$ (H_ψ being a field thermodynamically conjugate to the order parameter), etc.:

$$k_B T \chi = N[\langle m^2 \rangle_T - \langle m \rangle_T^2], \qquad k_B T^2 C = N[\langle E^2 \rangle_T - \langle E \rangle_T^2],$$
$$k_B T \chi_\psi = N[\langle \psi^2 \rangle_T - \langle \psi \rangle_T^2] \tag{25}$$

Note that χ translates into the derivative $\partial\theta/\partial\mu$ of the lattice gas, and hence measures the increase in coverage as the adsorbate gas pressure increases, assuming there is equilibrium between the layer and surrounding gas. The quantity $k_B T \chi_\psi$ is proportional to the peak intensity of the diffuse scattering at the Bragg position corresponding to the overlayer ordering described by the order parameter ψ.

As an example, Fig. 5 shows ψ as a function of temperature for the

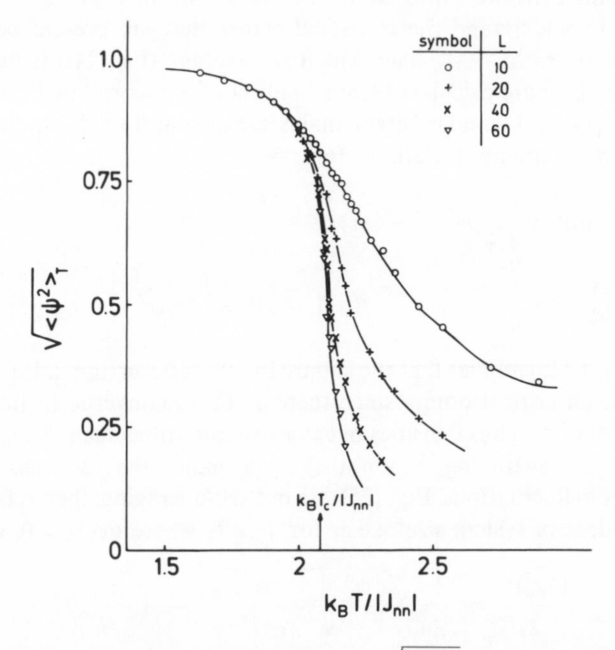

Fig. 5. Root mean square order parameter $\sqrt{\langle \psi^2 \rangle_T}$ plotted vs. temperature for the Ising model with nearest neighbor interaction $J_{nn} < 0$, and next nearest neighbor interaction $J_{nnn} = J_{nn}$, for different lattice sizes. The arrow shows the location of the critical temperature for $L \to \infty$. Note that this model orders in the (2×1) structure [cf. Fig. 4c], which has a two-component order parameter (ψ_1, ψ_2): ψ_1 is nonzero if the alternating rows of up spins and of down spins are oriented along the x-axis; ψ_2 is nonzero if the rows are oriented along the y-axis, and $\psi^2 \equiv \psi_1^2 + \psi_2^2$. (*The data shown in this figure were analyzed in Refs. 5, 68.*)

square lattice gas model, Eq. (2), with $\varphi_{\text{nn}} < 0$, $\varphi_{\text{nnn}} < 0$, $\varphi_{\text{t}} = 0$, and $R = \varphi_{\text{nnn}}/\varphi_{\text{nn}} = 1$[5,68]. While in the thermodynamic limit $L \to \infty \psi$ is expected to be singular near the critical temperature (Eq. (19)), in our finite lattices these singularities are rounded off: ψ has a pronounced 'tail' at $T > T_c$; similarly, the specific heat[5] and χ_ψ have only rounded maxima of finite heights somewhat offset from T_c. Sometimes there is interest to obtain properties for a particular finite size which corresponds to the linear dimension over which a real substrate is homogeneous (in adsorption on grafoil the effective sizes on which the homogeneous monolayers can be adsorbed, often is only of the order[69] of $60 \times 60\,\text{Å}^2$; in chemisorption at metal surfaces the effective size often is controlled by the density of steps appearing at the surface[70]). Thus finite size rounding of both first-order[69] and second-order transitions[70] was indeed observed. In most cases, however, one rather is interested in the properties that macroscopic ideal systems would have; the appropriate extrapolation to $L \to \infty$ by means of finite size scaling theories[71–78] will be discussed in section 2.3.

In obtaining Monte Carlo 'data' such as shown in Figs. 2, 3, 5, it is also necessary to understand the statistical errors that are present because the number of states $M - M_0$ over which we average (Eq. (24)) is finite. If the averages $\bar{m}$, $\bar{E}$, $\bar{\psi}$ are calculated from a subset of n uncorrelated observations $m(\vec{X}_\nu)$, $E(\vec{X}_\nu)$, $\psi(\vec{X}_\nu)$, standard error analysis applies and yields estimates for the expected mean square deviations, for $n \to \infty$,

$$\overline{(\delta \text{m})^2} \approx \frac{1}{n}\left[\langle m^2 \rangle_{\text{T}} - \langle m \rangle_{\text{T}}^2\right],$$

$$\overline{(\delta E)^2} \approx \frac{1}{n}\left[\langle E^2 \rangle_{\text{T}} - \langle E \rangle_{\text{T}}^2\right], \overline{(\delta \psi)^2} \approx \frac{1}{n}\left[\langle \psi^2 \rangle_{\text{T}} - \langle \psi \rangle_{\text{T}}^2\right] \tag{26}$$

Note that Eq. (25) implies that the square brackets occurring in Eq. (26) are of order $1/N$, off critical points since there χ, C, χ_ψ converge to finite values independent of N. Thus densities of extensive quantities such as $\langle m \rangle_{\text{T}}$, $\langle E \rangle_{\text{T}}$, $\langle \psi \rangle_{\text{T}}$ are 'self-averaging'[41]. On the other hand, the response functions sampled from fluctuations, Eqs. (25), are not self-averaging: their relative error is independent of system size[41]: e.g., for $T > T_c$ where $\langle \psi \rangle_{\text{T}} = 0$, we have

$$\frac{\overline{(\delta \chi_\psi^2)}^{1/2}}{\chi_\psi} = \frac{1}{\sqrt{n}} \frac{\sqrt{\langle \psi^4 \rangle_{\text{T}} - \langle \psi^2 \rangle_{\text{T}}^2}}{\langle \psi^2 \rangle_{\text{T}}} \xrightarrow[L \to \infty]{} \frac{1}{\sqrt{n}}\sqrt{2} \tag{27}$$

using the fact that for $N = L^2 \to \infty$ the distribution function $P_L(\psi)$ simply tends to a Gaussian[74],

$$P_L(\psi) \propto \exp\left[-\frac{\psi^2}{2k_B T \chi_\psi / L^2}\right] \tag{28}$$

Relations similar to Eqs. (27), (28) hold for C and χ as well.

Obviously, it is advantageous to choose n as large as possible since then all errors go down. However, if subsequent observations included in the subset for the average are separated by a time interval Δt over which 'time correlations' have not yet died out, errors are no longer given by Eq. (24) but rather by[2–4,79]

$$\overline{(\delta m^2)} \approx \frac{1}{n}[\langle m^2\rangle_T - \langle m\rangle_T^2](1 + 2\tau_m^{(l)}/\Delta t), \qquad n\Delta t \gg \tau_m^{(l)}, \tag{29}$$

where the 'linear' relaxation time $\tau_m^{(l)}$ of the magnetization is defined by

$$\tau_m^{(l)} = \int_0^\infty [\langle m(0)m(t)\rangle_T - \langle m\rangle_T^2]\,\mathrm{d}t/[\langle m^2\rangle_T - \langle m^2\rangle_T^2] \tag{30}$$

and similar expressions hold for other quantities. The 'time' t is associated with the label ν of subsequently generated configurations ν, as mentioned above, $t = \nu/N$: it turns out that it makes sense to call this variable 'time' since the Monte Carlo procedure can be interpreted as a realization of a master equation for the probability $P(\vec{X}, t)$ that state $\vec{X}$ occurs at time t[2–4,79],

$$\frac{\mathrm{d}P(\vec{X},t)}{\mathrm{d}t} = -\sum_{\vec{X}'} W(\vec{X}\to\vec{X}')P(\vec{X},t) + \sum_{\vec{X}'} W(\vec{X}'\to\vec{X})P(\vec{X}',t) \tag{31}$$

In the example discussed above, the transition $\vec{X}\to\vec{X}'$ stands simply for a single spin flip at a randomly chosen lattice site, and $W(\vec{X}\to\vec{X}') = 1$ if $\delta\mathscr{H} < 0$ while $W(\vec{X}\to\vec{X}') = \exp(-\delta\mathscr{H}/k_BT)$ for $\delta\mathscr{H} > 0$ should be interpreted as transition probability per unit time. Note that other choices for W would also be possible provided they satisfy the principle of detailed balance[2–4],

$$P_{\mathrm{eq}}(\vec{X})W(\vec{X}\to\vec{X}') = P_{\mathrm{eq}}(\vec{X}')W(\vec{X}'\to\vec{X}) \tag{32}$$

Equation (32) ensures that $P(\vec{X}, t\to\infty)$ according to Eq. (31) must relax towards $P_{\mathrm{eq}}(\vec{X})$, and thus the correct equilibrium distribution must be reached, as anticipated above.

Note also that the choice of what the move $\vec{X}\to\vec{X}'$ from one phase space point to the next means microscopically depends on the type of problem that one wishes to study; e.g., for a simulation of surface diffusion in the framework of the lattice gas model[80] (see section 4.2), this move may mean a hop of a randomly chosen adatom to a randomly chosen nearest neighbor site (and $W \equiv 0$ if this latter site is already taken).

From this dynamic interpretation of the Monte Carlo averaging we can obtain a formal estimate of the number of steps M_0 that have to be omitted at the beginning of the averaging. Usually, the order parameter ψ is the slowest relaxing quantity and then

$$M_0/N \gg \tau_\psi^{(nl)}, \quad \tau_\psi^{(nl)} \equiv \int_0^\infty \mathrm{d}t[\langle\psi(t)\rangle_T - \langle\psi(\infty)\rangle_T]/[\langle\psi(0)\rangle_T - \langle\psi(\infty)\rangle_T] \tag{33}$$

In this nonlinear relaxation time, $\langle \psi(0) \rangle_T$ is the order parameter for the initial condition, in which the Monte Carlo run is started, and $\langle \psi(\infty) \rangle_T = \langle \psi \rangle_T$ is simply the equilibrium value of the order parameter.

2.3. Finite size effects on phase transitions

If a temperature-driven phase transition occurs at a temperature $T_c(\infty)$ in a macroscopically large system, in a finite $L \times L$ geometry this transition will be smeared out over a temperature region $\Delta T(L)$ around a shifted 'effective' transition temperature $T_c(L)$,

$$\Delta T(L) \propto L^{-\theta}, \qquad |T_c(L) - T_c(\infty)| \propto L^{-\lambda}, \tag{34}$$

with the exponents θ, λ defining the rounding and the shift of the transition. In addition, the specific heat $C(T)$ which at a second-order transition diverges according to a power law (Eq. (19)) and at a first-order transition exhibits a delta-function singularity due to the latent heat, will only reach a finite maximum value $C_{\max}(L)$,

$$C(T_c(L)) \equiv C_{\max}(L) \propto L^{\alpha_m} \tag{35}$$

and similarly the susceptibility maximum at a second-order transition $\chi_\psi^{\max} \propto L^{\gamma_m}$ (note that in general the temperatures where $C(T)$ and $\chi_\psi(T)$ have their maxima are different). This finite size behavior must be understood, if extrapolation of finite lattice data to $L \to \infty$ is desired; in particular, knowledge of the exponents θ, λ, α_m, γ_m is required.

For a second-order transition, this problem is conveniently studied in terms of the order parameter distribution function, $P_L(\psi)$. Finite size scaling theory implies that near the critical point $P_L(\psi)$ no longer depends on the three variables L, ψ, $1 - T/T_c$ separately but rather is a scaled function of two variables $\psi L^{\beta/\nu}$, $(1 - T/T_c)L^{1/\nu}$ only[74], where ν is the critical exponent of the correlation length ξ, $\xi = \hat{\xi}|1 - T/T_c|^{-\nu}$ as $T \to T_c$[24]. Denoting this function by $\tilde{P}$, one obtains[74]

$$P_L(\psi) = L^{\beta/\nu}\tilde{P}\{\psi L^{\beta/\nu}, (1 - T/T_c)L^{1/\nu}\} \tag{36}$$

The prefactor $L^{\beta/\nu}$ in Eq. (36) is understood from the normalization condition, $\int_{-1}^{+1} d\psi P_L(\psi) = 1$, which must hold at all temperatures. From Eq. (36) one immediately obtains finite-size scaling relations for the order parameter $\langle |\psi| \rangle_T$ and ordering susceptibility χ_ψ by taking suitable moments of the distribution (note $P_L(\psi)$ is symmetric around $\psi = 0$ in the absence of symmetry-breaking fields and thus $\langle \psi \rangle \equiv 0$):

$$\langle |\psi| \rangle_T = \int_{-1}^{+1} d\psi |\psi| P_L(\psi) = L^{-\beta/\nu}\tilde{\psi}\{(1 - T/T_c)L^{1/\nu}\}, \tag{37}$$

where $\tilde{\psi}(z)$ is another scaling function, whose argument z is basically $(L/\xi)^{1/\nu}$, expressing the fact that finite size scaling simply means to compare lengths: 'L scales with ξ'[72]. Similarly, for $T > T_c$ we have from Eq. (25) (remember $N = L^2$ for the $L \times L$ geometry considered here)

$$k_B T \chi_\psi = L^2 \langle \psi^2 \rangle_T = L^{2-2\beta/\nu} \tilde{\chi}\{(1 - T/T_c)L^{1/\nu}\} = L^{\gamma/\nu} \tilde{\chi}\{(1 - T/T_c)L^{1/\nu}\} \quad (38)$$

In the last step, the hyperscaling exponent relation $d\nu = \gamma + 2\beta$ ($d = 2$ here) was used. From Eqs. (37), (38) one sees $\langle |\psi| \rangle_{T_c} \propto L^{-\beta_m}$ with $\beta_m = \beta/\nu$, $\gamma_m = \gamma/\nu$ and similarly one derives $\alpha_m = \alpha/\nu$ (Eq. (35) holds only if $\alpha > 0$). The effective critical temperature $T_c(L)$ corresponds to the scaling variable $z = (1 - T/T_c)L^{1/\nu}$ taking a critical value z_c, which immediately yields $|T_c(L) - T_c| \propto L^{-1/\nu}$, i.e. the shift exponent, Eq. (34), becomes $\lambda = 1/\nu$. Since Eqs. (37), (38) for $L \to \infty$ must smoothly tend towards the bulk critical behaviour, Eq. (19), we conclude that $\tilde{\psi}(z \gg 1) \propto z^{\beta}$, $\tilde{\chi}(z \gg 1) \propto z^{-\gamma}$. Obviously, strong deviations of χ_ψ and $\langle |\psi| \rangle_T$ from their bulk limit occur if $z \lesssim 1$, and hence the width of the rounding regime ΔT is estimated putting $z_{\Delta T} = 1$, which yields the rounding exponent θ in Eq. (34) which is also $1/\nu$.

Figures 6–9 illustrate the use of these finite size scaling relations for the square lattice gas with repulsion between both nearest and next nearest neighbors. In Fig. 6 the 'raw data' of Fig. 5 are replotted in scaled form, as suggested by Eq. (37). Note that neither $T_c = T_c(\infty)$ nor the critical exponents are known in beforehand – the phase transition of the (2×1) phase falls in the universality class of the 'XY model with uniaxial anisotropy'[81,82], which has nonuniversal exponents depending on R. Clearly, it is desirable to estimate T_c without being biased by the choice of the critical exponents. This is possible

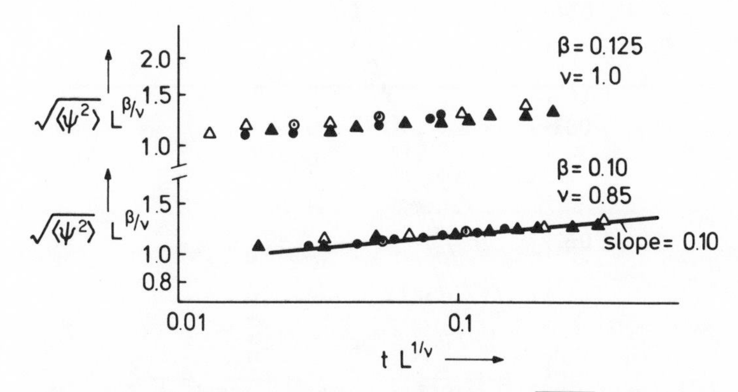

Fig. 6. Finite size scaling plot of the order parameter $\sqrt{\langle \psi^2 \rangle_T}$ of the square lattice gas (at coverage $\Theta = 1/2$), with repulsive interactions between both nearest and next nearest neighbors φ_{nn}, φ_{nnn} of equal strength ($R = \varphi_{nn}/\varphi_{nnn} = 1$). Upper part shows the choice of the 2-d Ising exponents[24] $\beta = 1/8$, $\nu = 1$; for this choice the 'raw data' fail to collapse on a single curve. Adjusting β and ν until an optimal 'data collapsing' is obtained yields $\beta \approx 0.10$, $\nu \approx 0.85$ as shown in the lower part of the figure. (*From Binder and Landau*[5].)

considering the reduced fourth-order cumulant U_L[73,74],

$$U_L = 1 - \langle \psi^4 \rangle_T / [3 \langle \psi^2 \rangle_T] = \tilde{U}\{(1 - T/T_c)L^{1/\nu}\} \tag{39}$$

since this quantity has no power law prefactor in front of the scaling function $\tilde{U}(z)$. In the disordered phase, where Eq. (28) holds for $L \to \infty$s we have $U_L \to 0$; similarly, deep in the ordered phase we can write (considering for the moment the simplest case of a scalar one-component order parameter)

$$P_L(\psi) \propto \exp\left[-\frac{(\psi - \langle \psi \rangle_T)^2}{2k_B T \chi_\psi / L^2}\right] + \exp\left[-\frac{(\psi + \langle \psi \rangle_T)^2}{2k_B T \chi_\psi / L^2}\right] \tag{40}$$

i.e. the distribution function just is a sum of two Gaussians centered at $\psi = \pm \langle \psi \rangle_T$. Equation (40) implies $U_L \xrightarrow[L \to \infty]{} 2/3$ (this also holds for more-component order parameters). Thus Eq. (39) implies that plotting U_L vs. T for different L yields a family of curves starting at 2/3 for low T and decreasing to zero at high T, intersecting at T_c in a common intersection point $U^* \equiv \tilde{U}(0)$. Thus T_c can be located from this common intersection point. Alternatively, we may plot $U_{L'}$ versus U_L for various choices of U_L (Fig. 7), where T is a

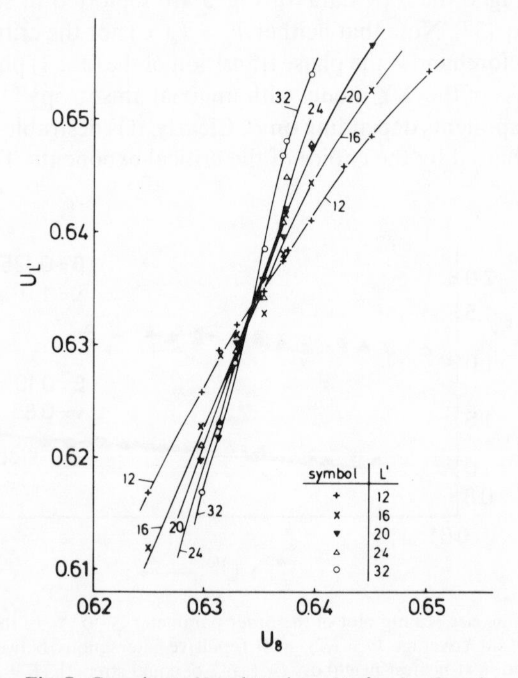

Fig. 7. Cumulants $U_{L'}$ plotted vs. U_L for $L = 8$ and different choices of L' for the same model as in Figs. 5, 6. (*The data shown in this figure were analyzed in Ref. 66.*)

parameter of the curves. From the slope at the intersection point one can obtain estimates for the exponent ν via $[b \equiv L'/L]$,

$$\nu^{-1} = \ln\{\partial U_{bL}/\partial U_L\}_{U^*}; \tag{41}$$

similarly, from the second moment of Eq. (36) one finds

$$2\beta/\nu = \ln\{\langle\psi^2\rangle_{bL}/\langle\psi^2\rangle_L\}/\ln b. \tag{42}$$

The advantage of this analysis is twofold: estimates for T_c, ν, and β/ν are obtained from independent pieces of information and are hence not as correlated as in a fit such as shown in Fig. 6; in addition, corrections to finite size scaling can be taken into account systematically. In fact, Eqs. (34)–(42) are expected to hold only for $L \to \infty$; for finite L correction terms described by correction exponents x_c and amplitudes f_c are present, e.g.

$$\langle\psi^2\rangle_L \propto L^{-2\beta/\nu}\{1 + f_c L^{-x_c} + \cdots\} \tag{43}$$

which implies that Eq. (42) gets modified to

$$\ln\{\langle\psi^2\rangle_{bL}/\langle\psi^2\rangle_L\}/\ln b = 2\beta/\nu - \frac{L^{-x_c} f_c}{b^{x_c} - 1}\frac{1}{\ln b} + \cdots \tag{44}$$

Figure 8 shows that a plot of the exponents $2\beta/\nu$, ν and T_c as function of $1/\ln b$

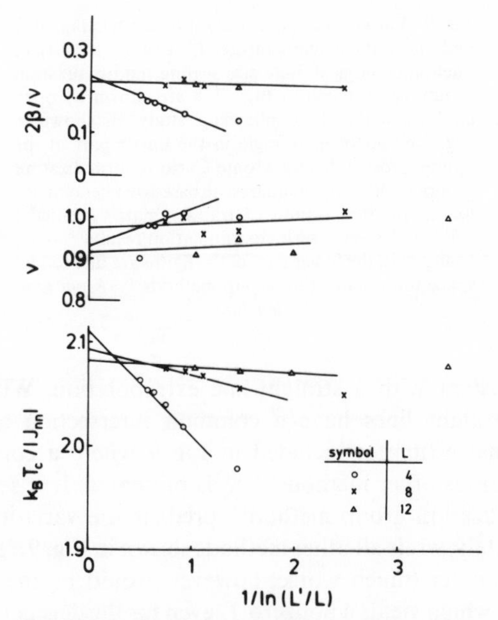

Fig. 8. Variation of critical parameters with $b = L'/L$ for the model of Figs. 5–7 using the analysis described in the text. (*The data shown in this figure were analyzed in Ref. 66.*)

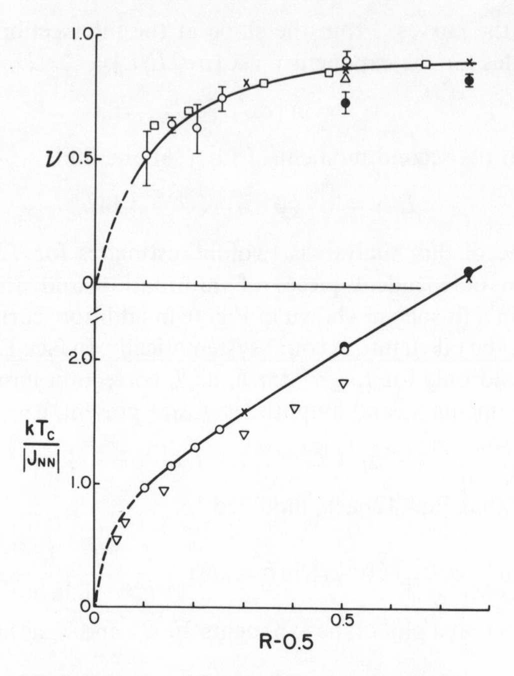

Fig. 9. Variation of the correlation length exponent ν and the critical temperature T_c with R. Results of phenomenological finite size scaling renormalization group as described in Figs. 7, 8 are shown by open circles and the 'data collapsing' study[5] as shown in Fig. 6 by an open triangle in the upper part of the figure. Crosses denote Monte Carlo renormalization group results[83], open squares are based on phenomenological renormalization of transfer matrix results[84]; solid circles show series extrapolation results[85]; open triangles in the lower part of the figure are due to real space renormalization group methods[86]. (*From Landau and Binder*[66].)

indeed is consistent with a straight-line extrapolation. Within reasonable errors these straight lines have a common intersection at the ordinate, yielding the final estimates included in Fig. 9 where a comparison of the results with various other methods[83–86] is presented. It is seen that the real space renormalization group method[86] predicts the variation of T_c with R rather inaccurately, while all other methods shown in Fig. 9 agree with respect to their T_c estimates (much worse, however, would be the molecular field approximation which yields a nonzero T_c even for the degeneracy point where the $c(2 \times 2)$ and (2×1) phases coexist in the groundstate; see e.g. Refs. 5, 68). However, with respect to the exponent ν also the series extrapolations of Ref. 85 clearly are less accurate than the methods based on phenomenological renormalization of either Monte Carlo 'data'[66], as described here, or transfer matrix results for finite strips[84], and the Monte Carlo renormalization group (MCRG) method[83]. Note, however, that the transfer matrix method would be rather cumbersome and less efficient if longer range interactions are present, which are still easily included in the Monte Carlo calculation[66]. While the MCRG method presumably is the most efficient way to estimate critical exponents, it must be supplemented by standard MC methods as described here, if a more complete information on the thermodynamic properties of the model (e.g. adsorption isotherms, Fig. 2; ordering energies, Fig. 3; order parameters, Fig. 5; specific heats[5,68]; etc.) over a wide range of parameters is desired.

We now briefly consider finite size effects at first-order phase transitions. The easiest case is transitions driven by the field H_ψ conjugate to the order parameter ψ in systems at $T < T_c$. Then Eq. (40) is easily generalized by introducing Boltzmann weight factors for the two states $\pm\langle\psi\rangle_T$ according to their Zeemann energies $\pm H_\psi \langle\psi\rangle_T L^d$ [76,77]:

$$P_L(\psi) \propto \exp\left\{\frac{\langle\psi\rangle_T H_\psi L^d}{k_B T}\right\}\exp\left[-\frac{(\psi - \langle\psi\rangle_T - \chi_\psi H_\psi)^2}{2k_B T\chi_\psi/L^d}\right]$$
$$+\exp\left\{-\frac{\langle\psi\rangle_T H_\psi L^d}{k_B T}\right\}\exp\left[-\frac{(\psi + \langle\psi\rangle_T - \chi_\psi H_\psi)^2}{2k_B T\chi_\psi/L^d}\right] \quad (45)$$

In Eq. (45) we have accounted for the fact that for $H_\psi \neq 0$ the Gaussian peaks are centered at $\pm\langle\psi\rangle_T + \chi_\psi H_\psi$, respectively.

While for $L \to \infty$, $\langle\psi\rangle_T$ at $H_\psi = 0$ has a discontinuous jump and χ_ψ hence exhibits a delta-function singularity, for finite L this singularity is rounded in a finite peak:

$$\chi_\psi(T, L) \equiv L^2 \frac{\langle\psi^2\rangle_T - \langle\psi\rangle_T^2}{k_B T} = \chi_\psi + \frac{\langle\psi\rangle_T^2}{k_B T} L^2\left[\cosh^2 \frac{H_\psi \langle\psi\rangle_T L^2}{k_B T}\right] \quad (46)$$

This implies that the exponents Θ and γ_m defined above are $\Theta = \gamma_m = 2(=d)$ for a first-order transition. Since the symmetry around $H_\psi = 0$ is preserved for finite L, there is no shift of the transition. This feature is different, however, if we consider temperature-driven first-order transitions[78], since there is no symmetry between the disordered high-temperature phase and the ordered low-temperature phase. In order to understand the rounding of the delta-function singularity of the specific heat, which measures the latent heat for $L \to \infty$, it now is useful to consider the energy distribution, for which again a double Gaussian approximation applies[78]:

$$P_L(E) \propto \frac{a_+}{(C_+)^{1/2}} \exp\left\{\frac{-[E - (E_+ + C_+\Delta T)]^2}{2k_B T^2 C_+/L^2}\right\}$$
$$+\frac{a_-}{(C_-)^{1/2}} \exp\left\{\frac{-[E - (E_- + C_-\Delta T)]^2}{2k_B T^2 C_-/L^2}\right\}. \quad (47)$$

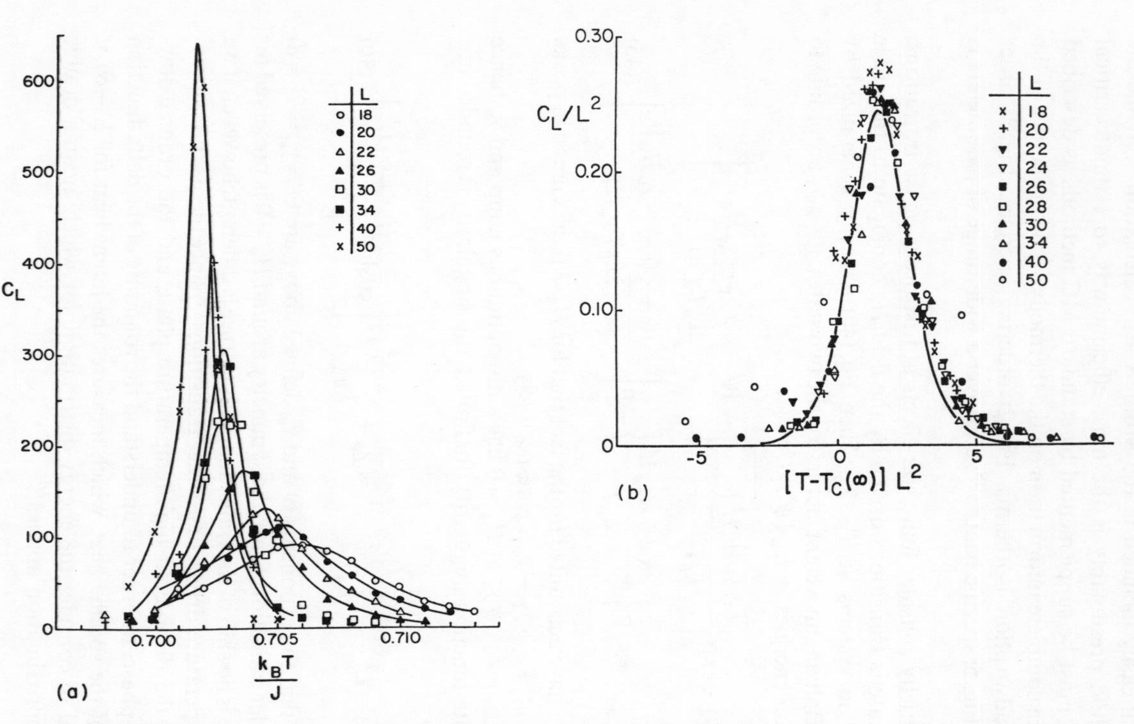

Fig. 10. (a) Temperature variation of the specific heat of the q-state Potts model with $q = 10$ for various lattice sizes as indicated in the figure. (b) Scaling representation of the specific heat data, $C_L(L)/L^2$ plotted vs. $[T - T_c(\infty)]L^2$. The solid curve is the theoretical scaling function resulting from Eq. (47), namely $C_L/L^2 = \{(E_+ - E_-)^2 q(C_-/C_+)^{1/2}\}/\{k_B T_c^2 [e^x + e^{-x} q(C_-/C_+)^{1/2}]^2\}$ with C_-/C_+ fitted as 0.7, all other parameters $\{E_+, E_-, T_c(\infty)\}$ being taken from the exact solution[87].

Here E_+, E_- are the energies of the two phases coexisting at the first-order transition, and C_+, C_- their specific heats, and $\Delta T = T - T_c$. Therefore the Gaussians are centered at $E = E_+ + C_+ \Delta T$ and $E = E_- + C_- \Delta T$, respectively. The weight factors a_+, a_- are given by

$$a_+ = C_+^{1/2} \exp\left[-\frac{\Delta F L^2}{2k_B T}\right], \qquad a_- = C_-^{1/2} q \exp\left[\frac{\Delta F L^2}{2k_B T}\right], \tag{48}$$

q being the degeneracy of the ordered low-temperature phase, and ΔF the free energy difference between both states, $\Delta F \cong -(E_+ - E_-)\Delta T/T_c$. From these results it is easy to show that the rounded peak of the specific heat occurs at $T_c(L)$,

$$\{T_c(L) - T_c\}/T_c = k_B T_c \ln [q(C_-/C_+)^{1/2}]/\{(E_+ - E_-)L^2\} \tag{49}$$

and has a maximal height

$$C^{\max} \equiv C(T_c(L)) \approx (C_+ + C_-)/2 + \{(E_+ - E_-)^2 L^2/[4k_B T_c^2]\} \tag{50}$$

Since this peak is rounded over a region $\Delta T \approx 2k_B T_c^2/[(E_+ - E_-)L^2]$, we conclude that in this case $\Theta = \lambda = \alpha_m = 2$.

These ideas have been tested by calculations on the ferromagnetic Ising model[77] and the 10-state Potts model[78], where exact results on T_c, E_+, E_- are available[87], and excellent agreement with the theoretical predictions Eqs. (45)–(50) was found. As an example, Fig. 10 shows the specific heat for finite 10-state Potts lattices and the scaling representation of these data resulting from Eq. (47). Also experimental data on rounded specific heats of O_2 adsorbed on grafoil[69] can be accounted by Eq. (47) quantitatively[8].

2.4. Free energy and the thermodynamic integration method

While the finite size scaling method for first-order phase transitions as described above has the advantage of being a first-principles approach, it is very time-consuming since the time needed to 'jump' from one Gaussian peak to the other one increases exponentially with L. Often it is more convenient to simulate a very large lattice, where no such jumps between the phases occur near T_c, and finite size effects can be disregarded. However, in this case pronounced hysteresis inevitably occurs (Fig. 11): starting from an ordered state and sweeping the temperature up, metastable superheated ordered states occur, while one similarly finds metastable supercooled disordered states when one sweeps the temperature down from disordered initial states. Since then in a certain temperature region both phases are found to be stable or at least metastable, the precise location of the transition becomes a problem. For doing this, one requires the free energies of both phases which are not a direct output of the Monte Carlo simulation, however (in the importance sampling method, Eq. (24), the denominator is not the partition function

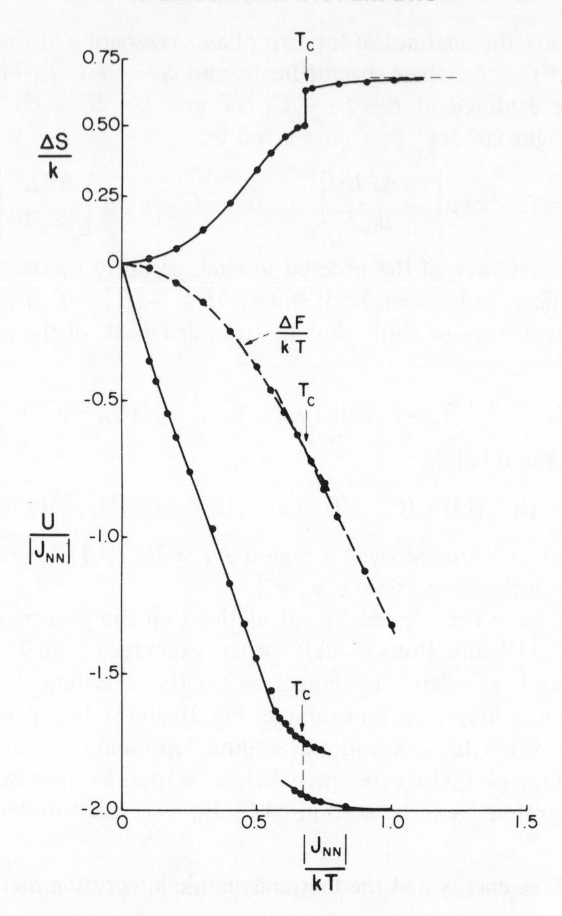

Fig. 11. Determination of the location of the first-order transition of the Ising square lattice with nearest (J_{NN}), next nearest (J_{NNN}) and third nearest (J_{3NN}) neighbor interactions at zero field, from comparison of free energies. Data shown refer to $R = J_{NN}/J_{NNN} = 1/2$, $R' = J_{3NN}/J_{NN} = 1$, for which the ground state is a (4×4) 'checkerboard' structure ((2×2) cells of up spins alternate with (2×2) cells of down spins). Lower part is a plot of the internal energy versus inverse temperature. Upper part shows the entropy difference ΔS and middle part the free energy difference ΔF, obtained via thermodynamic integration (Eqs. (51), (52)). (*From Landau and Binder*[66].)

Z but simply the number of configurations included in the average, and methods to sample Z, or the entropy S approximately are difficult and not straightforward[2, 3]).

A simple way round this difficulty is the thermodynamic integration of the internal energy E (see e.g. Ref. 89).

$$S/k_B = S(T=\infty)/k_B + E/k_BT - \int_0^{(1/k_BT)} E\,d(1/k_BT) \tag{51}$$

$$F/k_BT = -S(T=\infty) + \int_0^{1/k_BT} E\,d(1/k_BT) \tag{52}$$

At a first-order transition, the jump in the internal energy $(E_+ - E_-)$ also gives rise to a jump in the entropy $\Delta S = (E_+ - E_-)/k_BT_c$. The transition temperature T_c must then be adjusted (in a trial-and-error procedure!) until the total entropy resulting from the integration is equal to its exact value. (In Ising models $S(T=\infty)/k_B = \ln 2$, and if the ground state has no entropy, $\Delta S = S(T=\infty) - S(T=0) = \ln 2/k_B$. If T_c is chosen too low or too high, ΔS will be too large or too small, respectively.) Alternatively, one may integrate the specific heat from a $T=0$ reference state[89], or one may also perform isothermal integrations varying the field (using the state where all spins are up or all spins are down as a reference state), etc. In fact, any path connecting two states in the space of thermodynamic variables that is convenient may be chosen for a thermodynamic integration to estimate the free energy difference of the two states[89]. Figure 11 shows that this method indeed gives good results with modest effort (i.e. a moderately large number of temperatures where E needs to be sampled).

2.5. Translation from the grand-canonical to the canonical ensemble

For many problems of practical interest (see sections 3.1, 3.2), one is interested in studying phase diagrams in the T–Θ plane. Now the coverage Θ is the density of an extensive thermodynamic variable (namely the number N_a of adsorbed atoms), and for extensive variables it is clear that first-order transitions show up as two-phase coexistence regions. It is then not possible to study such phase diagrams by direct simulations at fixed coverage, however: if the system state falls inside such a two-phase region, an initially homogeneous state would need to decompose into two macroscopic phases, to reach thermal equilibrium (the amounts of these phases are given by the lever rule). Consider e.g. the lattice gas model on the square lattice, with interactions $J_{nn} = J$, J_{nnn} between nearest and next nearest neighbors but $R = J_{nn}/J_{nnn} = -1$: this model orders in the $c(2\times 2)$ phase, but for $T_t \approx 2.5|J|/k_B$ a tricritical point appears, and then the transition from the disordered phase to the $c(2\times 2)$ phase is first order. Figure 12b shows that there occurs for $T < T_t$ a wide two-phase region, where lattice gas (with Θ very small at low T) coexists with 'islands' of ordered $c(2\times 2)$ phase (with Θ close to 0.5 at low T). If we simulated a system inside this region, we would observe the formation and growth of these islands, which is an extremely slow process. This is a problem

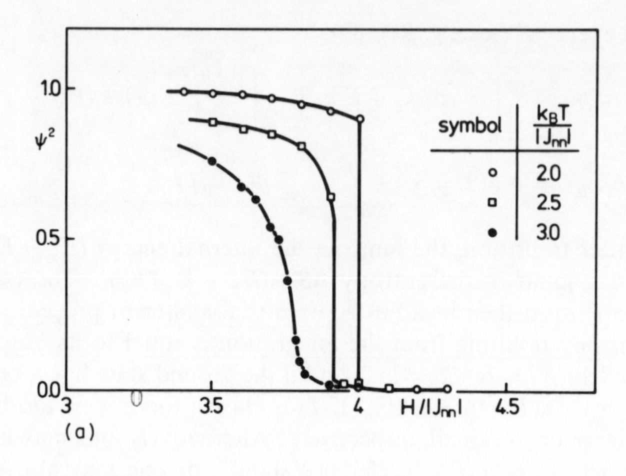
1.0
ψ^2
0.5
0.0
3
3.5
4
$H/|J_{nn}|$
4.5
(a)
symbol
$\frac{k_BT}{|J_{nn}|}$
2.0
2.5
3.0

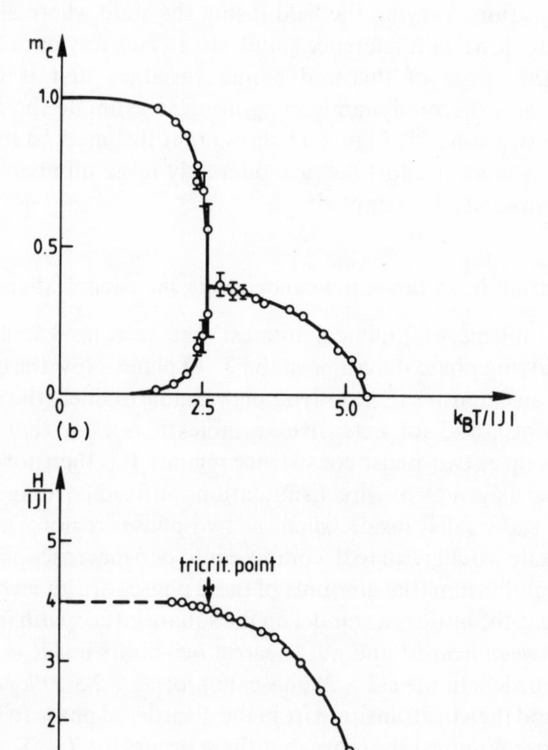
m_c
1.0
0.5
0
2.5
5.0
$k_BT/|J|$
(b)
$\frac{H}{|J|}$
5
4
3
2
1
0
tricrit. point
2.5
5.0
$k_BT/|J|$
(c)

in its own right (see section 4), but it would be extremely hard to follow this process until full equilibrium would be established (ideally one then must have one 'island' of the minority phase on the background 'sea' of the majority phase). Besides the fact that simulations of island formation and growth can hardly ever be carried out long enough to reach this equilibrium state, its properties in a finite lattice geometry are strongly affected by non-negligible energy contributions from the 'coastline' of the island.

Thus, what one actually does is a simulation in the grand-canonical ensemble (T, μ) (or (T, H) in Ising magnet-terminology), and there data are 'translated' into the canonical ensemble (T, Θ) (or (T, m), respectively). For the model of Fig. 12, we perform scans at constant T varying $\mu(H)$, and locate the transition from an analysis of the order parameter (Fig. 12a). This yields the critical field $H_c(T)$ at each temperature (Fig. 12c). From the magnetization process $m(H)$ (which is nothing but the adsorption isotherm, Fig. 2, in the lattice gas language) we may read off at H_c (or μ_c, respectively) the two values m_c^-, m_c^+ [or θ^-, θ^+] for the magnetization (coverage) jump. Plotting these values in the (T, m) [(T, θ)] plane then yields the phase diagram in the canonical ensemble. Due to uncertainties in the precise values of H_c (or μ_c) and due to statistical fluctuations (particularly along the second-order line) the error bars shown in Fig. 12b result. All phase diagrams discussed in section 3 have been constructed in this way.

3. SOME STUDIES OF STATIC CRITICAL BEHAVIOR AND OF PHASE DIAGRAMS

3.1. Square lattice gases with two- and three-body interactions: a model for the adsorption of hydrogen on Pd(100)?

Behm *et al.*[19] have measured LEED diffraction intensities for H monolayers on Pd(100) surfaces as function of temperature at different coverages (Fig. 13b). Taking the temperature $T_{1/2}$ where the intensity has dropped to 50% of its low-temperature value as estimated for $T_c(\Theta)$, they constructed the phase diagram shown by crosses in Fig. 13a. (Alternatively using the inflection points of the I vs T curves (Fig. 13b)[19] yields similar results.)

Fig. 12. (a) Square of the order parameter ψ (staggered magnetization) of the square lattice gas model with $R = \varphi_{nn}/\varphi_{nnn} = -1$ plotted versus magnetic field at three temperatures. Highest temperature corresponds to a second-order transition while for the two lower temperatures the transition is of first order. (b) Critical magnetization m_c plotted versus temperature. Two values m_c^-, m_c^+ for m_c indicate the magnetization jump at the transition, which translates into the two-phase coexistence region. (c) Critical magnetic field (for $R = -1$) plotted versus temperature. The transition is second order for temperatures higher than the tricritical temperature T_t while for $T < T_t$ it is of first order. (*From Binder and Landau*[20].)

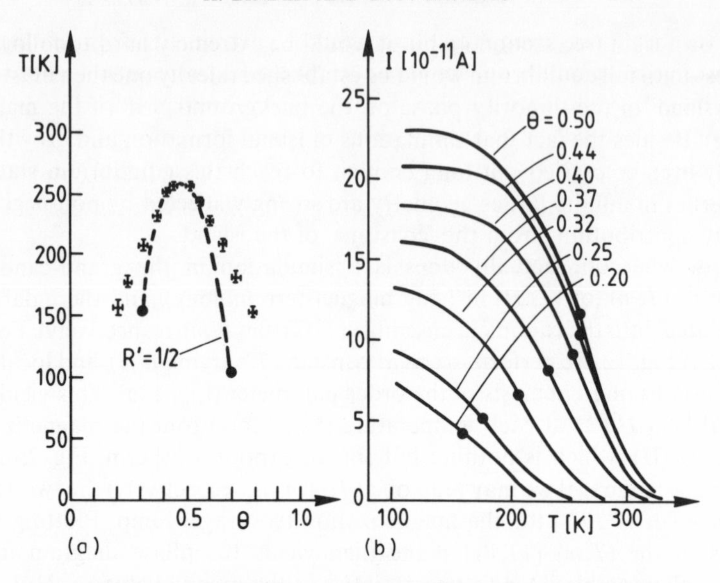

Fig. 13. (a) Experimental phase diagram for H adsorbed on Pd(100). Crosses denote the points $T_{1/2}$ where the LEED intensities have dropped to one-half of their low-temperature values. Dashed curve is a theoretical phase diagram obtained from[20] for $R_t = \varphi_t/\varphi_{nnn} = 1/2$. (b) LEED intensities plotted versus temperature at various coverages as indicated in the figure. (*From Behm et al.*[19].)

While in this system the lattice gas model is believed to be a good approximation of reality, the fact that the maximum transition temperature occurs for $\Theta_c \approx 0.48$ instead of $\Theta = 1/2$ shows that a model with strictly pairwise interaction is not adequate. Calculations with a reasonable value for the strength of the trio-interaction φ_t in Fig. 1 ($R_t = \varphi_t/\varphi_{nn} = \frac{1}{2}$) yield agreement with the experimental phase diagram at temperatures near the maximum transition temperature, but fail to reproduce the apparent widening of the ordered phase regime at lower temperatures.

Is this an indication that the lattice gas model is bound to fail? We believe not, for the following reason: due to the limited resolution of the LEED technique, Bragg scattering in the ordered region and critical scattering in the disordered phase close to the transition are not really distinguished. Therefore the intensity vs T curves look so smooth. Consequently their inflection points do not yield the true phase boundary accurately. We may model this limited resolution by finite system size, and construct also curves where the order parameter is plotted vs. temperature at constant coverage (Fig. 14a[20]). These curves in fact look strikingly similar to their experimental counterparts in Fig. 13b. Taking also the temperature $T_{1/2}$ as a criterion for drawing an apparent phase diagram, one indeed observes an apparent widening of the regime of ordered phase at low temperatures, similar to the experimental observations. Thus the conclusion of Ref. 20 was that a lattice gas model with nearest neighbor repulsion, some (weaker) next nearest neighbor attraction and a bit of trio interaction is a reasonable model for H on Pd(100), though at present the comparison of experiment and simulation does not allow one to make an unique precise choice of these interaction parameters. In any case, this study illustrates one advantage of the Monte Carlo simulations: one

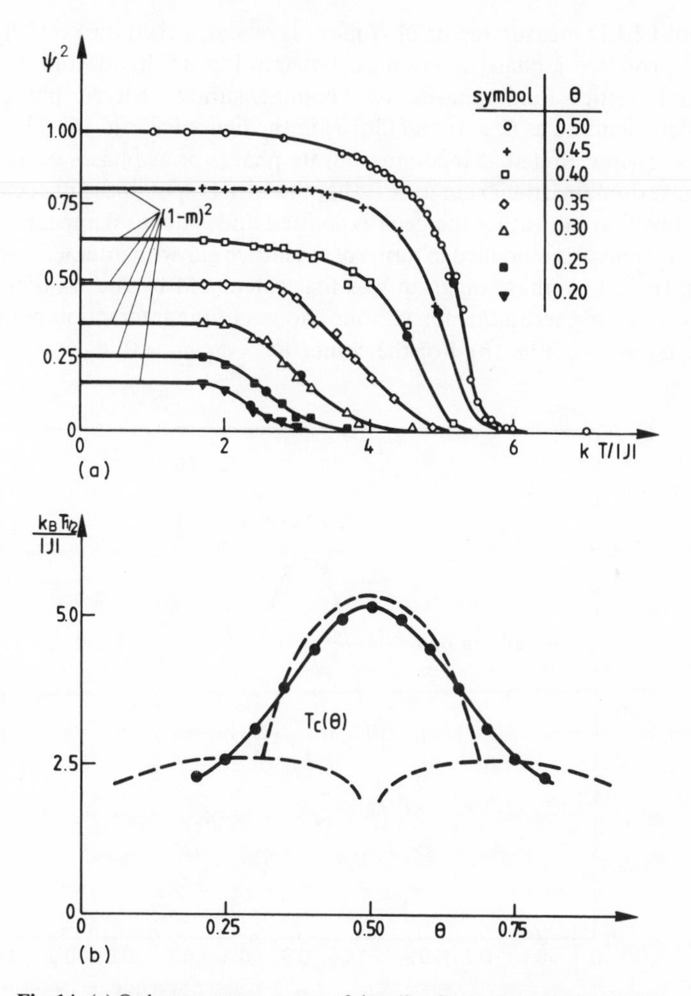

Fig. 14. (a) Order parameter square of the $c(2 \times 2)$ structure plotted versus temperature at constant coverage, as obtained from reconstructing constant field data as shown in Fig. 12a. Solid circles denote temperatures $T_{1/2}$ where the order parameter square has dropped to 50% of its low-temperature value. These data are for $R = -1$, $R_t = 0$. (b) 'Phase diagram' as derived from $T_{1/2}$ (dots and full curve) in comparison with the correct phase diagram (broken curves). (*From Binder and Landau*[20].)

can consider also experimental limitations such as limited resolution (or substrate inhomogeneity, etc.) rather directly. We shall return to this aspect in section 3.4.

3.2. Centered rectangular lattice gases: a model for the adsorption of H on Fe(110)?

From LEED measurements of H monolayers adsorbed on Fe(110) Imbihl *et al.*[21] proposed a phase diagram as shown in Fig. 15. In addition to 'lattice gas' and 'lattice fluid' phases, two commensurate ordered phases were identified, denoted as (2×1) and (3×1) in the figure (cf. Fig. 16). The shaded regions are interpreted as incommensurate phases or as phases composed of antiphase domains; their signature is that the LEED spot does not occur at the Bragg position but rather the peak is splitted and 'satellites' appear (Fig. 17).

This system was modelled in terms of the lattice gas with interactions shown in Fig. 1b[22]. The phase diagram was first calculated by the transfer matrix finite size scaling technique for various choices of the interaction parameters $\varphi_1, \varphi_2, \varphi_3$ and φ_t (Fig. 1b). For the choice $R_1' = \varphi_1/\varphi_2 = 0$, $R_3' = \varphi_3/\varphi_2 = 1/3$

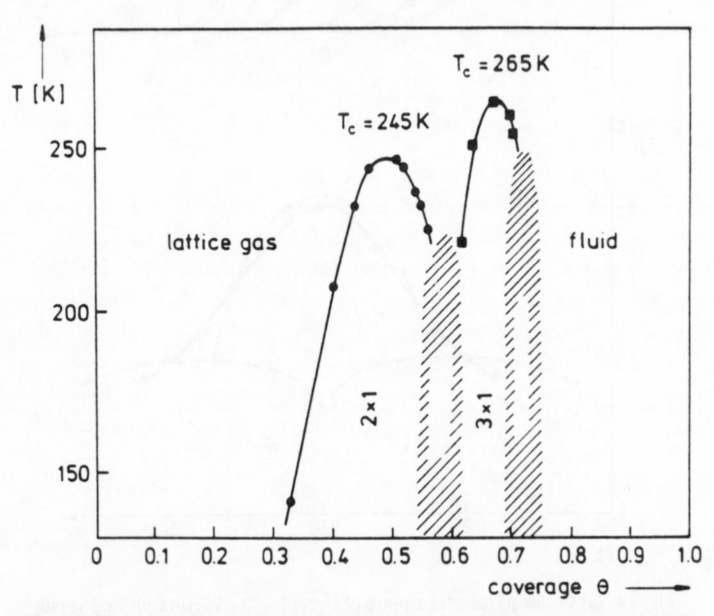

Fig. 15. Phase diagram of the H/Fe(110) system, as determined from LEED intensities. Full dots represent experimentally determined data points; shaded areas correspond to incommensurate or 'antiphase domain' regions. A possible interpretation for the ordered (2×1) and (3×1) phases is indicated in Fig. 16, assuming that the adsorption sites form a centered rectangular lattice as shown in Fig. 1b. (*From Imbihl et al.*[21].)

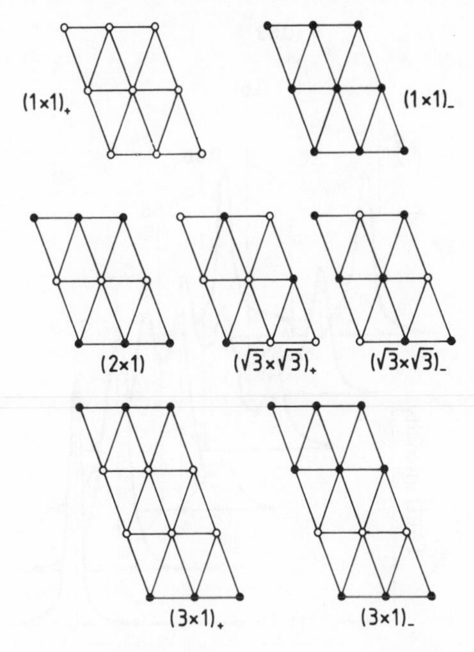

Fig. 16. Several ground state structures of the lattice gas model on the centered rectangular lattice. Solid circles represent occupied sites, open circles represent empty sites. (*From Kinzel et al.*[22].)

and $R'_t = \varphi_t/\varphi_2 = -1/3$ the phase diagram has also been checked by Monte Carlo simulation (Fig. 18), since this case seemed to have qualitative similarity with the experiment (Fig. 15): the maximum transition temperature of the (2×1) phase and the (3×1) phase with $\Theta \approx 2/3$ are comparable, while the transition temperature of the (3×1) phase with $\Theta \approx 1/3$ is much smaller (due to the strong trio-interaction): the initial interpretation was that in the real system (Fig. 15) this phase was not seen at all because at low temperatures ($T \lesssim 140$ K) the chemisorbed layer can no longer be equilibrated, since surface diffusion is frozen out.

In this model a study of the short-range order in the disordered phase of the lattice gas is also very interesting. This short-range order is described by a correlation function

$$\langle S_i S_j \rangle_T \propto \exp[-|\vec{r}_i - \vec{r}_j|/\xi] \cos[|\vec{r}_i - \vec{r}_j|\phi + \Delta\phi] \tag{53}$$

where ξ is the correlation length which diverges at T_c, and the angle ϕ determines the oscillations of the large distance correlations and we also have allowed for a 'phase shift' $\Delta\phi$. For the (2×1) structure one has $\phi = \pi$, $\Delta\phi = 0$, while for the (3×1) and $(\sqrt{3} \times \sqrt{3})$ structures (Fig. 16) one has $\phi = 2\pi/3$. It

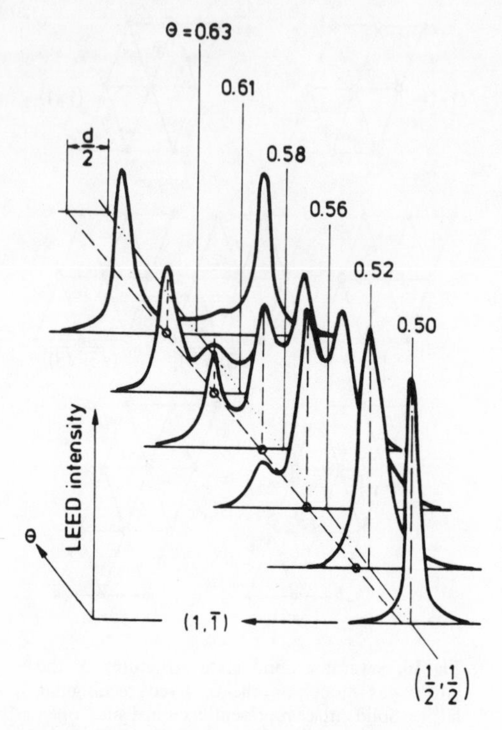

Fig. 17. Angular LEED beam profiles plotted for various values of the coverage for H on Fe(110) at a temperature $T = 200$ K. (*From Imbihl et al.*[21].)

turns out that for the model of Fig. 18 the angle ϕ sticks at its commensurate value $\phi = \pi$ for temperatures T less than the disorder line temperature T_D (dashed curves in Fig. 18), while for $T > T_D$, ϕ depends on both T and ϕ[22].

A related behavior is also found for the structure factor $S(q)$, which for a $N \times M$ lattice is

$$S(\vec{q}) = \frac{1}{(NM)^2} \sum_{i,j} \langle S_i S_j \rangle_T \exp[i\vec{q} \cdot (\vec{r}_i - \vec{r}_j)] \tag{54}$$

Figure 19 shows $S(\vec{q})$ for $\vec{q}$ being chosen in the direction of the φ_3-interaction, at a temperature $k_B T/|J_2| = 1.4$. It is seen that for $\Theta \lesssim 0.53$ the peak occurs at the commensurate position ($\phi = \pi$ corresponds to $q = 30$), while for $\Theta \gtrsim 0.53$ the peak splits and these peaks move to $q = 20, 40$ ($\phi = 2\pi/3$) with increasing coverage. Although this is somewhat similar to the experimental structure factor, Fig. 17, a situation where $S(q)$ exhibits a split peak together with two side peaks is not found.

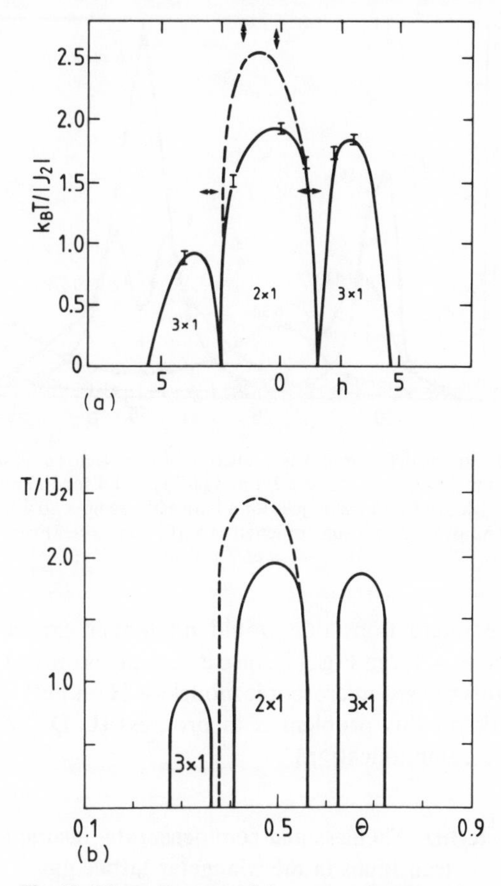

Fig. 18. Phase diagram of the centered rectangular lattice gas model with $\varphi_1 = 0$, $\varphi_3/\varphi_2 = 1/3$, $\varphi_t/\varphi_2 = -1/3$ plotted in the temperature–field plane (a) and in the temperature–coverage plane (b). The solid and dashed lines give the critical temperatures T_c and the disorder temperature T_D, as obtained from transfer matrix finite-size scaling (strips of width $N = 2$ and $N = 4$ are used). The error bars and arrows indicate T_c and T_D from Monte Carlo simulations. (*From Kinzel et al.*[22].)

Although this model calculation has many features in common with the real system H on Fe(110), one should not conclude too hastily that the data are explained uniquely by the model. In fact, recent work[90] has revealed that the actual adsorption sites for H on Fe(110) are not those shown in Fig. 1b; in reality there are twice as many adsorption sites: thus the coverage axis in Fig. 17 is labelled incorrectly, the '(2 × 1) phase' really occurs at $\Theta = 1/4$, and the '(3 × 1) phase' occurs at $\Theta = 1/3$ rather than at $\Theta = 2/3$. Thus even with

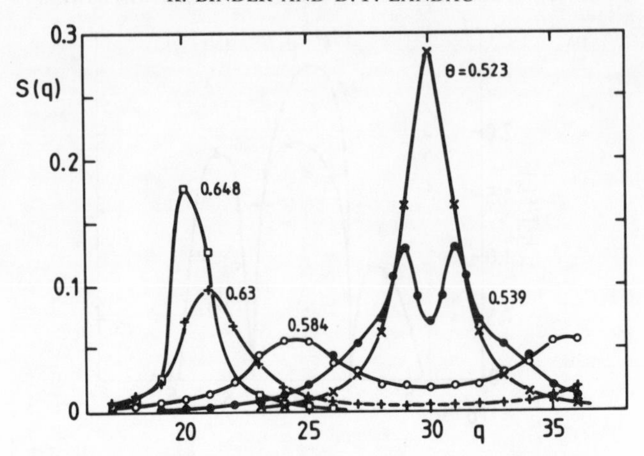

Fig. 19. Structure factor $S(q)$ as a function of wavevector q (in units of $2\pi/60$) for coverages $1/2 < \theta < 2/3$ at $k_B T/|J_2| = 1.4$ for the model of Fig. 18. Monte Carlo techniques for a lattice of size 60×20 were used, averaging over several runs of length 3×10^4 steps/site. (*From Kinzel et al.*[22].)

strictly pairwise interactions one would no longer expect any symmetry around the value $\Theta = 1/2$ in Fig. 17, since it actually is only $\Theta = 1/4$. A Monte Carlo study of this revised microscopic model for H on Fe(110) would be very interesting; work on this problem is in progress (L. D. Roelefs and G. C. Gumars, private communication).

3.3. Kosterlitz–Thouless and commensurate–incommensurate transitions in the triangular lattice gas

In the model of the previous section, we have encountered the situation that outside the regime bounded by the disorder line the correlation function, Eq. (53), is characterized by an angle ϕ which is incommensurate with the lattice. Now it is well known that in $d = 2$ dimensions, incommensurate long-range order (which would be described by $\langle S_i S_j \rangle_T = \psi^2 \cos[|\vec{r}_i - \vec{r}_j|\phi + \Delta\phi]$ instead of Eq. (53)) is unstable against thermal fluctuations (this is a two-component ordering – amplitude ψ, phase $\Delta\phi$ – similar to the XY model). Nevertheless, it is possible to have a phase transition where ξ in Eq. (53) diverges, but for $T < T_c$ one then has a power-law decay of the correlation function

$$\langle S_i S_j \rangle_T \propto |\vec{r}_i - \vec{r}_j|^{-\eta(T)} \cos[|\vec{r}_i - \vec{r}_j|\phi + \Delta\phi] \tag{55}$$

with $\eta(T \to 0) \to 0$, $\eta(T_c) = 1/4$[36]. The Kosterlitz–Thouless transition[36] from the disordered phase (where Eq. (53) holds) to the 'floating phase' (where Eq. (55) holds) is characterized by an exponent $1/\nu = 0$, i.e. there is an exponential divergence of the correlation length, namely[36]

$$\ln \xi \propto (T/T_c - 1)^{-1/2} \tag{56}$$

It is not clear whether in the centered rectangular lattice gas of section 3.2 such a Kosterlitz–Thouless transition occurs, or whether the disordered phase extends, though being incommensurate, down to the commensurate (3 × 1) phase (then this transition is believed to belong to a new 'chiral' universality class[38]), or whether there is another disorder line for (3 × 1) correlations. However, Kosterlitz–Thouless type transitions have been found for various two-dimensional models: the XY ferromagnet[36,30,91,92], the Coulomb gas[93],

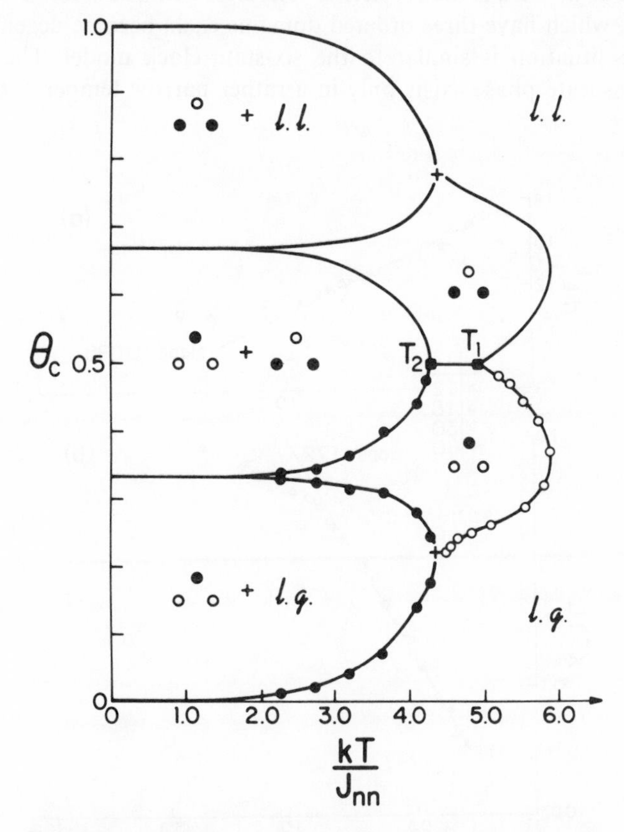

Fig. 20. Phase diagram of the triangular lattice gas model with nearest–neighbor repulsion and next-nearest neighbor attraction, for $J_{nnn}/J_{nn} = -1$, in the coverage–temperature plane. For $\theta = 0.5$ a Kosterlitz–Thouless transition occurs at T_1 and a commensurate–incommensurate transition at T_2. Two commensurate $\sqrt{3} \times \sqrt{3}$ phases (with ideal coverages of 1/3 and 2/3, respectively) occur, as well as several two-phase regions, as indicated in the figure. Here l.g. stands for 'lattice gas' and l.l for 'lattice liquid'. (*From Landau*[94].)

the anisotropic Heisenberg antiferromagnet in a field[43], the ANNNI model[32–35], and the triangular lattice gas with nearest neighbor repulsion and next-nearest neighbor attraction[94], the six-state clock model[95], etc. As an example, we describe here briefly the work on the triangular lattice gas[94] and reproduce the phase diagram for $J_{nnn}/J_{nn} = -1$ in Fig. 20[94]. At high temperatures there occur second-order transitions from the disordered phase to commensurate $\sqrt{3} \times \sqrt{3}$ phases (cf. Fig. 16); these transitions belong to the 3-state Potts universality class[26,27]. These transition lines end at low coverage (and high coverage, respectively) at tricritical points, where two-phase coexistence regions open up. The temperature T_1 where both second-order lines meet at $\Theta = 1/2$ is the Kosterlitz–Thouless transition: for $\Theta = 1/2$ both orderings, which have three ordered domains each, become degenerate and hence the situation is similar to the six-state clock model. The resulting incommensurate phase exists only in a rather narrow temperature interval

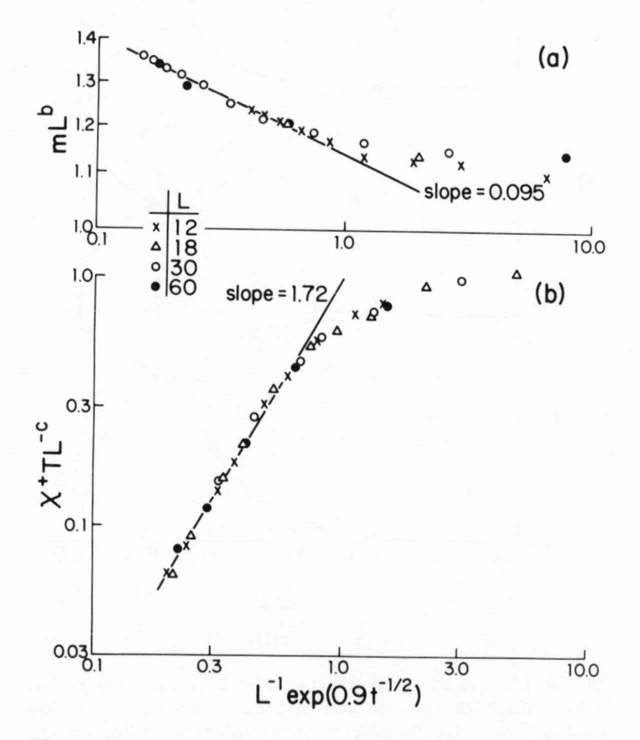

Fig. 21. Finite-size scaling plot of: (a) order parameter m using $T_c = T_2 = 4.26\ J_{nn}/k_B$, (b) the ordering susceptibility χ^+ using $T_c = T_1 = 4.89\ J_{nn}/k_B$. Here $t = (T_c - T)/T_c$, and the exponents $b = 0.095$, $c = 1.72$. (*From Landau*[94].)

($T_2 < T < T_1$), while for $T < T_2$ one has phase coexistence of commensurate $\sqrt{3} \times \sqrt{3}$ phases only.

In the finite size scaling analysis Eqs. (36)–(38) need modification at Kosterlitz–Thouless transitions[43]: the argument $(1 - T/T_c)L^{1/\nu}$ or alternatively $(L/\xi)^{1/\nu}$ must be replaced by L/ξ itself since $1/\nu = 0$, and Eq. (56) must be taken into account explicitly. The same fact is true for the commensurate–incommensurate transition. Figure 21 shows that then a reasonable data collapsing is obtained. In the incommensurate phase, Eq. (55) is also verified (Fig. 22) and the temperature dependence of $\eta(T)$ can be estimated.

Particularly interesting is the fact that in the incommensurate phase one can identify local defect configurations (triangles corresponding to the $(1 \times 1)_+$, $(1 \times 1)_-$ structure in Fig. 16), which are the analogues of the vortex (antivortex) cores in the XY model[91,92,43]. While for $T < T_1$ only few tightly bound 'vortex–antivortex' pairs occur (Fig. 23a), for $T > T_1$ also 'unbound' pairs of such excitations are identified (Fig. 23b). The direct analysis of system configurations generated by the Monte Carlo averaging in this case gives a very clear evidence in favor of the Kosterlitz–Thouless theory[36].

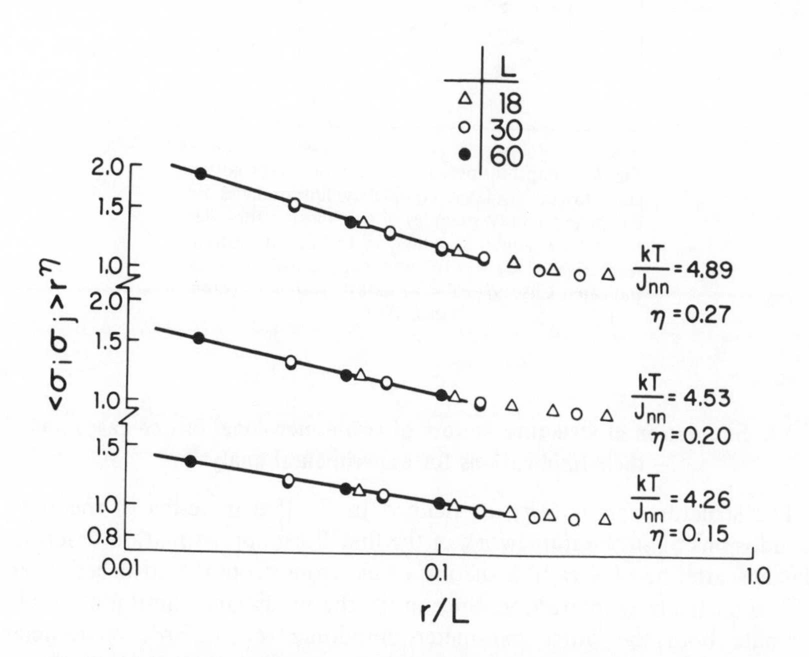

Fig. 22. Finite-size scaling plot of the order parameter correlation function at $\theta = 1/2$ and temperatures in the regime $T_2 < T < T_1$, for the model of Fig. 20. Resulting estimates $\eta(T)$ are indicated. (*From Landau*[94].)

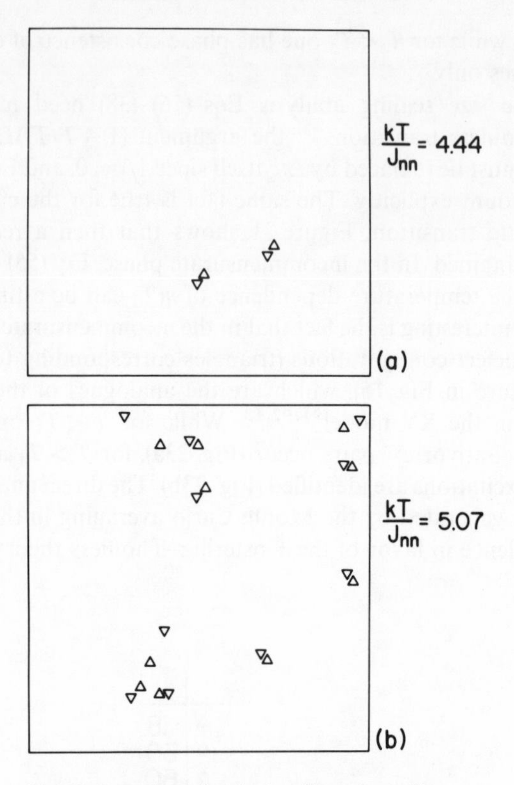

Fig. 23. Snapshots of typical configurations of 'vortex core' defects (isolated completely full triangles or completely empty triangles of neighboring sites, denoted by triangles standing on top or on bottom, respectively), for $\theta = 1/2$ and two temperatures as indicated. Case (a) to $T < T_1$, case (b) to $T > T_1$. (*From Landau*[94].)

3.4. Simulation of structure factors of two-dimensional lattice gases and their implications for experimental analysis

The structure factor $S(\vec{q})$, as defined in Eq. (54) in terms of the Ising pseudospins S_i, in the framework of the first Born approximation describes elastic scattering of X-rays, neutrons, or electrons, from the adsorbed layer. $S(\vec{q})$ is particularly interesting, since in the thermodynamic limit it allows to estimate both the order parameter amplitude ψ, the order parameter susceptibility χ_ψ and correlation length ξ, since for $\vec{q}$ near the superstructure Bragg reflection $\vec{q}_B$ we have ($\vec{k} \equiv \vec{q} - \vec{q}_B$)

$$S(\vec{q}) \propto \{\psi^2 \delta(\vec{k}) + \chi_\psi/(1 + k^2\xi^2)\}, \qquad k\xi \lesssim 1. \tag{57}$$

While in the ideal case $S(\vec{q})$ according to Eq. (57) clearly reflects the singular behavior at T_c, due to substrate inhomogeneity and/or limited resolution the actual behavior of scattering data is quite smooth; see Fig. 13b for an example. A detailed analysis of finite resolution effects on the structure factor of two-dimensional lattice gas models has been presented by Bartelt *et al.*[96–102]. If the instrumental resolution is characterized by a wave-vector k_I, the intensity for $\vec{q} = \vec{q}_B$ satisfies a scaling form ($t = 1 - T/T_c$)

$$I(k_I, T) = |t|^{2\beta} \tilde{I}_{\pm}(k_I \xi), \tag{58}$$

where $\tilde{I}_{\pm}(z)$ is a scaling function. In the limit of small z ($k_I \to 0$), one picks up just the terms expected from Eq. (57), namely the order parameter $\psi^2 = |t|^{2\beta} \tilde{I}_-(0)$ for $T < T_c$ and the ordering susceptibility χ_ψ for $T > T_c$, where $\tilde{I}_+(0) \equiv 0$ and hence $I(k_I, T) \propto k_I^2 \xi_0^2 |t|^{2\beta - 2\nu} \propto |t|^{-\gamma}$. However, for large z ($k_I > 1/\xi$) the LEED intensity is just due to a finite sum of correlation functions, all of which have an energy-like singularity, and hence also I itself is dominated by the energy-like singularity.

$$I(k_I, T) \approx I_0 - I_1 t \mp B_{\pm} |t|^{1-\alpha} + \ldots \tag{59}$$

where α is the exponent of the specific heat, and B_+ (B_-) are corresponding critical amplitudes for $T > T_c (T < T_c)$, and I_0, I_1 are constants.

While it is clear that Eq. (59) must hold for $t \to 0$ in the thermodynamic limit, in a realistic situation this singularity will be rounded off also (e.g. because of finite size of the substrate, quenched impurities, etc.) and thus it is not always clear that there is a significant range of t where Eq. (59) holds (for large t further correction terms come into play). Bartelt *et al.*[96,97] present model calculations for various structures to test Eq. (59). Figure 24 presents, as an example, the structure factor of a $\sqrt{3} \times \sqrt{3}$ overlayer in the triangular lattice gas with nearest-neighbor repulsive interaction, obtained from Monte Carlo calculations of lattices containing 3888 sites, and the resulting temperature dependence of the intensity integrated over 2.3% of the surface Brillouin zone (Fig. 25). Compelling evidence that the temperature dependence of the integrated structure factor coincides with the temperature dependence of the internal energy is in fact obtained. However, at the same time it is demonstrated that the analysis in terms of Eq. (59) is notoriously difficult, since Eq. (59) contains too many adjustable parameters. If the linear term $I_1 t$ is omitted, exponents $\alpha \gtrsim 1/2$ are obtained, although in the 3-state Potts model universality class[26,27] $\alpha = 1/3$ has been established[31]. A similar difficulty occurs with the study of the energy in the 3-state Potts model as well[103]. With this linear term $I_1 t$ included, the estimate for α drops to a value in between 0.27 and 0.39, depending on the temperature range and the k-vector range k_I included in the fit. Similar difficulties are encountered for other structures. Although the idea of distinguishing the various universality classes for adsorbed monolayers by use of Eq. (59) is rather attractive in principle, since

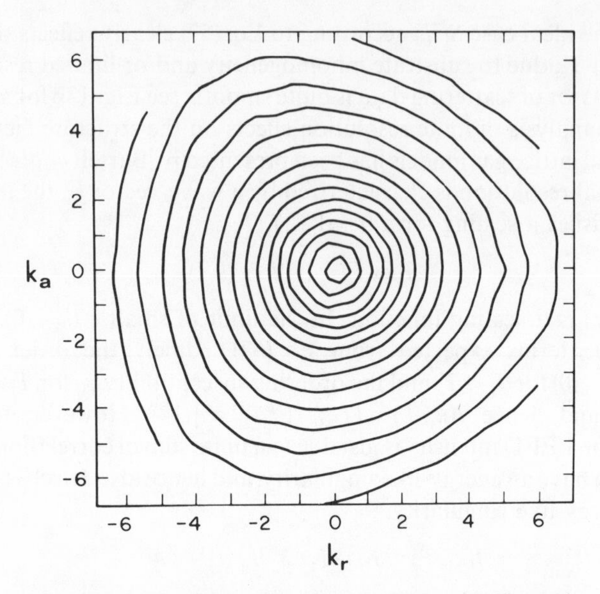

Fig. 24. Contour plot of the structure factor (the kinematic LEED intensity) of a $\sqrt{3} \times \sqrt{3}$ monolayer in a triangular lattice gas with nearest-neighbor repulsion, at a temperature $k_B T/\varphi_1 = 0.355$ (about 5% above T_c) and a chemical potential $\mu = 1.5\ \varphi_1$ ($\theta_c \cong 0.336$ at the transition temperature.) Contour increments are in a (common) logarithmic scale separated by 0.1, starting with 3.2 at the outermost contour. Center of the surface Brillouin zone is to the left; k_r and k_a, the radial and azimuthal components of $k\vec{k}$, are in units of $\pi/27a$, a being the lattice spacing. Data are based on averages over 2×10^5 Monte Carlo steps per site. (*From Bartelt et al.*[96].)

clear differences in α occur (α(Ising) = 0, α(3-state Potts) = 1/3, α(4-state Potts) = 2/3[31]), in practice there is the added complication that these exponents can only be observed in a chemisorbed layer, where the coverage rather than the chemical potential μ is fixed, if one works at the maximum critical coverage Θ_c for which $\mathrm{d}T_c(\Theta)/\mathrm{d}\Theta|_{\Theta_c} = 0$: if $\mathrm{d}T_c(\Theta)/\mathrm{d}\Theta \neq 0$, Fisher renormalization[104] of critical exponents occurs, and then the exponents to be observed are (for $\alpha > 0$) $\alpha_r = -\alpha/(1-\alpha)$, $\beta_r = \beta/(1-\alpha)$, $\gamma_r = \gamma/(1-\alpha)$. Then the singular term in Eq. (59) is very hard to identify, since it vanishes more strongly for $\alpha > 0$ than the linear term.

Studying the temperature dependence of structure factor 'data' as shown in Fig. 24, Bartelt *et al.*[101] also extract 'effective exponent' estimates for β, γ, and ν, working in the limits $k_l\xi < 1$ and using Eq. (57). They find that above T_c the exponents are within of order 10% of their expected values, while below T_c reliable estimates are more difficult to obtain. This work gives some justification for the notion of identifying T_c with the inflection point of the

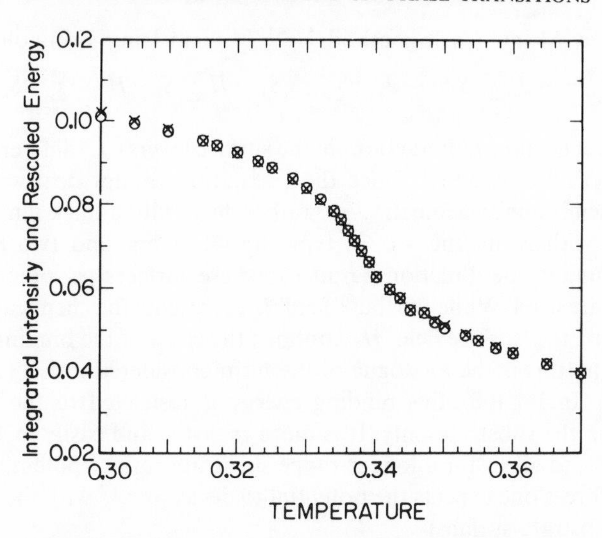

Fig. 25. Structure factor integrated over 2.3% of the surface Brillouin zone (radius of 5 mesh lengths in Fig. 24) vs. T(circles), plotted with the rescaled energy (crosses) for the $\sqrt{3} \times \sqrt{3}$ overlayer on the triangular lattice. Rescaling involves multiplication by a negative number and shifting by a constant. Temperature is measured in units of φ_1. (*From Bartelt et al.*[96].)

intensity vs T curve (for $k = 0$). A caveat about the implications of this work for experiment is that Ref. 101 only considers periodic boundary conditions, while in reality size effects are more pronounced due to various boundary perturbations and less regular shapes of the system (in Figs. 24, 25 a hexagonal regular parallelepiped is used). Bartelt and Einstein[98] consider this problem for the case of the nearest neighbor square Ising lattice, comparing size effects on the structure factor for free boundary conditions in the 1×1, 2×1 and circular geometry.

3.5. Simulation of wetting and layering transitions

So far we only have considered adsorption phenomena in the submonolayer range. It is well known, however, that for certain substrate–adsorbate partners it is possible to observe multilayer adsorption phenomena[15,54,56,57] or adsorption of fluid films which may undergo 'wetting transitions'[105–107] from a microscopic to a macroscopic thickness.

These problems are also readily modelled in terms of lattice gas systems, which now need to be semi-infinite three-dimensional lattices rather than two-dimensional ones, with an appropriate boundary condition at the 'free' surface to model the effects due to the substrate. A model which has been studied intensively by Monte Carlo methods[59,61–64] is, in Ising terminology,

$$\mathscr{H} = -J \sum_{\text{bulk}} S_i S_j - J_s \sum_{\text{surfaces}} S_i S_j - H \sum_i S_i - H_1 \sum_{\text{surfaces}} S_k \tag{60}$$

where nearest neighbor interactions in the surface layers (J_s) different from the bulk ones (J) are assumed. Since the simulation cannot deal with a truly infinite or semi-infinite geometry, one rather deals with a thick film geometry $L \times L \times D$, with a thickness D of typically 40 layers, and two free $L \times L$ surfaces, while in the direction parallel to these surfaces periodic boundary conditions are used. While the 'bulk field' H represents the chemical potential in the system, the 'surface field' H_1 contains the effect of the binding forces to the substrate (this is the analogue of the term considered in Eq. (10) for the monolayer). In Eq. (60), this binding energy is restricted to the first layer adsorbed on the substrate only. It is more realistic, and easily included in a simulation[55], to allow for a longer-range substrate surface potential (for van der Waals forces one expects the potential to decay as z^{-3} with the distance z from the substrate surface).

The Monte Carlo simulation of wetting and layering phenomena is a field of very great activity at the moment[58,59,61–64], as is the related field of critical phenomena at surfaces near phase transitions in the bulk[108–111]. A full account of this work is outside of the scope of this chapter. We only mention that the model, Eq. (60), allows the study of both wetting and layering phenomena, depending on whether one works above or below the interfacial roughening temperature T_R[112], which in the Ising model is estimated to occur at $T_R/T_c \approx 0.54 \pm 0.02$[113,114]. As an example, Fig. 26 shows the surface excess internal energy U_s and the surface excess magnetization m_s for layering, as well as the variation of m_s with H for second-order wetting, when H_1 has its critical value H_{1c}. The excess density n_s adsorbed at the substrate surface, $n_s = \int_0^\infty \{n(z) - n_{\text{bulk}}\} \, dz$, due to Eqs. (7), (12) is simply related to m_s as $n_s = m_s/2$. Since $H = 0$ corresponds to the chemical potential μ_{coex} for gas–liquid coexistence in the lattice gas, the logarithmic variation $m_s \propto -\ln H$ seen in Fig. 26a thus is evidence for the expected law $n_s \propto |\ln(\mu_{\text{coex}} - \mu)|$[15,107]. In contrast the stepwise adsorption isotherm of Fig. 26b is a clear signature of multilayer adsorption.

While Eq. (2) models submonolayer order–disorder transitions and Eq. (60) model multilayer adsorption, it is of course possible to formulate a combined model which considers the competition between order–disorder phenomena in the first layer and adsorption of further layers[58]. Then instead of Eqs. (2), (60) we write, for the simple cubic lattice,

$$\mathscr{H} = -\sum_{\text{surface}} \varepsilon_i c_i - \sum_{i \neq j} \varphi_{ij} c_i c_j,$$

$$\varepsilon_i = \varepsilon + \frac{1}{2} u_1 \left[\cos\left(\frac{\pi}{a} x\right) + \cos\left(\frac{\pi}{a} y\right) \right] + u_2 \cos\left(\frac{\pi}{a} x\right) \cos\left(\frac{\pi}{a} y\right) \tag{61}$$

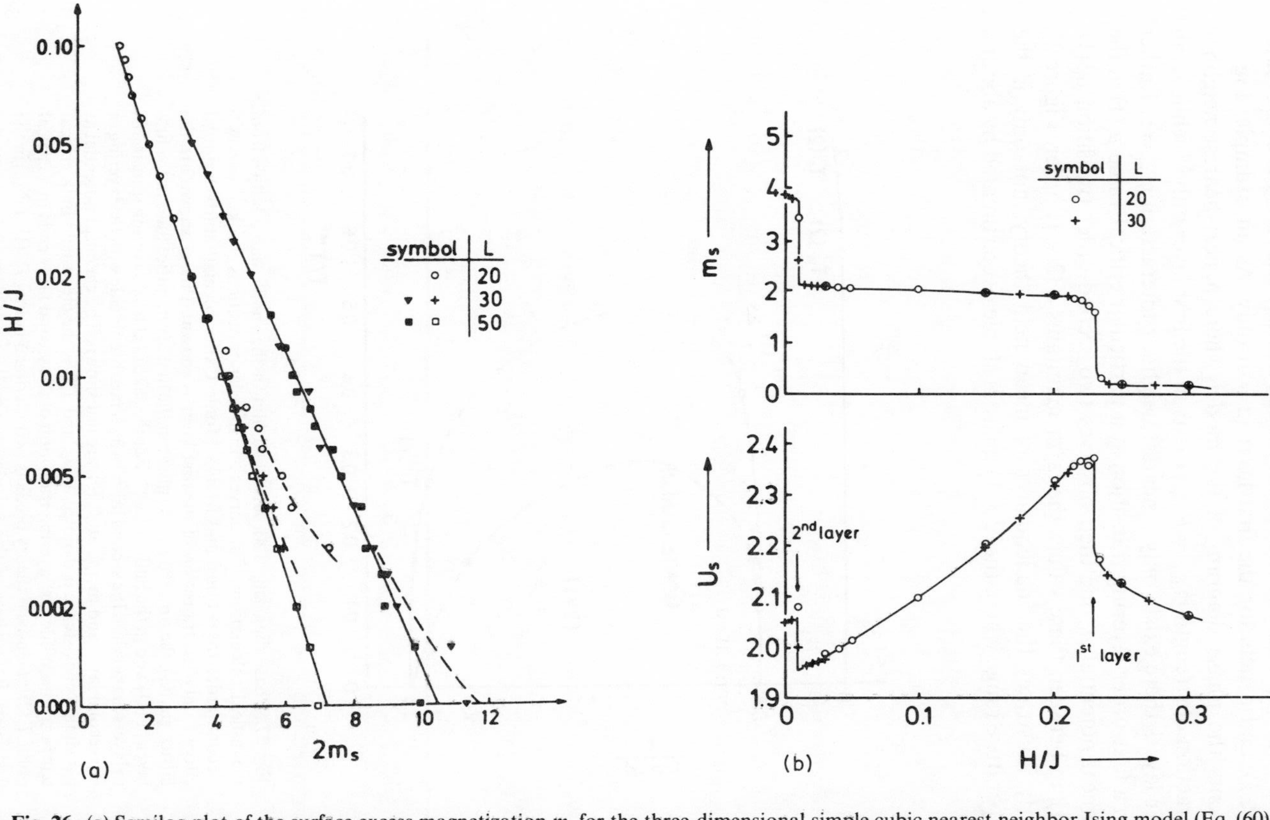

Fig. 26. (a) Semilog plot of the surface excess magnetization m_s for the three-dimensional simple cubic nearest neighbor Ising model (Eq. (60)) with $J_S/J = 1$ plotted vs. H/J (semi/log scales!) for $J/k_BT = 0.25$ and $H_1 = H_{1c} = -0.55$ J (solid circles and squares) and for $J/k_BT = 0.35$ and $H_1 = H_{1c} = -0.89$ J (open circles and squares). Various linear dimensions are included to show the onset of size effects for very small field. $D = 40$ is used throughout. (b) Surface excess magnetization m_s (upper part) and surface excess internal energy U_S (lower part), plotted vs. bulk field H at $J/k_BT = 0.45$, $J_S = 1$, and $H_1 = -1.2$ J. (*From Binder and Landau*[59].)

where u_1, u_2 represent amplitudes of a corrugation potential which has the periodicity of twice the lattice spacing a, and a lattice site i has the coordinates (x, y, z), and ε_i acts for the first layer $(z = a)$ only. As an example, Fig. 27 presents the phase diagram of the model where a next-nearest neighbor interaction of strength $\varphi_{\text{nnn}} = R\varphi_{\text{nn}}$ exists in the first layer only[58] while in all other layers there exists only a nearest neighbor interaction $\varphi_{\text{nn}} = v$. Rather than a first-order layering transition at a particular critical value $\mu_c(1)$ of the chemical potential μ, one then observes two second-order transition $\mu_c^1(1)$, $\mu_c^2(1)$; in between these values the layer maintains a (2×1) order. Figure 27 clearly illustrates the inadequacies of mean field theory: not only is the temperature range where the (2×1) phase is stable overestimated by about a

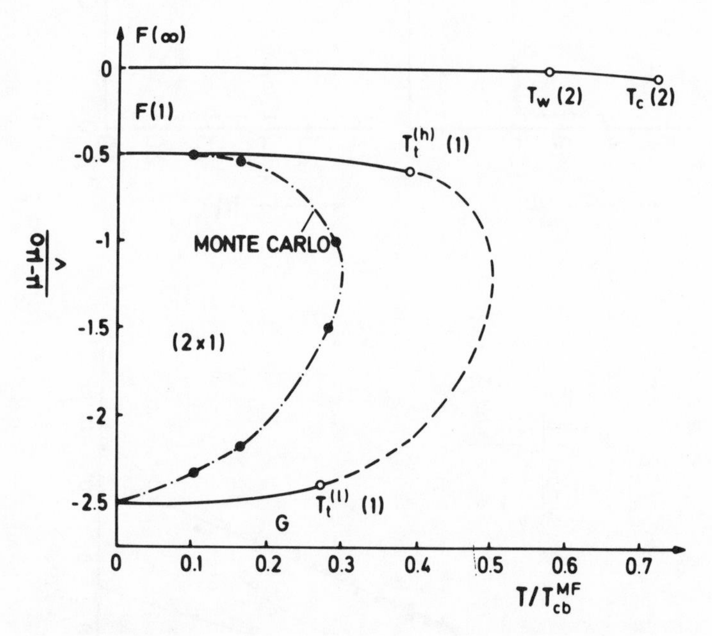

Fig. 27. Phase diagram of an adsorbed film in the simple cubic lattice from mean-field calculations (full curves – first-order transitions, broken curves – second-order transitions) and from a Monte Carlo calculation (dash-dotted curve – only the transition of the first layer is shown). Phases shown are the lattice gas (G), the ordered (2×1) phase in the first layer, lattice fluid in the first layer $\{F(1)\}$ and in the bulk $\{F(\infty)\}$. For the sake of clarity, layering transitions in layers higher than the second layer (which nearly coincide with the layering of the second layer and merge at $T_W(2)$, are not shown. The chemical potential at gas–liquid coexistence is denoted as μ_0, and T_{cb}^{MF} is the mean-field bulk critical temperature. While the layering transition of the second layer ends in a critical point $T_c(2)$, mean-field theory predicts two tricritical points $T_t^{(l)}(1)$, $T_t^{(h)}(1)$ in the first layer. Parameters of this calculation are $R = -0.75$, $\varepsilon = 2.5\,v$, $u_2 = u_1/^2 = v/2$, $D = 20$, and L varied from 6 to 24. (*From Wagner and Binder*[58].)

factor of two, but also the order of the transition at low temperatures erroneously is predicted to be first order.

Finally we return to monolayers at surfaces again and consider wetting phenomena at interfaces separating ordered domains[115–117]. For example, in the 3×1 phase at the centered rectangular lattice of section 3.2 at coverage $\Theta = 2/3$ we have three ordered domains A, B, C: introducing in Fig. 1b three

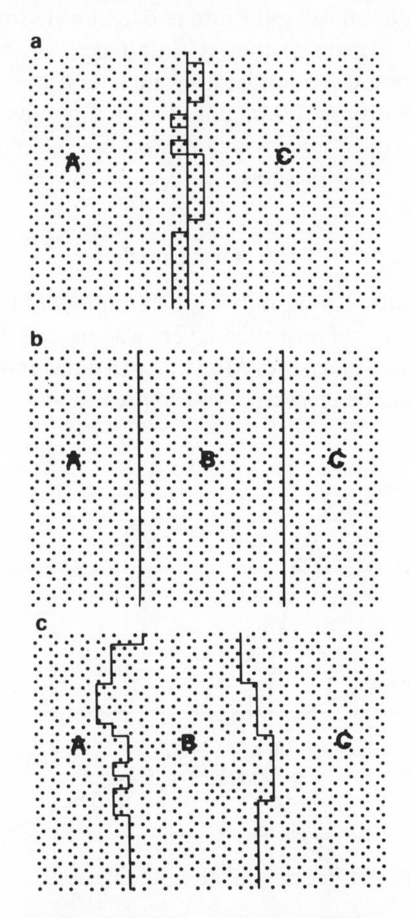

Fig. 28. Typical Monte Carlo equilibrium arrangements of atoms adsorbed on the centered rectangular lattice illustrating the wetting transition $A||C \rightarrow A|B|C$. Systems of sizes $L \times M = 20 \times 60$ are used at (a) $k_B T/|J_2| = 1.4$, $H/|J_2| = 2.8$; (b) $k_B T/|J_2| = 1.4$, $H/|J_2| = 2.3$; and (c) $k_B T/|J_2| = 2.8$. The solid lines show the borders of the A and C domains. Case (c) is rather close to the transition to the disordered phase. (*From Sega et al.*[115].)

one-dimensional sublattices by choosing subsequent rows in the y-direction, domain A has sublattices 1, 2 filled, sublattice 3 empty; domain B has sublattice 2 empty, while domain C has sublattice 1 empty. Now a domain boundary AC involves a 'heavy wall' (three successive filled rows), while a domain boundary AB or BC involves a 'light wall' each (three successive rows where only the middle one is filled). Now an interfacial wetting transition can occur where a heavy wall AC splits into two light walls (of type AB and BC), by adsorption of a B-domain at the AC interface (Fig. 28). A similar wetting transition occurs for a CA boundary, where a light wall splits into two heavy walls CB, BA by adsorbing B-domains at a CA interface. The location of these interfacial wetting transitions, which always are of second order, have been found with Monte Carlo methods[115].

3.6. Other models

In the previous subsections, only a small number of typical applications of the Monte Carlo study of adsorbed layers was treated in detail to show how one proceeds with the analysis of Monte Carlo data, and to illustrate the type of questions that can be answered. In the present subsection, we give a brief

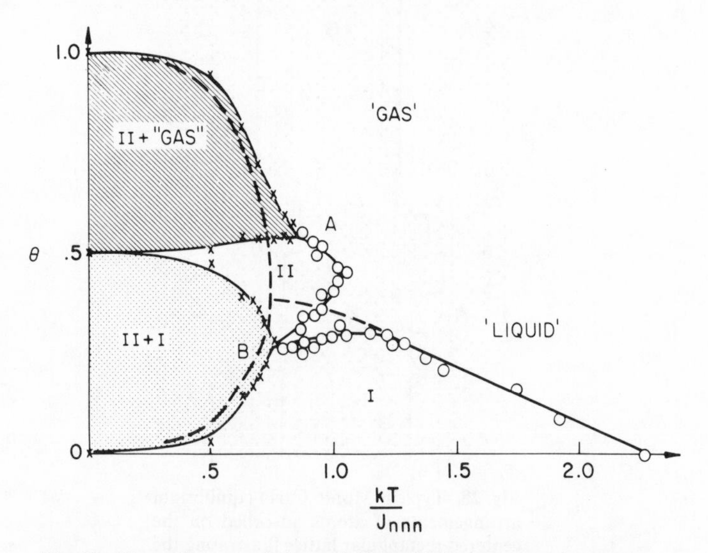

Fig. 29. Phase diagram of the model Eq. (22) for coadsorption of two kinds of atoms in the temperature–coverage space. Circles indicate a second-order phase transition, while crosses indicate first-order transitions. Point A is believed to be a tricritical point and point B a bicritical point. The dashed curve shows the boundary from the Blume–Capel model on a square lattice with a nearest-neighbor coupling equal to J_{nnn} in the present model (for $K_{nn} = 0$ Eq. (22) reduces to this model), only the ordered phase I then occurs. (*From Lee and Landau*[52].)

enumeration of various other applications which we find rather interesting, yet without aiming at completeness.

In Fig. 29 we return to the model for coadsorption of two different kinds of atoms (Eq. (22)); the expected ordered structures of this model have been shown in Fig. 4. Figure 29 now shows the phase diagram in the coverage–temperature space[52], for the special case $K_{nn}/J_{nnn} = 1/4$ (another case has been treated in Ref. 118). Again the finite size scaling technique (section 2.3) has been instrumental in mapping out the details of this rather complicated phase diagram. Further work on the multicritical behavior of this model should be rather interesting.

A particular complex problem has been the modelling of Si/W(110): Amar *et al.*[119] have included pairwise interactions up to the sixth nearest neighbor shell, as estimated experimentally from field-ion microscopic studies[120]. The predicted phase diagram (Fig. 30) exhibits (5×1), (6×1) and $p(2 \times 1)$ commensurate phases, as well as a broad regime of an incommensurate phase. In contrast to the ANNNI model[33, 34], the present model does seem to have a finite-temperature Lifshitz point, where the incommensurate, commensurate

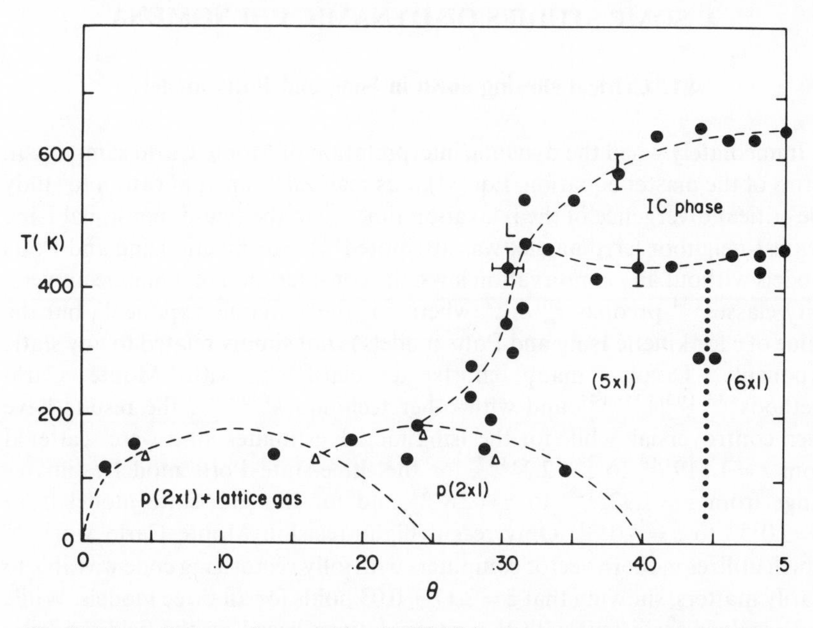

Fig. 30. Phase diagram of a model for Si/W(110) in the temperature versus Θ plane. Experimentally determined interactions $J_1, J_2, \ldots, J_6$ are used. Full dots are from Monte Carlo calculations, while triangles are based on transfer matrix finite size scaling using strip widths of 8 and 12. The point labelled L indicates approximate location of Lifshitz point. The dotted line indicates the transition region between the (5×1) and (6×1) phases. (*From* *Amar et al.*[119].)

and disordered phases meet. Experimental checks of this predicted phase diagram would be most illuminating.

There have been numerous other studies of phase diagrams of lattice gas models on the square lattice (e.g. Refs. 121–123), the triangular lattice (e.g. Refs. 124–128), the honeycomb lattice (e.g. Ref. 101), etc. Some of these studies have helped to understand phase transitions in real systems such as Se adsorbed on Ni(100)[123] or Cl on Ag(100)[122] or O on Ni(111)[129]. Some studies are done also with the motivation to understand phase transitions in graphite intercalation compounds[128]. Finally we again draw attention to models with continuous instead of discrete degrees of freedom, such as the ϕ^4 model on the square lattice[39–42], Eq. (20), which exhibits a phase transition in the Ising universality class, and the anisotropic Heisenberg antiferromagnet in a field[43], Eq. (21), which exhibits both an Ising transition to a simple antiferromagnetic structure, and a Kosterlitz–Thouless transition to a spin–flop-like phase. Particularly interesting is also the triangular XY antiferromagnet[130–132], which possesses a very complicated phase diagram with both Ising and Kosterlitz–Thouless orderings occurring.

4. SOME STUDIES OF DYNAMIC PHENOMENA

4.1. Critical slowing down in Ising and Potts models

Immediately when the dynamic interpretation of Monte Carlo sampling in terms of the master equation, Eq. (31), was realized[79], an application to study the critical divergence of the relaxation time $\tau_m^{(l)}$ in the two-dimensional Ising nearest-neighbor ferromagnet was attempted[133]. For kinetic Ising and Potts models without any conservation laws, the consideration of dynamic universality classes[134] predicts $\tau_m^{(l)} \propto \xi^z$, where z is the 'dynamic exponent', but the value of z for kinetic Ising and Potts models is not simply related to any static exponents. Despite many efforts to clarify z with Monte Carlo methods[133, 103, 135–141] and with other techniques[142–151], the results have been controversial: while for the Ising model, estimates for z were scattered from $z = 1.819$[148] to $z = 2.24$[147], for the three-state Potts model estimates range from $z = 1.922$[148] to $z = 2.8$[151] and for the four-state model from $z = 2.0$[148] to $z = 4.0$[151]. Only recent high-precision Monte Carlo work[141] which utilizes modern vector computers with fully vectorizing code was able to clarify matters, showing that $z = 2.17 \pm 0.03$ holds for all three models. While this result is consistent with the extrapolations based on the field-theoretic renormalization of the kinetic Ising model[146], the Domany conjecture[151] is clearly ruled out. As expected, real space renormalization[147–150] is not a reliable approach to critical dynamics either.

While a high-resolution inelastic neutron scattering study[152] has attempted to estimate z for the quasi-two-dimensional Ising-like antiferromagnet Rb_2CoF_4 (yielding $z = 1.69 \pm 0.05$, however), we are not aware of any experimental work on dynamic critical phenomena in adsorbed layers. As pointed out in Refs. 153, 154, 11, 41, a promosing method to do this could be quenching experiments, where the initially disordered chemisorbed layer is suddenly quenched to $T = T_c$ and one monitors the time evolution of the equal-time structure factor $S(q, t)$ (defined by Eq. (54)) as function of time t after the quench. The prediction of Sadiq and Binder[153,154] (see also Refs. 155, 135) is expressed in terms of a scaling function $\tilde{S}$

$$S(k, t) = [l(t)]^{\gamma/\nu} \tilde{S}\{kl(t)\}, \qquad l(t) \underset{t \to \infty}{\propto} t^{1/z} \tag{62}$$

while for $T > T_c$ the time-dependent characteristic length $l(t)$ would saturate at the correlation length ξ, and $S(k, t)$ would saturate at the scaling form of the structure factor in equilibrium[24], $S(k) = \xi^{\gamma/\nu} \tilde{\tilde{S}}(k\xi)$, $\tilde{\tilde{S}}$ being another scaling function. Quenching experiments to $T < T_c$, by which one can study the growth kinetics of ordered domains at surfaces, will be mentioned in section 4.3 together with the corresponding simulations.

4.2. Surface diffusion in ordered monolayers

In this subsection, we are not concerned with simulations which study the motion of single adsorbate atoms for realistic choices of the corrugation potential, but again restrict attention to simplified lattice gas models, where diffusion events are modelled by stochastic hops of adatoms from one lattice site to the next[80,156–162]. The description of the dynamics hence again is done in terms of the master equation, Eq. (31): but now the step $\vec{X} \to \vec{X}'$ is not a single-spin flip of the corresponding kinetic Ising model, but rather a move where the concentration variable c_i is interchanged with the concentration variable c_{li} at a randomly chosen nearest neighbor site l_i of i. The transition probability $W(\vec{X} \to \vec{X}')$ then becomes

$$W(c_i \to c_{li}, c_{li} \to c_i) = f_T(i) c_i (1 - c_{li})/\Theta \tag{63}$$

where the factor $f_T(i)$ has to be chosen in accord with the detailed balancing principle, Eq. (32). Explicit choices are[80] $f_T(i) = \{\exp(\delta\mathcal{H}/k_B T) + 1\}^{-1}$ or[158,159] $f_T(i) \propto \exp\{\sum_j \varphi_{ij} c_j / k_B T\}$ for a pairwise Hamiltonian $\mathcal{H} = \sum_{\langle i,j \rangle} \varphi_{ij} c_i c_j$. The factor $c_i(1 - c_{li})$ in Eq. (63) means that only occupied sites i are considered for a move, and double occupancy of a lattice site is forbidden. Now the selfdiffusion coefficient D_t of adsorbate atoms is defined from the mean square displacement of 'tagged' particles, $D_t = \lim_{t \to \infty} \sum_i \langle r_i^2(t) \rangle_T / 4N_a t$. In addition, it is also of interest to consider the 'collective' diffusion constant D which measures how a local excess of coverage spreads out.

Order–disorder transitions of the adsorbed monolayers now are reflected in a complicated dependence of both D_t and D on coverage Θ and temperature T. A rather extensive model study has been made in (Sadiq and Binder[80], for the square lattice gas with repulsive interactions between nearest and next-nearest neighbors (and the special case $\varphi_{nn} = \varphi_{nnn}$). For Θ near 1/2, this model orders in the (2 × 1) structure (see section 2.3 for a discussion of static properties of this model). For Θ near 1/4 and near 3/4 additional orderings occur, which are of (2 × 2) type if degeneracies of these structures are disregarded[5]. Figures 31, 32 show the collective diffusion constant and the self-diffusion constant for this model as a function of coverage for several temperatures. The data on D were obtained[80] with a linear response technique, while D_t is obtained straightforwardly from mean-square displacements. It is seen that D_t has distinct minima in the stoichiometric structures (2 × 1, (2 × 2): due to nearly perfect order, each adatom is nearly 'locked' in its position, while off stoichiometry the inevitable disorder makes self-diffusion more easy. In contrast, the collective diffusion is found to have maxima near $\Theta \approx 0.2$, 0.8 and a minimum near $\Theta = 1/2$. This behavior results from a competition between the coverage dependence of the jump rate Γ and the static 'susceptibility' χ defined as $\chi = (\partial\Theta/\partial\mu)_T$. It is

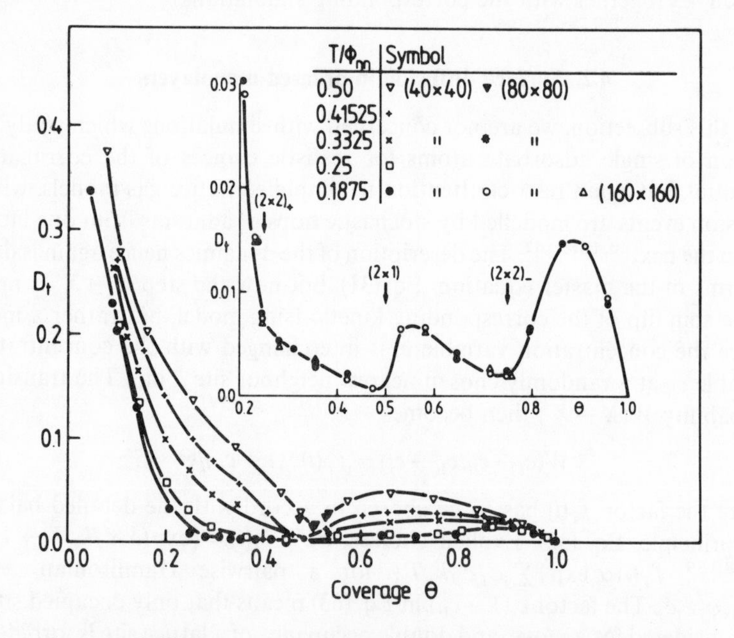

Fig. 31. Self-diffusion constant D_t plotted versus coverage at several temperatures for the square lattice gas with repulsive interactions φ_{nn}, φ_{nnn} between nearest and next nearest neighbor interactions ($\varphi_{nn} = \varphi_{nnn}$) At all temperatures shown the system orders in the (2 × 1) structure; the lowest temperature, where also $(2 \times 2)_+$, $(2 \times 2)_-$ orderings at $\Omega = 1/4$, 3/4 appear is shown in the inset. Various lattice sizes were used, as indicated by different symbols. (*From Sadiq and Binder*[80].)

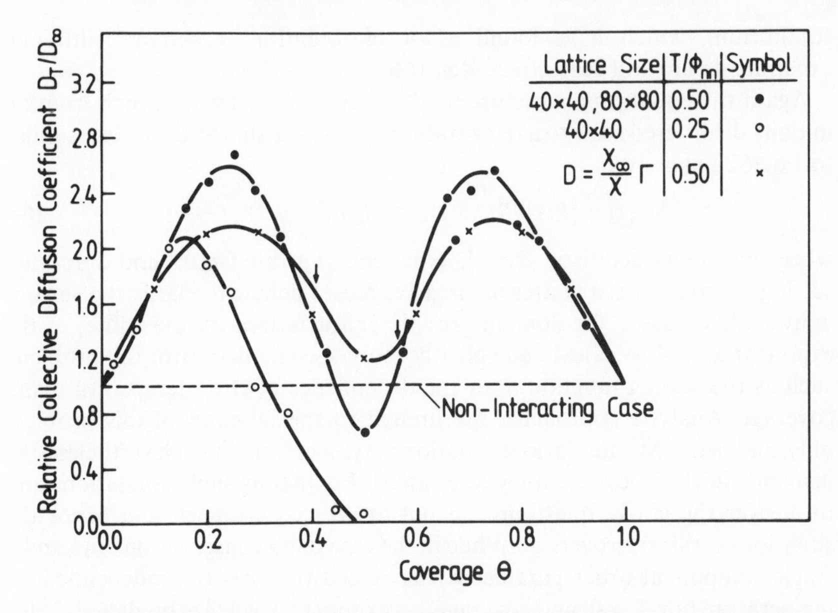

Fig. 32. Collective diffusion constant D plotted versus coverage at two temperatures where the square lattice gas of Fig. 31 is ordered near $\Omega = 1/2$ (arrows denote the coverages at which the order–disorder transitions occur for $k_B T/\varphi_{nn} = 0.5$). The mean-field approximation for D, $D = (\chi_\infty/\chi)\Gamma, \chi_\infty \equiv \chi(T = \infty)$, is shown for comparison, but using the correct values of susceptibility χ and jump rate Γ obtained from the same simulation. (*From Sadiq and Binder*[80].)

notable that neither D nor D_t have a detectable singularity at the coverages Θ_c where the order–disorder transitions occur.

A rather strong variation of D with coverage has been observed in the system O/W(110)[163,164], where a $p(2 \times 1)$ structure occurs for Θ near 1/2. An interesting attempt to model diffusion in this system by Monte Carlo methods is due to Tringides and Gomer[62], but only qualitative agreement with experiment is obtained. Clearly, the work shown in Figs. 31, 32[80] rather is a feasibility study, and the particular choice of $f_T(i)$ which yields a Θ-dependence of Γ symmetric, around $\Theta = 1/2$, is not expected to be realistic for adsorbate monolayers. In addition, the conversion of the 'time' unit of the Monte Carlo simulation to real times also involve unknown factors (reflecting the energy barriers involved in a simple diffusion jump for an isolated adatom). This may account for some of the problems encountered in comparing the simulation results[162] to the O/W(110) data.

4.3. Growth of ordered domains at surfaces; growth of wetting layers

In this section, we very briefly turn to phenomena far from thermal equilibrium, which have found a lot of attention recently[165] although experiments are still scarce (e.g. Ref. 166).

Again the equal-time structure function at time t after a quench from an initially disordered state to a temperature $T < T_c$ satisfies a scaling law similar to Eq. (62), namely

$$S(k,t) = [l(t)]^2 \hat{S}\{kl(t)\}, \qquad l(t)/\xi = (\Omega t)^x, t \to \infty \tag{64}$$

where x is a characteristic growth exponent, Ω a rate factor, and $\hat{S}$ another scaling function. It is a matter of intensive recent debate to clarify the various 'universality classes' for domain growth, characterized by the values of the exponent x, and to calculate explicitly more specific details of the problem, such as the scaling function $\hat{S}$, or the dependence of Ω on temperature and coverage. Analytic approaches are limited to special cases of this problem only, and hence Monte Carlo simulation plays a unique role. Nevertheless, the demands in statistical accuracy to establish Eq. (64) by such simulations and to answer the above questions are not at all easy to meet, and hence the situation is still controversial. While in the case of no conservation laws and a single-component order parameter it is agreed that $x = 1/2$, independent of temperature (for $T > 0$) and coverage, an exponent $x = 1/3$ is predicted[165] for transitions involving unmixing (i.e. a dynamics with 'conserved one-component order parameter'). For more complicated situations, other exponent values have also been suggested[165] but are not established beyond doubt. Clearly substantially larger computational effort will be needed to resolve these problems. In a pioneering experiment[167] on the O/W(110) system first experimental evidence for Eq. (64) was recently obtained.

Another very complicated problem where the approach to equilibrium with time after a quenching experiment is described by an asymptotic law is the growth of wetting layers, in a situation where thermal equilibrium would require the surface to be coated with a macroscopically thick film, but is initially nonwet. For a short-range surface potential as discussed in section 3.5, analytical theories[168,169] predict for a non-conserved density a growth of the thickness of the layer according to a law $l(t) \propto \ln t$, and this has in fact been observed by simulations[62,170]. In the case where the surface potential decays with distance z from the surface as z^{-p}, the prediction for the thickness $l(t)$ is $l(t) \propto t^{1/(p+2)}$ for the nonconserved case[168] and $l(t) \propto t^{1/2(p+2)}$ for the case of conserved density[171], which grows by diffusion of adatoms from the gas phase to the surface. The first of these predictions has already been verified for several choices of the potential exponent p, by Monte Carlo simulation of a solid on solid model[170].

Finally we draw attention to simulations of the growth of adsorbate islands for models of chemical reactions at surfaces, such as $A(a) + B(a) \to AB(g)$ where two reactants (A, B) are adsorbed at the surface (a) while the reaction product AB is rapidly desorbing to the gas phase (g)[172,173]. A well-known system exhibiting such behavior is $O(a) + CO(a) \to CO_2(g)$ on metal surfaces[174–177].

In the Monte Carlo modeling[172] it was assumed that the $\tau_B \ll \tau_R$, τ_A, where $\tau_A(\tau_B)$ are the diffusion times (reciprocal jump frequencies) of A(B) and τ_R is the reaction time of an AB pair. The system is modelled by a lattice gas mixture, with nearest-neighbor interaction (typically assumed to be nonzero for AA pairs only). At time $t = 0$, A atoms are adsorbed randomly at the surface with an initial coverage Θ_A^0. Subsequent equilibration at a chosen temperature is modelled by simulating surface diffusion as described in section 4.2. At some later time the adsorption of B-atoms is started, and chemical reactions take place according to some specified reaction probabilities if A, B are nearest neighbors. Following experimental procedures, the temperature is changed linearly with time in this reaction phase and the total reaction rate per site, i.e. $-\,\mathrm{d}\Theta_A/\mathrm{d}t$, is 'measured'.

Clearly this is a very interesting problem and of great practical relevance, very well suited to Monte Carlo simulation. At the same time, simulations of such problems have just only begun. In the context of crystal growth kinetics, models where evaporation–condensation processes compete with surface diffusion processes have occasionally been considered before[178]. But many related processes can be envisaged which have not yet been studied at all.

5. DISCUSSION

The Monte Carlo method has now been used for about 15 years to answer questions about phase transitions in adsorbed layers. Many models have been studied in the context of putting theoretical predictions to a critical test as well as from the perspective of modeling experiments on real systems. It has been clearly shown that the method is a useful and versatile tool: it is applicable to models with both continuous and discrete degrees of freedom, it allows inclusion of both pairwise additive and non-additive interactions and it may be used on systems with interactions extending over very different length scales, etc. Information can be obtained on all quantities of interest – ordering energies and specific heats, adsorption isotherms, order parameters, diffuse scattering intensities, etc. Both homogeneous and inhomogeneous systems may be treated and thus systems with interfaces, multiple coexisting phases, or with imperfections such as steps or impurities may be considered without difficulty. The information produced by simulations is not restricted to bulk quantities, and one can also look at typical configurations of the system and study the excitations on an atomic scale including how they evolve in time. Time-dependent studies, of course, rely on the stochastic dynamics inherent in Monte Carlo simulations which to some extent simulate random evaporation–condensation events (if one works in the grand-canonical ensemble where the chemical potential of the adsorbed layer is held fixed) or surface diffusion events (if one works in the canonical ensemble where the coverage is held fixed). Monte Carlo simulations are well suited to the study of dynamics of fluctuations near thermal equilibrium as well as to the investig-

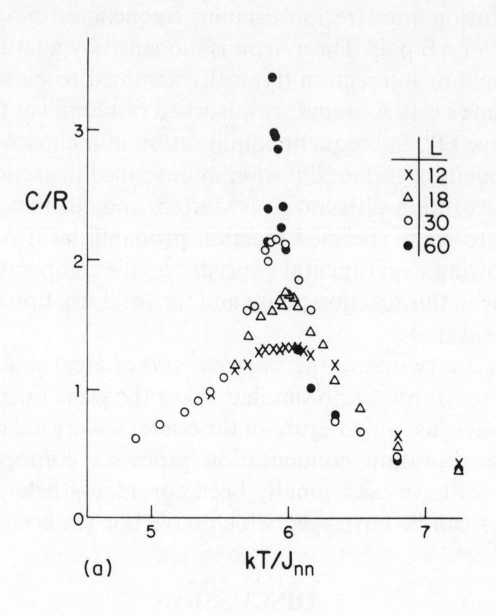

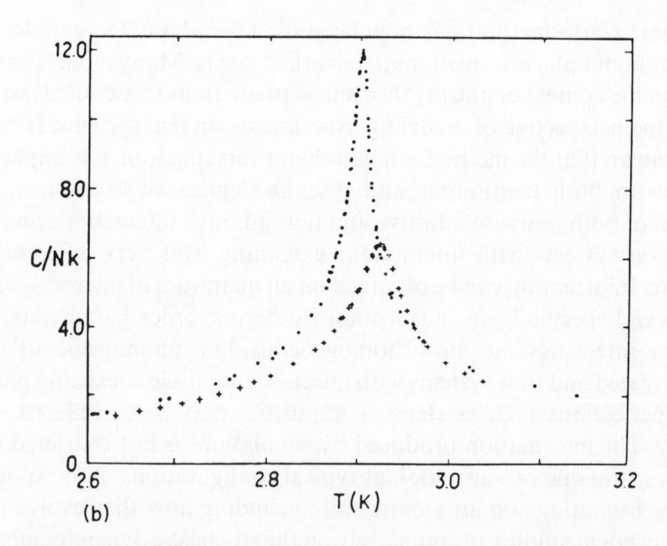

Fig. 33. (a) Specific heat of the triangular lattice gas vs. temperature for $H/J_{nn} = 2.43$, $J_{nnn}/J_{nn} = -1$. (*From Landau*[94].) (b) Specific heat of He^4 adsorbed on grafoil (plus signs) and on UCAR-ZYX (dots). (*From Bretz*[25].)

ation of relaxation phenomena far from equilibrium (e.g. domain growth in surface layers which are initially disordered, nucleation phenomena, etc.).

In this review we have tried to illustrate most of the above points by giving a few representative examples. To avoid excessive length we have concentrated on static phenomena in lattice gas models and have dealt with other topics only rather superficially. These examples by themselves illustrate that the Monte Carlo method has already made significant contributions to the understanding of adsorbed monolayers at surfaces: it has stimulated the development of analytical theories, and for many theoretical approximations the comparison to a Monte Carlo simulation has become *the* 'yardstick' with which its quality is most reliably assessed. Simulations have been also very useful for experiments: e.g. the finite size rounding and shifting of specific heat peaks seen in simulations (Fig. 33a) closely resembles analogous phenomena seen in real experiments (Fig. 33b), which are due to the limited size of the regions over which the substrate is homogeneous. Thus the simulations directly can contribute to better understand such size effects. In addition, simulations may provide useful guidance for future experiments: e.g. in a case where either from electronic structure calculations or from the experiments one knows already the effective interaction parameters, one can predict the nature of the ordered phases and the phase diagram (see Fig. 30 for the Si/W(110) model as an example), in cases where measurements of the phase diagram do not yet exist. We expect that simulations will become increasingly helpful in directing experimentalists to 'interesting' regions of thermodynamic parameter space where Monte Carlo data suggest novel or unusual behavior in relevant model systems. Also simulations can stimulate experimentalists to perform new types of experiments – the observation of the dynamics of domain growth via the time dependence of LEED intensities in the O/W(110) system[167] is an example.

An important topic which we have not dealt with at all is the simulation of adsorption on non-ideal surfaces – such as surfaces containing steps[179], point impurities[180], etc.

Finally, we wish to emphasize that much remains to be done in this field. With the development of faster (and cheaper!) computers it has become possible to perform some of the simpler calculations outlined in this chapter on a personal computer or workstation; conversely, with the investment of substantial computing time on supercomputers, the Monte Carlo simulations of very realistic models for adsorbed layers (e.g. using empirical pair potentials and corrugation potentials and removing the restriction to a lattice) now seem feasible. More theoretical effort indeed seems worthwhile as experimental progress is made in obtaining 'better' surfaces with larger defect-free regions. Thus, results have recently been obtained for $p(2 \times 2) - \mathrm{O}/Ru(0001)$[181] which are in reasonable agreement with the predicted 4-state Potts exponents.

Acknowledgements

Part of the work reviewed here has been performed together with M. S. S. Challa, W. Kinzel, H. H. Lee, A. Milchev, A. Sadiq, I. Sega, W. Selke and P. Wagner; it is a pleasure to thank them for their fruitful collaboration. We also have greatly benefitted from stimulating interaction with T. L. Einstein, G. Ertl and J. D. Gunton.

References

1. Metropolis, N., Rosenbluth, M. N., Rosenbluth, A., Teller, A., and Teller, E., *J. Chem. Phys.*, **21**, 1087 (1953).
2. Binder, K. (Ed.), *Monte Carlo Methods in Statistical Physics*, Springer, Berlin, 1979.
3. Binder, K. (Ed.), *Applications of the Monte Carlo Method in Statistical Physics*, Springer, Berlin, 1984.
4. Heermann, D. W., *An Introduction to Computer Simulation Methods in Theoretical Physics*, Springer, Berlin, 1986.
5. Binder, K., and Landau, D. P., *Phys. Rev.*, **B21**, 1941 (1980).
6. Landau, D. P., in Ref. 2, Chapter 9, and in Ref. 3, Chapter 3.
7. Binder, K., *Adv. Coll. Interface Sci.*, **7**, 297 (1977).
8. Binder, K., Kinzel, W., and Landau, D. P., *Surface Sci.*, **117**, 232 (1982).
9. Roelofs, L. D., in *Chemistry and Physics of Solid Surfaces* IV, Springer Series in Chemical Physics, Vol. 20 (Eds. R. Vanselow and R. Howe), p. 219, Springer, Berlin, 1982.
10. Selke, W., Binder, K., and Kinzel, W., *Surface Sci.*, **125**, 74 (1983).
11. Binder, K., *Ber. Bunsenges. Phys. Chem.*, **90**, 257 (1986).
12. Dash, J. G., and Ruvalds, J. (Eds.), *Phase Transitions in Surface Films*, Plenum, New York, 1979.
13. Sinha, S. (Ed.), *Ordering in Two Dimensions*, North-Holland, Amsterdam, 1980.
14. Dash, J. G., *Phys. Repts.*, **38C**, 177 (1978).
15. Dash, J. G., *Films on Solid Surfaces*, Academic, New York, 1975.
16. Bauer, E. in Ref. 12, p. 267.
17. Stoop, L. C. A., *Thin Solid Films*, **103**, 375 (1983).
18. Chin, K. K., and Landau, D. P., *Phys. Rev.*, **B36**, 275 (1987); Blöte, H. W. J., Compagner, A., Cornelissen, P. A. M., Hoogland, A., Mallezie, F., and Vanderzander, C., *Physica*, **139A**, 387 (1986).
19. Behm, R. J., Christmann, K., and Ertl, G., *Surface Sci.*, **99**, 320 (1980).
20. Binder, K., and Landau, D. P., *Surface Sci.*, **108**, 503 (1981).
21. Imbihl, R., Behm, R. J., Christmann, K., Ertl, G., and Matsushima, T., *Surface Sci.*, **117**, 257 (1982).
22. Kinzel, W., Selke, W., and Binder, K., *Surface Sci.*, **121**, 13 (1982).
23. Milchev, A., and Binder, K., *Surface Sci.*, **164**, 1 (1985).
24. Fisher, M. E., *Rev. Mod. Phys.*, **46**, 597 (1974).
25. Bretz, M., *Phys. Rev. Lett.*, **38**, 501 (1977).
26. Alexander, S., *Phys. Lett.*, **54A**, 353 (1975).
27. Domany, E., Schick, M., Walker, J. S., and Griffiths, R. B., *Phys. Rev.*, **B18**, 2209 (1978).
28. Potts, R. B., *Proc. Cambridge Phil. Soc.*, **48**, 106 (1952).
29. Wu, F. Y., *Rev. Mod. Phys.*, **54**, 235 (1982).

30. José, J. V., Kadanoff, L. P., Kirkpatrick, S., and Nelson, D. R., *Phys. Rev.*, **B16**, 1217 (1977).
31. Nijs, den M. P. M., *J. Phys.*, **A12**, 1857 (1979); Nienhuis, B., Riedel, E. K., and Schick, M., *J. Phys.*, **A13**, L 189; Pearson, R. B., *Phys. Rev.*, **B22**, 2579 (1980).
32. Selke, W., and Fisher, M. E., *Z. Physik*, **B40**, 71 (1980).
33. Selke, W., *Z. Physik*, **B43**, 335 (1981).
34. Rujan, P., Selke, W., and Uimin, G. V., *Z. Phys.*, **B53**, 221 (1983).
35. Selke, W., *Surface Sci.*, **144**, 89 (1984).
36. Kosterlitz, J. M., and Thouless, D. J., *J. Phys.*, **C6**, 1181 (1973).
37. Selke, W., and Yeomans, J. M., *Z. Physik*, **B46**, 311 (1982); Ostlund, S., *Phys. Rev.*, **B2**, 398 (1981).
38. Huse, D. A., and Fisher, M. E., *Phys. Rev.*, **B29**, 239 (1984); *Phys. Rev. Lett.*, **49**, 793 (1982).
39. Roelofs, L. D., Hu, G. Y., and Ying, S. C., *Phys. Rev.*, **B28**, 6369 (1983).
40. Milchev, A., Heermann, D. W., and Binder, K., *J. Statist. Phys.*, **44**, 749 (1986).
41. Milchev, A., Binder, K., and Heermann, D. W., *Z. Physik*, **B63**, 521 (1986).
42. Bruce, A. D., *J. Phys.*, **A18**, L873 (1985).
43. Landau, D. P., and Binder, K., *Phys. Rev.*, **B24**, 1391 (1981).
44. Schneider, T., and Stoll, E., *Phys. Rev.*, **B13**, 1216 (1976).
45. Lecante, J., Riwan, R., and Guillot, C., *Surface Sci.*, **35**, 271 (1973).
46. Anderson, J., and Estrup, P. J., *J. Chem. Phys.*, **46**, 563 (1967).
47. Felter, T. E., and Estrup, P. J., *Surface Sci.*, **54**, 179 (1976).
48. Jona, F., Legg, K. C., Shih, H. D., Sepsen, D. W., and Marcus, P. M., *Phys. Rev. Lett.*, **40**, 1466 (1978).
49. Conrad, H., Ertl, G., Küppers, J., and Latta, E. E., *Surface Sci.*, **50**, 296 (1975).
50. Conrad, H., Ertl, G., and Latta, E. E., *J. Catalysis*, **35**, 363 (1974).
51. Riwan, R., Guillot, C., and Paigne, J., *Surface Sci.*, **47**, 183 (1975).
52. Lee, H. H., and Landau, D. P., *Phys. Rev.* **B20**, 2893 (1979).
53. Blume, M., Emergy, V. J., and Griffiths, R. B., *Phys. Rev.*, **A4**, 1070 (1971).
54. de Oliviera, M. J., and Griffiths, R. B., *Surface Sci.*, **71**, 687 (1978).
55. Kim, I. M., and Landau, D. P., *Surface Sci.*, **110**, 415 (1981).
56. Pandit, R., and Wortis, M., *Phys. Rev.*, **B25**, 3226 (1982).
57. Pandit, R., Schick, M., and Wortis, M., *Phys. Rev.*, **B25**, 5112 (1982).
58. Wagner, P., and Binder, K., *Surface Sci.*, **175**, 421 (1986).
59. Binder, K., and Landau, D. P., *Phys. Rev.*, **B37**, 1745 (1988).
60. Nakanishi, H., and Fisher, M. E., *Phys. Rev. Lett.*, **49**, 1565 (1982).
61. Binder, K., and Landau, D. P., *J. Appl. Phys.*, **57**, 3306 (1985).
62. Mon, K. K., Binder, K., and Landau, D. P., *Phys. Rev.*, **B35**, 3683 (1987).
63. Mon, K. K., Landau, D. P., and Binder, K., *J. Appl. Phys.*, **61**, 4409 (1987).
64. Binder, K., Landau, D. P., and Kroll, D. M., *Phys. Rev. Lett.*, **56**, 2272 (1986).
65. Landau, D. P., *Phys. Rev.*, **B13**, 2997 (1976).
66. Landau, D. P., and Binder, K., *Phys. Rev.*, **B31**, 5946 (1985).
67. Binder, K., *Repts. Progr. Phys.*, **50**, 783 (1987).
68. Landau, D. P., *Phys. Rev.*, **B21**, 1285 (1980).
69. Marx, R., *Phys. Repts.*, **125**, 1 (1985).
70. Wendelken, F. J., and Wang, G.-C., *Phys. Rev.*, **B32**, 7542 (1985).
71. Fisher, M. E., and Barber, M. N., *Phys. Rev. Lett.*, **28**, 1516 (1972).
72. Barber, M. N., in *Phase Transitions and Critical Phenomena*, Vol. 8 (Ed. C. Domb and J. L. Lebowitz), Academic, New York, 1983, Chap. 2.
73. Binder, K., *Ferroelectrics*, **73**, 43 (1987).
74. Binder, K., *Z. Phys.*, **B43**, 119 (1981).

75. Imry, Y., *Phys. Rev.*, **B21**, 2042 (1980).
76. Privman, V., and Fisher, M. E., *J. Stat. Phys.*, **33**, 385 (1983).
77. Binder, K., and Landau, D. P., *Phys. Rev.*, **B30**, 1477 (1984).
78. Challa, M. S. S., Landau, D. P., and Binder, K., *Phys. Rev.*, **B34**, 1841 (1986).
79. Müller-Krumbhaar, H., and Binder, K., *J. Stat. Phys.*, **8**, 1 (1973).
80. Sadiq, A., and Binder, K., *Surface Sci.*, **128**, 350 (1983).
81. Krinsky, S., and Mukamel, D., *Phys. Rev.*, **B16**, 2313 (1977).
82. Domany, E., Schick, M., Walker, J. S., and Griffiths, R. B., *Phys. Rev.*, **B18**, 2209 (1978).
83. Swendsen, R. H., and Krinsky, S., *Phys. Rev. Lett.*, **43**, 177 (1979).
84. Nightingale, M. P., *Phys. Lett.*, **59A**, 486 (1977).
85. Oitmaa, J., *J. Phys.*, **A14**, 1159 (1981).
86. Nauenberg, M., and Nienhuis, B., *Phys. Rev. Lett.*, **33**, 941 (1977).
87. Baxter, R. J., *J. Phys.*, **C6**, L445 (1973).
88. Marx, R. Preprint.
89. Binder, K., *Z. Physik*, **B45**, 61 (1981).
90. Moritz, W., Imbihl, R., Behm, R. J., Ertl, G., and Matsushima, T., *J. Chem. Phys.*, **83**, 1959 (1985).
91. Kawabata, C., and Binder, K., *Solid State Comm.*, **22**, 705 (1977).
92. Tobochnik, I., and Chester, G. V., *Phys. Rev.*, **B20**, 3761 (1979).
93. Saito, Y., and Müller-Krumbhaar, H., *Phys. Rev.*, **B23**, 308 (1981).
94. Landau, D. P., *Phys. Rev.*, **B27**, 5604 (1983).
95. Challa, M. S. S., and Landau, D. P., *Phys. Rev.*, **B33**, 437 (1986).
96. Bartelt, N. C., Einstein, T. L., and Roelofs, L. D., *Phys. Rev.*, **B32**, 2993 (1985).
97. Bartelt, N. C., Einstein, T. L., and Roelofs, L. D., *Surf. Sci.*, **149**, L47 (1985).
98. Bartelt, N. C., and Einstein, T. L., *J. Phys.*, **A19**, 1429 (1986).
99. Bartelt, N. C., Einstein, T. L., and Roelofs, L. D., *J. Vac. Sci. Technol.*, **A3**, 1568 (1985).
100. Bartelt, N. C., Einstein, T. L., and Roelofs, L. D., in *The Structure of Surfaces-I* (Ed. S. Y. Tong and M. A. Van Hove), Springer, Berlin, 1985, p. 101.
101. Bartelt, N. C., Einstein, T. L., and Roelofs, L. D., *Phys. Rev.*, **B35**, 1776 (1987).
102. Bruce, A. D., *J. Phys.*, **C14**, 193 (1981).
103. Binder, K., *J. Stat. Phys.*, **24**, 69 (1981).
104. Fisher, M. E., *Phys. Rev.*, **176**, 257 (1968); For an experimental application, see Campbell, J. H., and Bretz, M., *Phys. Rev.* **B32**, 2861 (1985).
105. Cahn, J. W., *J. Chem. Phys.*, **66**, 3667 (1977).
106. Ebner, C., and Saam, W. F., *Phys. Rev. Lett.*, **38**, 1486 (1977).
107. For recent reviews, see Sullivan, D. W., and Telo, da Gama, M. M., in *Fluid Interfacial Phenomena* (Ed. C. A. Croxton), Wiley, New York, 1986, Hauge, E. H., in *Fundamental Problems in Statistical Physics VI* (Ed. E. G. D. Cohen), North-Holland, Amsterdam, 1985; deGennes, P. G., *Rev. Mod. Phys.* **55**, 825 (1985); S. Dietrich, S., in *Phase Transitions and Critical Phenomena*, Vol. 12 (Ed. C. Domb and J. L. Lebowitz), Academic, New York, 1987.
108. Binder, K., and Landau, D. P., *Phys. Rev. Lett.*, **52**, 318 (1984).
109. Binder, K., and Landau, D. P., Surface Sci. **151**, 409 (1985).
110. Landau, D. P., and Binder, K., *J. Appl. Phys.* **63**, 3077 (1988); and preprint in preparation.
111. For recent reviews, see Binder, K., in *Phase Transitions and Critical Phenomena*, Vol. 8 (Ed. C. Domb and J. L. Lebowitz), Academic, New York, 1983; H. W. Diehl *ibid.*, Vol. 1, Academic, New York; 1986; Binder, K., in *Polarized Electrons from Surfaces* (Ed. R. Feder), World Scientific, Singapore, 1985.

112. For a review on the roughening transition, see Weeks, J. D., in *Ordering in Strongly Fluctuating Condensed Matter Systems* (Ed. T. Riste), Plenum, New York, 1980, p. 293.
113. Bürkner, E., and Stauffer, D., *Z. Phys.*, **B55**, 241 (1983).
114. Mon, K. K., Landau, D. P., Wansleben, S., and Binder, K., *Phys. Rev. Lett.* **60**, 708 (1988).
115. Sega, I., Selke, W., and Binder, K., *Surface Sci.*, **154**, 331 (1985).
116. Selke, W., in *Lecture Notes in Physics*, Vol. 206 (Ed. A. Pekalski and J. Sznajd), Springer, Berlin, 1984, p. 191.
117. Selke, W., Huse, D. A., and Kroll, D. M., *J. Phys.*, **A17**, 3019 (1984).
118. Tanaka, M. and Kawabe, T., *J. Phys. Soc. Japan*, **55**, 1873 (1986).
119. Amar, J., Katz, S., and Gunton, J. D., *Surface Sci.*, **155**, 667 (1985).
120. Casanova, R., and Tsong, T. T., *Thin Solid Films* **93**, 41 (1982).
121. Sadiq, A., and Yaldran, K., *Surface Sci.* (1988).
122. Taylor, D. E., Williams, E. D., Park, R. L., Bartelt, N. C., and Einstein, T. L., *Phys. Rev.*, **B32**, 4653 (1985).
123. Bak, P., Kleban, P., Unertl, W. N., Ochab, J., Akinci, G., Bartelt, N. C., and Einstein, T. L., *Phys. Rev. Lett.*, **54**, 1539 (1985).
124. Saito, Y., *Phys. Rev.*, **B35**, 6652 (1981).
125. Glosli, J., and Plischke, M., *Can. J. Phys.*, **61**, 1515 (1983).
126. Bartelt, N. C., Einstein, T. L., and Roelofs, L. D., *J. Vac. Sci. Technol.*, **A1**, 1217 (1983).
127. Roelofs, L. D., Bartelt, N. C., and Einstein, T. L., *Phys. Rev. Lett.*, **47**, 1348 (1981).
128. Cai, Z.-X., and Mahanti, S. D., *Phys. Rev.*, **B36**, 6928 (1987).
129. Roelofs, L. D., Kortan, A. R., Einstein, T. L., and Park, R. L., *Phys. Rev. Lett.*, **46**, 1465 (1981).
130. Lee, D. H., Joannopoulos, J. D., Negele, J. W., and Landau, D. P., *Phys. Rev.*, **B33**, 450 (1986).
131. Lee, D. H., Joannopoulos, J. D., Negele, J. W., and Landau, D. P., *Phys. Rev. Lett.*, **52**, 433 (1984).
132. Miyashita, S., and Shiba, H., *J. Phys. Soc. Japan*, **53**, 1145 (1984).
133. Stoll, E., Binder, K., and Schneider, T., *Phys. Rev.*, **B8**, 3266 (1973).
134. Hohenberg, P. C., and Halperin, B. I., *Revs. Mod. Phys.*, **49**, 435 (1977).
135. Kretschmer, R., Binder, K., and Stauffer, D., *J. Stat. Phys.*, **15**, 267 (1976).
136. Chakrabarty, B. K., Baumgaertel, H. G., and Stauffer, D., *Z. Phys.*, **B44**, 333 (1981).
137. Miyashita, S., and Takano, H., *Progr. Theor. Phys.*, **73**, 1122 (1985).
138. Williams, J. K., *J. Phys.*, **A18**, 49 (1985).
139. Williams, J. K., *J. Phys.*, **A18**, 1781 (1985).
140. Wansleben, S., and Landau, D. P., *J. App. Phys.*, **61**, 3968 (1987).
141. Tang, S. Y., and Landau, D. P., *Phys. Rev.*, **B36**, 567 (1987).
142. Aydin, M., and Yalabik, M. C., *J. Phys.*, **A18**, 174 (1985).
143. Tobochnik, J., and Jayaprakash, C., *Phys. Rev.*, **B35**, 4893 (1981).
144. Yalabik, M. C., and Gunton, J. D., *Progr. Theor. Phys.*, **62**, 1573 (1979).
145. Yahata, H. and Suzuki, M., *J. Phys. Soc. Japan*, **27**, 1421 (1969); Racz, Z., and Collins, M. F., *Phys. Rev.*, **B13**, 3074 (1976).
146. Bausch, R., Dohm, V., Janssen, H. K., and Zia, R. K. P., *Phys. Rev. Lett.*, **47**, 1837 (1981).
147. Achiam, Y., *J. Phys.*, **A13**, 1355, (1980).
148. Forgacs, G., Chui, S.-T., and Frisch, H. L., *Phys. Rev.*, **B22**, 415 (1980).
149. Mazenko, G. F., and Valls, O. T., *Phys. Rev.*, **B24**, 1419 (1981).

150. Mazenko, G. F., and Valls, O. T., *Phys. Rev.*, **B31**, 1565 (1985).
151. Domany, E., *Phys. Rev. Lett.*, **52**, 871 (1984).
152. Hutchings, M. T., Ikeda, H., and Janke, E., *Phys. Rev. Lett.*, **49**, 386 (1982).
153. Sadiq, A., and Binder, K., *Phys. Rev. Lett.*, **51**, 674 (1983).
154. Sadiq, A., and Binder, K., *J. Stat. Phys.*, **35**, 617 (1984).
155. Fisher, M. E., and Racz, Z., *Phys. Rev.*, **B13**, 5039 (1976).
156. Bowker, M., and King, D. A., *Surface Sci.*, **71**, 583 (1978); **72**, 208 (1978).
157. Sadiq, A., *Phys. Rev.*, **B9**, 2299 (1981).
158. Murch, G. E., and Thorn, R. J., *Phil. Mag.*, **A35**, 493 (1977); **A36**, 529 (1977).
159. Murch, G. E., *Phil. Mag.*, **A43**, 871 (1981); *Solid State Ionics*, **5**, 117 (1981).
160. Reed, L. A., and Ehrlich, G., *Surface Sci.*, **105**, 603 (1981).
161. Zwerger, W., *Z. Physik*, **B42**, 333 (1981).
162. Tringides, M., and Gomer, R., *Surf. Sci.*, **145**, 121 (1984).
163. Butz, R., and Wagner, G., *Surface Sci.*, **63**, 448 (1977).
164. Chen, J. R. and Gomer, R., *Surface Sci.*, **79**, 413 (1979); **94**, 456 (1980).
165. For recent reviews on the kinetics of domain growth in two dimensions, see Ref. 11 and Binder, K. and Heermann, D. W., in *Scaling Phenomena in Disordered Systems* (Ed. R. Pynn and A. Skjeltorp), Plenum, New York, 1985, p. 207; Furukawa, H., *Advances in Physics*, **34**, 703 (1985); and Binder, K., Heermann, D. W., Milchev, A., and Sadiq, A., in *Glassy Dynamics and Optimization* (Ed. J. L. van Hemmen and I. Morgenstern), Springer, Berlin, 1987.
166. Wang, G. C., and Lee, T. M., *Phys. Rev. Lett.*, **50**, 2014 (1983).
167. Tringides, M. C., Wu, P. K., and Lagally, M. G., *Phys. Rev. Lett.*, **56**, 315 (1987).
168. Lipowsky, R., *J. Phys.*, **A18**, L585 (1985).
169. Schmidt, I., and Binder, K., *Z. Phys.*, **B67**, 369 (1987).
170. Jiang, Z., and Ebner, C., *Phys. Rev.*, **B36**, 6976 (1987).
171. Lipowsky, R., and Huse, D. A., *Phys. Rev. Lett.*, **52**, 353 (1986).
172. Silverberg, M., and Ben-Shaul, A., *Chem. Phys. Lett.*, **134**, 491 (1987).
173. Silverberg, M., Ben-Shaul, A., and Rebentrost, F., *J. Chem. Phys.*, **83**, 6501 (1985).
174. Engel, T., and Ertl, G., *Advanc. Catal.*, **28**, 1 (1979).
175. Gland, J. L., and Kollin, E. B., *J. Chem. Phys.*, **78**, 963 (1983).
176. Behm, R. J., Thiel, P. A., Norton, P. R., and Bindner, P. E., *Surface Sci.*, **147**, 143 (1984).
177. Stuve, E. M., Madix, R. J., and Brundle, C., *Surface Sci.*, **146**, 155 (1984).
178. Saito, Y., and Müller-Krumbhaar, H., *J. Chem. Phys.*, **70**, 1078 (1979).
179. Albano, E. V., and Martin, H. O., *Phys. Rev.*, **B35**, 7820 (1987).
180. Novotny, M. A., and Landau, D. P., *Phys. Rev.*, **B32**, 5874 (1985).
181. Piercy, P., and Pfnür, H., *Phys. Rev. Lett.*, **59**, 1124 (1987); Piercy, P., Maier, H., and Pfnür, H., in *The Structure of Surfaces-II* (Ed. J. F. van der Veen and M. A. van Hove), Springer, Berlin, Heidelberg, New York (1988).

Advances in Chemical Physics
Edited by K. P. Lawley

MODEL STUDIES OF SURFACE CATALYZED REACTIONS

A. G. SAULT and D. W. GOODMAN

Surface Science Division, Sandia National Laboratories, Albuquerque, NM 87185, USA

CONTENTS

Abstract

The study of high pressure catalytic reactions on metal single crystal surfaces is reviewed. Examples of structure-insensitive reactions are provided which demonstrate the relevance of single crystal studies for modeling the behavior of high surface area supported catalysts. Surface analysis following reaction provides information relating to surface composition and structure during the reaction. For structure-sensitive reactions, model single crystal studies allow correlations to be drawn between surface structure and catalytic activity. High pressure studies on single crystal surfaces also allow the kinetics of single elementary steps in a catalytic reaction to be studied, as demonstrated for alkane dissociation on nickel surfaces. In addition to studies of catalytic reactions on clean metal surfaces, the effects of both electropositive and electronegative additives on the catalytic activity of single crystal surfaces are reviewed. Finally, studies of model bimetallic catalysts, prepared by deposition of one metal on a single crystal surface of a second metal, are summarized.

1. INTRODUCTION

The widespread use of ultrahigh vacuum (UHV) surface science techniques for studying the interactions of organic molecules with metal single crystal surfaces has furthered our understanding of the organometallic chemistry of metal surfaces[1]. In general, the motivation for these UHV studies has been to understand the catalytic chemistry of metal surfaces on a molecular level. Unfortunately, the conditions under which the analytical techniques of surface science can be applied are highly idealized. The very low pressures required for surface science studies are typically many orders of magnitude below the pressures used in practical catalytic processes. In addition, the catalysts used in high pressure catalytic processes often consist of very small metal particles supported on high surface area materials such as silica and alumina. These metal particles almost certainly expose many crystal planes of the metal and probably have high defect concentrations. In contrast, the metal single crystals used in surface science are smooth on an atomic scale, exposing only a single crystal plane of the metal. Because of the highly idealized nature of the systems studied using UHV techniques, the relevance of UHV studies for modeling the chemistry of practical catalytic processes must be established. In order to assess the relevance of UHV techniques for studying catalytic reactions, several laboratories have developed experimental systems which combine a high pressure reactor system with a UHV analysis chamber[2–7]. The high pressure reactor allows the kinetics of catalytic reactions to be measured on single crystal metal surfaces, while analysis of the structure and composition of the surface both before and after reaction can be accomplished in the UHV analysis chamber. This combined approach not only allows direct comparison of reaction rates measured on single crystal surfaces with those measured on more realistic supported metal catalysts, but also allows detailed study of structure sensitivity, the effects of promoters and inhibitors on catalytic activity, and, in certain cases, identification of reaction intermediates by post-reaction surface analysis.

In this chapter, the work carried out in our laboratory using a combined UHV–high pressure reactor system will be reviewed. In addition to studies of hydrogenolysis on Ni surfaces and CO methanation which have already been reviewed[4], more recent results regarding CO oxidation, alkane activation, hydrogenolysis on Ir surfaces, and the catalytic activity of strained metal overlayers are summarized.

2. EXPERIMENTAL

The studies to be discussed were carried out utilizing the specialized apparatus shown in Fig. 1 and described in Refs. 3, 8. This device consists of two distinct regions, a surface analysis chamber and a microcatalytic reactor. The custom-built reactor, contiguous to the surface analysis chamber, employs a retraction belows that supports the metal single crystal and allows translation of the catalyst in vacuo from the reactor to the surface analysis region. Both regions are of ultrahigh vacuum construction, bakeable, and capable of ultimate pressures of less than 2×10^{-10} torr. Auger spectroscopy (AES) is used to characterize the sample before and after reaction. A second chamber equipped with Auger spectroscopy, low energy electron diffraction (LEED) and a mass spectrometer for temperature programmed desorption (TPD) was used to characterized the metal overlayers discussed in section 3.3. The single crystal catalysts, ~1 cm diameter × 1 mm thick, were aligned within 1/2° of the desired orientation. Thermocouples were spotwelded to the edges of the crystals for temperature measurement. Details of sample mounting, cleaning procedures, and preparation techniques are given in the

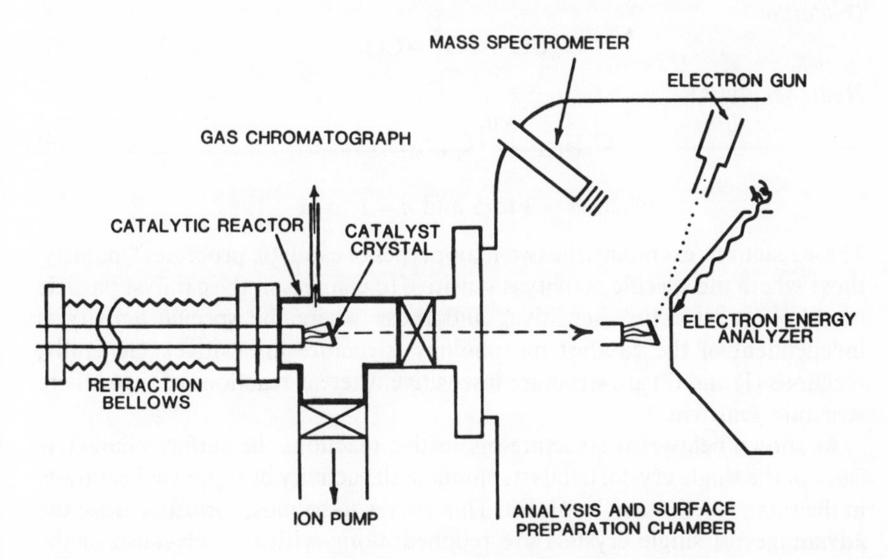

Fig. 1. An ultrahigh vacuum apparatus for studying single crystal catalysts before and after reaction at elevated pressures.

references accompanying the relevant data. All reactants were initially of high purity; however, further purification procedures were generally used to improve the gas quality. These typically included multiple distillations for condensables and/or cryogenic scrubbing using a low conductance glass wool-packed trap at 80 K. The kinetic data presented were obtained under steady-state reaction conditions and at low conversions (typically $< 5\%$). Products were measured by a gas chromatograph equipped with a flame ionization detector (FID). A methanizer was used to convert CO and CO_2 to methane, thereby allowing detection of these products with the FID. Throughout this review, catalytic activity will be expressed in terms of specific activity, i.e. the catalytic rate normalized to the number of exposed metal atoms. Specific activity is usually expressed as a turnover frequency (TOF), which is simply the number of molecules of product produced per metal atom site per second.

3. DISCUSSION

3.1. Correlations between surface structure and catalytic properties

The apparatus described above has been used to study the reactions in Eqs.(1)–(3), over single crystals of Ni, Ru, Rh, Pd, and Ir.

Methanation

$$3H_2 + CO \rightarrow CH_4 + H_2O \tag{1}$$

Oxidation

$$CO + \tfrac{1}{2}O_2 \rightarrow CO_2 \tag{2}$$

Hydrogenolysis

$$C_xH_{(2x+2)} \xrightarrow{H_2} C_{(x-n)}H_{[2(x-n)+2]} \tag{3}$$

where $x = 2$ to 5 and $n = 1$ to $(x - 1)$

These reactions exemplify the two major types of catalytic processes[9], namely, those where the specific activity is sensitive to changes in the catalyst particle morphology (structure sensitive), and those where the specific activity is independent of the catalyst morphology (structure insensitive). Generally, reactions (1) and (2) are structure insensitive whereas reactions of type (3) are structure sensitive.

As shown below, for structure-insensitive reactions the surface characteristics of the single crystal catalysts simulate the activity of supported catalysts in the same reactant environment. This proves to be most fortunate since the advantages of single crystals are retained along with the relevance of the measurements. Moreover, the use of single crystals allows the assessment of the crystallographic dependence of structure-sensitive reactions.

3.1.1. *Structure-insensitive reactions*

(a) Methanation

The methanation reaction[10,11] (reaction (1)) has a critical role in the production of synthetic natural gas from hydrogen-deficient carbonaceous materials. In addition, this reaction is an obvious starting point in studies of fuel and chemical synthesis from carbon sources. Historically this reaction has been considered to be structure insensitive[10] in that changes in catalyst morphology generally produce, at most, small changes in the catalytic activity. For methanation on nickel it has been possible to correlate kinetic parameters measured on single crystal catalysts with those found for supported high-area catalysts[3,11]. The data in the Arrhenius plot of Fig. 2a represent steady-state, specific methanation rates on both the Ni(111) and Ni(100) surfaces[12]. The atomic configurations of the Ni(100) and Ni(111) surfaces are shown in Figs. 2b and 2c, respectively. The similarity between the close-packed (111) and the more open (100) crystal plane of Ni is evident in both the value of the specific rates and activation energy (103 kJ/mole). In the temperature range of 450–700 K, methane production rates vary by almost five orders of magni-

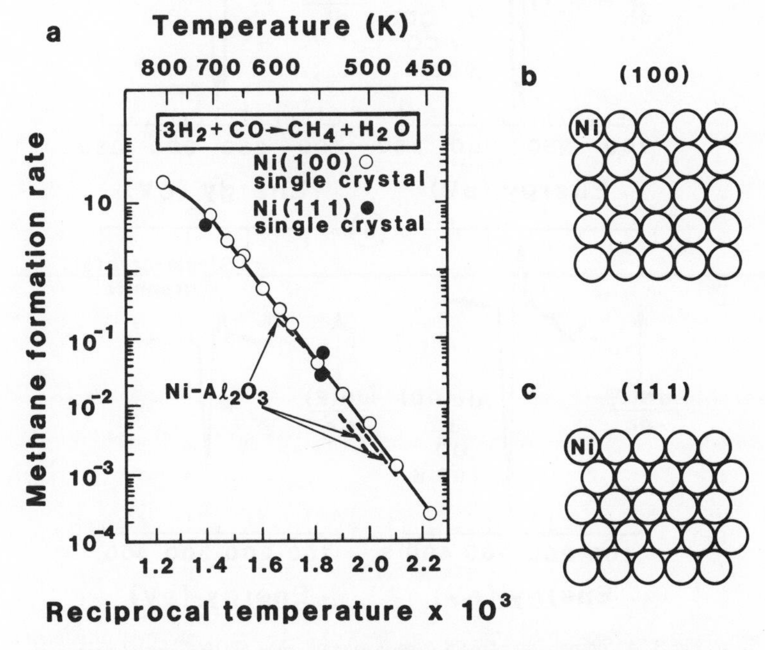

Fig. 2. (a) A comparison of the rate of methane synthesis over single crystal nickel catalysts and supported Ni/Al_2O_3 catalysts at 120 torr total reactant pressure. (*From Ref. 12.*) (b) Atomic configuration of a Ni(100) surface. (c) Atomic configuration of a Ni(111) surface.

tude. It should be pointed out that measurements over such a wide temperature range are difficult, if not impossible, with supported high surface area catalysts due to heat and mass transfer limitations at high temperatures.

The single crystal results are compared in Fig. 2 with three sets of data taken from Ref. 13 for nickel supported on alumina, a high surface area catalyst. This comparison shows extraordinary similarities in kinetic data taken under nearly identical conditions. Thus, for the H_2–CO reaction over nickel, there is no significant variation in the specific reaction rates or the activation energy as the catalyst changes from small metal particles to bulk single crystals. These data provide convincing evidence that the methanation reaction rate is indeed structure insensitive on nickel catalysts.

For the methanation reaction surface characterization subsequent to

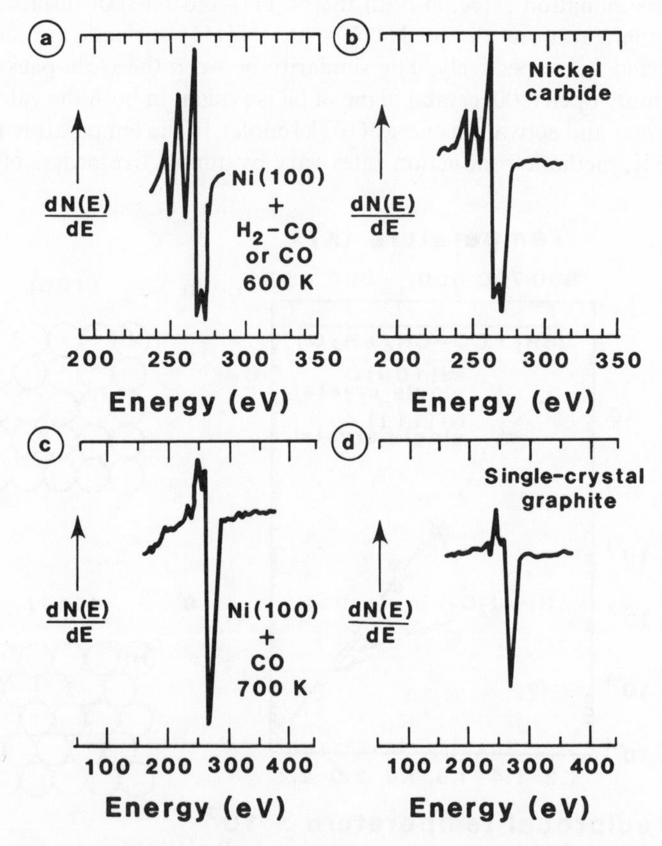

Fig. 3. A comparison of AES carbon signals on a Ni(100) crystal with those from single crystal graphite and nickel carbide. (a) Following 1000 s heating at 600 K in 24 torr CO. (b) Nickel carbide. (c) Following 1000 s heating at 700 K in 24 torr CO. (d) Single crystal graphite. (*From Ref. 3.*)

reaction has provided data addressing the reaction mechanism. Auger spectroscopic analysis of an active nickel catalyst following reaction at atmospheric pressure[3,11] has been accomplished by evacuating the reactor and transferring the crystal to the surface analysis chamber. Auger analysis after such a procedure shows a carbonaceous species present on the surface at a concentration equivalent to approximately 10% of a monolayer (a monolayer, ML, in this context is defined as one atom per one substrate metal atom). The AES spectrum shown in Fig. 3c indicates this carbon to be a 'carbidic' form with a lineshape distinctive from that of graphitic carbon[3,14] shown in Fig. 3d. This carbidic-type carbon can be readily removed by heating the crystal to 600 K in atmospheric H_2, with the product formed being methane. A carbon species with these same characteristics can be produced by heating the Ni crystal in CO in the absence of hydrogen. Figure 3 shows the AES carbon signal[3] measured after heating a Ni(100) crystal in several torr of CO at 600 K (Fig. 3a) and at 700 K (Fig. 3c). These carbon peaks are compared with those observed with single crystal graphite (Fig. 3d) and with bulk nickel carbide (Fig. 3b). Based on this comparison, the active carbon form has been designated 'carbidic'. The deposition of an active carbon and the absence of oxygen on the nickel surface following heating in pure CO is consistent with a well-known disproportionation reaction, the Boudouard reaction,

$$2CO \rightarrow C + CO_2 \tag{4}$$

Which has been studied on supported Ni catalysts[15,16] and on Ni films[17]. Studies such as those described here show that methane can be catalytically synthesized over Ni by an active (carbidic) carbon formed via the Boudouard reaction and its subsequent hydrogenation to methane. However, to demonstrate that this surface carbon route is the major reaction pathway, kinetic measurements of both carbon formation from CO and its removal by H_2 were carried out[18].

In the first set of measurements[18] the rate of carbon build-up on a Ni(100) surface was measured at various temperatures as follows: (1) surface cleanliness was established by AES; (2) the sample was retracted into the reaction chamber and exposed to several torr of CO for various times at a given temperature; (3) after evacuation the sample was transferred to the analysis chamber; and (4) the AES spectra of C and Ni were measured. Two features of this study are noteworthy. First, two kinds of carbon forms are evident – a carbidic type which occurs at temperatures < 650 K and a graphite type at temperatures > 650 K. The carbide form saturates at 0.5 monolayers. Second, the carbon formation data from CO disproportionation indicates a rate equivalent to that observed for methane formation in a H_2/CO mixture. Therefore, the surface carbon route to product is sufficiently rapid to account for methane production with the assumption that kinetic limitations are not imposed by the hydrogenation of this surface carbon.

A second set of experiments[18] further supported the surface carbon route to methane. In these experiments a Ni(100) surface was precarbided by exposure to CO and then treated with hydrogen in the reaction chamber for various times. Steps (3) and (4) above were then followed to measure the carbide level. This study showed that the rate of carbon removal in hydrogen compared favorably to the carbide formation rate in CO and to the overall methanation rate in H_2/CO mixtures. Thus in a H_2–CO atmosphere the reaction rate is determined by a delicate balance of the carbon formation and removal steps and neither of these is rate determining in the usual sense.

According to this mechanism, if the surface coverage of atomic hydrogen is close to saturation, it is predicted that (1) further increases in hydrogen pressure would have almost no effect on the methane rate, and (b) a low surface carbon level will result. This is the condition believed to exist under the reaction conditions of Fig. 2a. However, if reaction conditions are altered such that the surface hydrogen concentration decreases (e.g. low H_2 pressure and high temperature) then the mechanism demands a correlation of decreasing methane yield with increasing surface carbide. This correlation between the rate of production of methane and the steady-state surface carbide con-

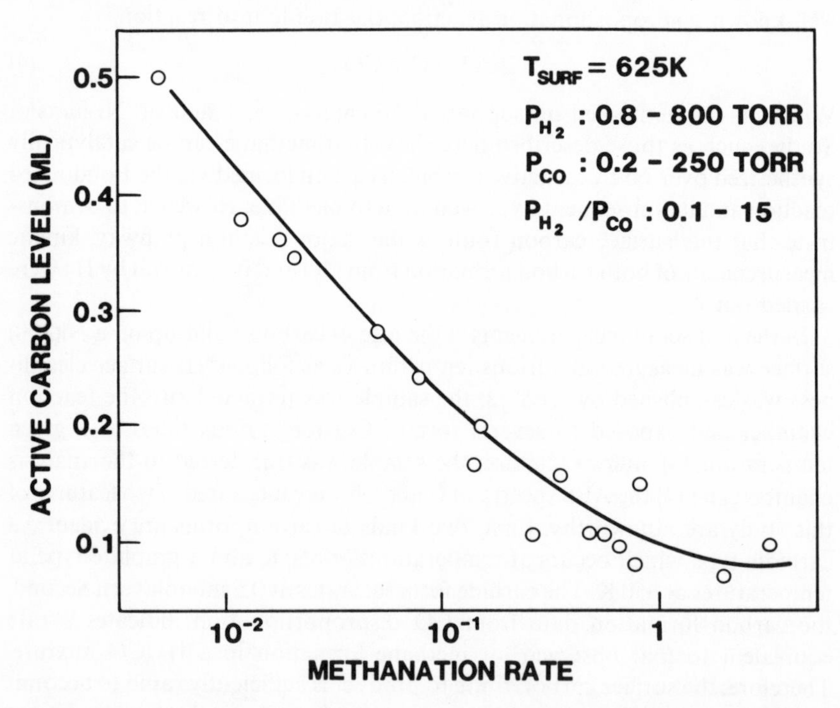

Fig. 4. Methane production rate at 625 K over a Ni(100) catalyst as a function of surface carbon coverage at various reaction conditions. (*From Ref. 12.*)

centration holds very well as evidenced by the data in Fig. 4[12]. Here all reaction rate data measured over a Ni(100) catalyst at 625 K lie on a smooth curve when plotted against the measured steady-state carbide level, regardless of H_2/CO ratio or total pressure.

Thus, the proposed reaction mechanism involving the dissociation of CO and the subsequent hydrogenation of the resulting carbon species (C_{ads}) accounts quite satisfactorily for the effect of pressure on the methanation rate, for the variation in the measured surface carbon level as reaction parameters are changed, and for the formation at a characteristic temperature and pressure conditions of a catalyst-deactivating graphitic carbon. Recent studies[19] using isotopically labeled CO have shown that the CO dissociation step is essentially unidirectional in that the rate of C_{ads} and O_{ads} recombination is insignificantly slow compared to the C_{ads} hydrogenation rate.

Methanation reactions over Ni, Ru[12,20], and Fe[5] show remarkable similarities in many critical parameters, suggesting that the three metals behave essentially the same catalytically. This conclusion finds support in studies with nickel, ruthenium, and cobalt high-surface-area supported catalysts[15,16,21,22] and with studies using Ni, Co and Ru films in UHV[23].

(b) Carbon monoxide oxidation

The oxidation of CO by O_2 over group VIII metal catalysts has been the subject of a large body of ultrahigh vacuum surface science and high pressure catalysis work due to its importance in pollution control[24]. Currently, the removal of CO as CO_2 from automobile exhaust is accomplished by catalytic converters which employ a supported Pt, Pd, and Rh catalyst. The importance of CO oxidation has led to numerous recent studies of the kinetics of this reaction on supported metal catalysts[25–34] and transient kinetic studies on polycrystalline foils[35–44], which have sought to identify and quantify the parameters of the elementary mechanistic steps in CO oxidation.

The relative simplicity of CO oxidation makes this reaction an ideal model system of a heterogeneous catalytic reaction. Each of the mechanistic steps (adsorption and desorption of the reactants, surface reaction, and desorption of products) has been probed extensively with surface science techniques, as has the interaction between O_2 and CO[45–62]. These studies have provided essential information necessary for understanding the elementary processes which occur in CO oxidation.

Recent reviews by Ertl and Engel have summarized most of the chemisorption and low pressure catalytic findings[63,64]. In general, the reaction proceeds through a Langmuir–Hinshelwood mechanism involving adsorbed CO and O atoms. Under reaction conditions typical in most high pressure, supported catalyst studies, and most low pressure UHV studies, the surface is almost entirely covered by CO, and the reaction rate is determined by the rate of desorption of CO. As first determined by Langmuir for Pt wire catalysts[65], the observed activation energy is close to the binding energy of adsorbed CO. Oxygen can only adsorb at sites where CO has desorbed, leading to first-order dependence in oxygen pressure, negative first order in CO partial pressure, and zero-order total pressure dependence. These features have allowed many of the reaction parameters determined in UHV to be applied directly to the kinetics at higher pressures[66].

Models based on chemisorption and kinetic parameters determined in surface science studies have been successful at predicting most of the observed high pressure behavior. Recently Oh *et al.*[66] have modeled CO oxidation by O_2 or NO on Rh using mathematical models which correctly predict the absolute rates, activation energy, and partial pressure dependence. Similarly, studies by Schmidt and coworkers on CO + O_2 on Rh(111)[67] and CO + NO on polycrystalline Pt[68] have demonstrated the applicability of steady-state measurements in UHV and relatively high (1 torr) pressures in determining reaction mechanisms and kinetic parameters.

Recently, the steady-state reaction kinetics of CO oxidation at high pressure over Ru[69], Rh[69–72], Pt, Pd, and Ir[73] single crystals have been studied in our laboratory. These studies have convincingly demonstrated the applicability and advantages of model single crystal studies, which combine UHV surface analysis techniques with high pressure kinetic measurements, in the elucidation of reaction mechanisms over supported catalysts.

Figure 5 compares the CO oxidation rates measured over single crystals of Rh, Pt, Pd, Ir, and Ru with the analogous supported catalysts[47]. The turnover frequencies for the supported catalysts were obtained by normalizing the measured reaction rates to the total number of surface metal atoms. Notice that the turnover frequencies for the single crystal catalysts traverse several orders of magnitude over a temperature range of 450–600 K. As mentioned in section 3.1.1(a), kinetic measurements over such a wide temperature range are not possible with supported catalysts due to heat and mass transfer limitations encountered at high temperatures. Thus, a direct comparison of the kinetic data between the two types of catalyst must necessarily be limited to a relatively small temperature range. Nevertheless, it is clear from Fig. 5 that there is excellent agreement between the single crystal and supported catalysts in both the specific reaction rates and apparent activation energies. This agreement indicates that CO oxidation is structure insensitive on Ru, Rh, Pt, Pd and Ir catalysts.

The Langmuir–Hinshelwood reaction between adsorbed CO and O atoms is well established as the dominant reaction mechanism for conditions where CO is the primary surface species[63,64]. This mechanism has been confirmed by numerous UHV studies of the coadsorption of the reactants[36,45,46,48,49,51,53,54,56,58,62], transient kinetic studies[74,37,39–42,44,46], and steady-state kinetics[38,52,55,59–61]. The reaction

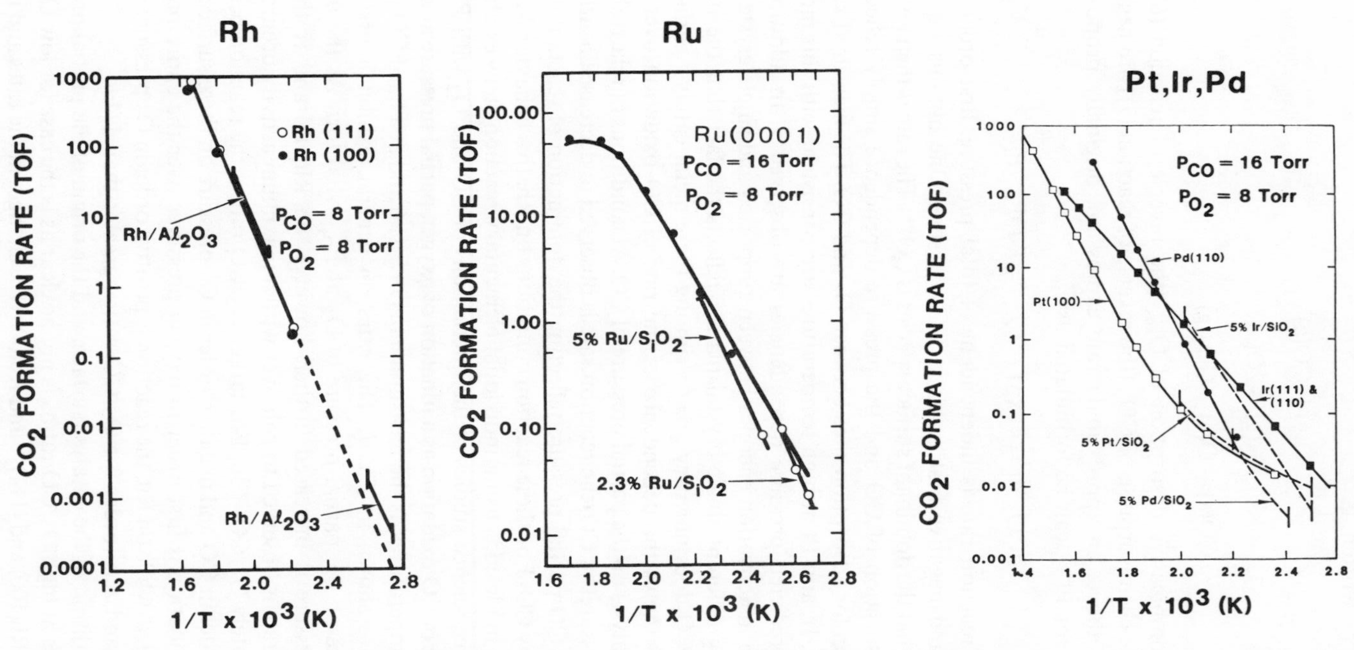

Fig. 5. Specific rates of CO oxidation (turnover frequencies) as a function of inverse temperature for single crystal and supported Rh, Ru, Pt, Ir, and Pd catalysts. (*From Refs. 69–73.*)

steps may be written as:

$$CO(g) \leftrightarrow CO(a) \tag{5}$$

$$O_2(g) \rightarrow 2O(a) \tag{6}$$

$$CO(a) + O(a) \rightarrow CO_2(g) \tag{7}$$

where recombinative desorption of O_{ads} (the reverse of reaction (6)) and dissociative chemisorption of CO_2 (the reverse of reaction (7)) are neglected. From the above, an approximate rate expression, originally proposed by Langmuir for Pt[65], can be formulated as:

$$dCO_2/dt = k\exp(-E_{des,CO}/RT)P_{O_2}/P_{CO} \tag{8}$$

Thus, the reaction rate is independent of total pressure, first order in O_2 pressure, and negative first order in CO pressure. One obtains Eq (8) by assuming that the dominant surface species is CO[66]. The rate is then governed by the desorption of CO and the pressure dependence simply reflects the competition for adsorption sites between O_2 and CO. The measured kinetics on Rh, Pd, Ir, and Pt at high temperatures are consistent with this model in that the predicted pressure dependencies are observed[73]. In addition, the correlation in activation energies between supported and single crystal data, and among different single crystal planes, reflects the fact that the binding energy of CO does not vary greatly among these metal catalyst surfaces.

Figure 6 shows the dependence of the rate of CO oxidation over single crystal catalysts on the partial pressure of CO. At conditions of relatively high partial pressures of CO, the reaction rate is observed to decrease linearly with increasing CO partial pressure reflecting the domination of reactant surface coverage by CO. For these reaction conditions on Rh, this behavior has been accurately modeled[66] using individual elementary reaction steps established from surface science studies of the interactions of CO and O_2 with Rh.

The rate of CO oxidation as a function of oxygen partial pressure at similar catalyst temperatures as the measurements in Fig. 6 and at a fixed CO partial pressure are shown in Fig. 7. The rates, in general, exhibit a first-order dependence on the partial pressure of O_2 at $P_{O_2} < 100$ torr. As the oxygen partial pressure is increased further, however, on Rh, Pd and Ir the CO oxidation rate is observed to roll over with a maximum activity occurring at approximately $O_2/CO = 30$. For larger O_2/CO ratios, the reaction becomes positive order in CO and negative order in O_2 pressure on these surfaces. For Pt, CO oxidation is first order in oxygen pressures over the entire pressure range studied, while on Ru, the reaction is positive order in O_2 pressure at low pressures, and zero order for O_2/CO ratios greater than 1/4.

The condition of the catalyst surface, and the origin of the partial pressure dependence at high O_2/CO ratios is not as clear as is the case for low O_2/CO ratios. On Rh(100) and (111)[70] and Ru(0001)[69], changes in the rates and partial

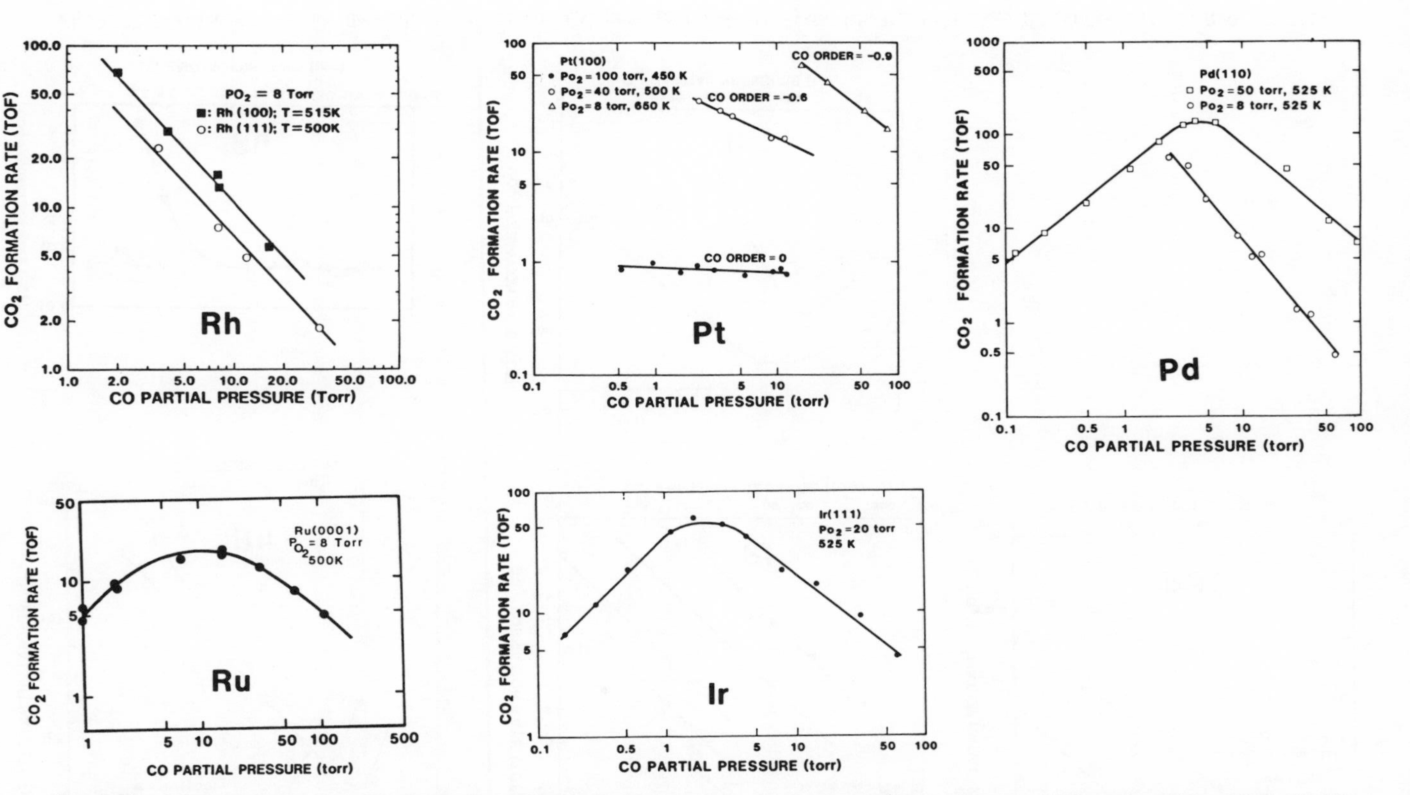

Fig. 6. The CO partial pressure dependence at constant oxygen pressure for single crystal Rh, Ru, Pt, Ir, and Pd catalysts. (*From Refs. 71, 73.*)

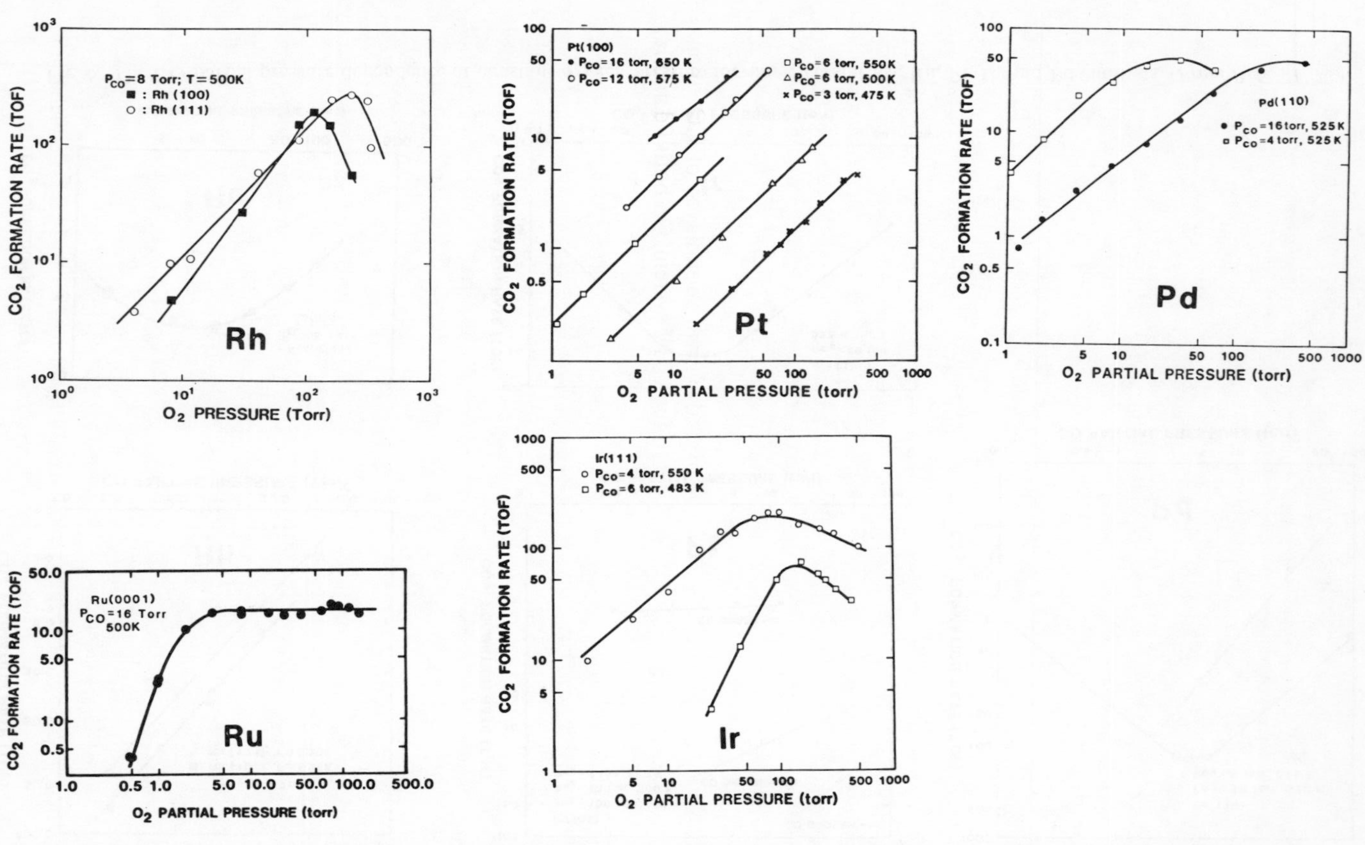

Fig. 7. The O_2 partial pressure dependence at constant CO pressure for single crystal Rh, Ru, Pt, Ir, and Pd catalysts. (*From Refs. 71, 73.*)

pressure dependencies have been directly correlated with the formation of a strongly bound oxygen species. On Ru, the oxide is substantially more active than the clean surface and the reaction order in oxygen pressure is close to three at low O_2 pressures, reflecting the gradual formation of the oxide layer as the O_2 pressure is increased. Once the oxide is fully formed, the reaction rate becomes zero order in oxygen pressure. On Rh(111) and (100) single crystal increases in the O_2/CO ratios result in an eventual decrease in rates, and a change from positive order in oxygen to negative order. This change has been correlated directly with the formation of an oxide-like species, characterized by AES and TPD subsequent to reaction[70], which is less active than the clean surfaces. Pd(110), and Ir(111), and Ir(110) exhibit partial pressure dependence and high oxygen pressure deactivation behavior which is very similar to Rh. In the case of Rh, the formation of a near surface oxide (probably Rh_2O_3[72] which is catalytically inactive) is responsible for the deactivation. The similar behavior on Pd and Ir suggests that oxide formation is also responsible for the deactivation on these metals. In contrast to Rh, however, the oxide species cannot be isolated on Pd and Ir. Instead, the strongly bound oxygen species reacts with CO and is detected as CO_2, which desorbs at high temperatures in post-reaction TPD. Thus, the oxide formed on Rh is resistant to reduction by CO in the post-reaction cool-down and transfer, while oxide reduction by CO is quite facile on Pd and Ir.

The metal surfaces are always covered with a monolayer of CO upon evacuation of the reactor and transfer to the UHV system. On both Pd and Ir the CO, which desorbs as CO_2 when reacted with the oxide species, desorbs at a much higher temperature than CO from the clean surface. This result implies that the oxide species forms an inactive complex with CO upon adsorption of CO under reaction conditions. While the presence of the oxide species reduces the overall rate of reaction, the activation energy is unchanged, suggesting that oxygen serves as a simple site blocker on the surface.

The differences between Pt, Pd, Ir, Ru and Rh may be explained by the ease of oxidation of these metals. The more easily oxidized Rh (and Ru) surfaces form oxides which are bound strongly enough to resist reaction when CO is flashed off the crystal prior to post-reaction AES analysis. Pd and Ir, which are less easily oxidized than Rh, form a less strongly bound oxide, possibly in a thinner layer, which is easily reduced. In a study by Savchenko and coworkers[75], it was shown that a reconstructive type of oxygen adsorption (i.e. the formation of an oxide) would only occur for metals with a heat of adsorption of oxygen above 220 kJ/mol. Pt falls well below this limit; Pd, Rh, and Ir are close to the borderline between types of oxides; while Ru falls well into the oxide range. This evaluation corresponds well with the observed oxidation behavior of these metals during CO oxidation.

For Pt(100) the lack of any rollover in the oxygen partial pressure behavior (Fig. 7) indicates that under our conditions no strongly bound, deactivating oxygen species is formed. In light of the trend from Ru to Rh to Pd and Ir, this is not surprising. In order to form an oxide species on Pt, much higher O_2 pressures and/or higher temperatures would be required.

It is interesting to compare our results on single crystal surfaces with those of Turner and coworkers[76] for Pt, Pd, and Ir. In this study wires of Pt formed less than one layer of 'oxide' under CO oxidation conditions. Considering that the Pt wires were known to have substantial Si impurities, which form subsurface oxides[77–79], it is not surprising that some oxide was formed. The absence of impurities on the rigorously cleaned, Pt single crystal surface used in this study precluded the formation of any oxides during CO oxidation.

3.1.2. Structure-sensitive reactions

(a) Dissociative adsorption of alkanes

The activation of alkanes on transition metal surfaces is an important step in many catalytic reactions. Hydrogenolysis, steam reforming and isomerization of alkanes all involve alkane dissociation. Thus, much interest exists in the mechanistic and kinetic aspects of alkane dissociation.

Relative to most other classes of organic molecules, alkanes are very unreactive at transition metal surfaces. While alkane dissociation is easily observable at high pressures on transition metal surfaces[80–82] the activation of alkanes in ultrahigh vacuum, where the analytical tools of surface science can provide detailed mechanistic and structural information, is quite difficult. In vacuum, low surface temperatures are required to stabilize molecular adsorption of alkanes[83–87]. Subsequent heating of the surfaces generally results in molecular desorption of the adsorbed alkanes without any detectable decomposition or reaction[83,88–90]. Two exceptions to this behavior are the Pt(110)–(1 × 2)[87] and Ir(110)–(1 × 2)[84,86] surfaces on which alkanes undergo dissociation at 200 K and 130 K, respectively. The high reactivity of these two surfaces for alkane decomposition has been related to the presence of low coordination metal atoms formed as a result of the (1 × 2) reconstruction[87]. The absence of dissociation on other single crystal transition metal surfaces indicates that the barriers to dissociation are generally too large for dissociation to compete effectively with desorption during TPD.

In order to overcome the barrier to dissociation in ultrahigh vacuum, several workers have used high energy molecular beams[91–98]. By far the most attention has been devoted to methane dissociation, although studies of the dissociation of higher alkanes on Ni(100)[93] and Ir(110)–(1 × 2)[92] have also been reported. While the use of molecular beams clearly enables one to observe alkane dissociation in vacuum, molecular beam techniques generally sample only the behavior of high energy molecules, providing relatively little information regarding the behavior of the low energy molecules which dominate the flux at surface at high pressures. Thus, it is not clear, *a priori*, that the dissociation behavior observed by molecular beam techniques accurately reflects the processes which occur at high pressures.

(i) Methane In order to determine if the results of the molecular beam studies can indeed be used to predict dissociation kinetics at high pressures, Beebe *et al.*[99] measured dissociative sticking coefficients of methane at 1.0 torr on the three low index surfaces of nickel, and compared their results to moelcular beam studies by Hamza and Madix[93] on Ni(100) and Lee *et al.*[95,96] on Ni(111). It was found that the activity of the nickel surfaces for methane dissociation increased in the order Ni(111) < Ni(100) < Ni(110), with activation energies of 12.6, 6.4, and 13.3 kcal/mole, respectively. The absolute magnitude of the sticking probabilities were quite small, falling in the range of 10^{-7} to 10^{-9} for all three surfaces. In addition, on Ni(100) and Ni(110), the effect of deuterium substitution on methane dissociation was found to be consistent with a mechanism involving quantum tunneling of a proton as the rate limiting step[99]. This finding is consistent with studies of methane activation on other transition metal surfaces[97,98,100], which all suggest a tunneling mechanism. The deuterium isotope effect was not measured for Ni(111).

As shown in Fig. 8, good agreement was obtained between the high pressure data[99] and the molecular beam data for Ni(111)[95,96]. The dissociative sticking probabilities measured at 1.0 torr were only a factor of four greater than predicted by the beam experiments, well within the combined uncertainty of the two studies. In contrast, very poor agreement was obtained between the high pressure data[99] and the molecular beam data[93] on the Ni(100) surface. As shown in Fig. 8, sticking probabilities measured at 1.0 torr were a factor of 100 to 200 lower than the values predicted from the molecular beam data. This disagreement is now believed to be due to differences in vibrational excitation of the methane molecules between the two studies[101]. In order to achieve the high translational energies necessary for methane activation, nozzle temperatures as high as 1000 K were used in the molecular beam study[102], resulting in significant vibrational excitation of the incident methane molecules. At the temperatures used in the high pressure study, relatively little vibrational excitation occurs. For both the W(110)[97,98] and Ni(111)[95,96] surfaces, vibrational excitation has been shown to be as effective as translational energy in promoting methane dissociation. Assuming that vibrational excitation also promotes dissociation on Ni(100), the presence of large amounts of vibrational energy in the beam experiments may have resulted in an overestimation of the dissociative sticking probabilities of methane at high pressures. In fact, Kay[101] has shown that by correcting the molecular beam data for the effects of vibrational excitation, good agreement can be obtained between the molecular beam study and the high pressure study.

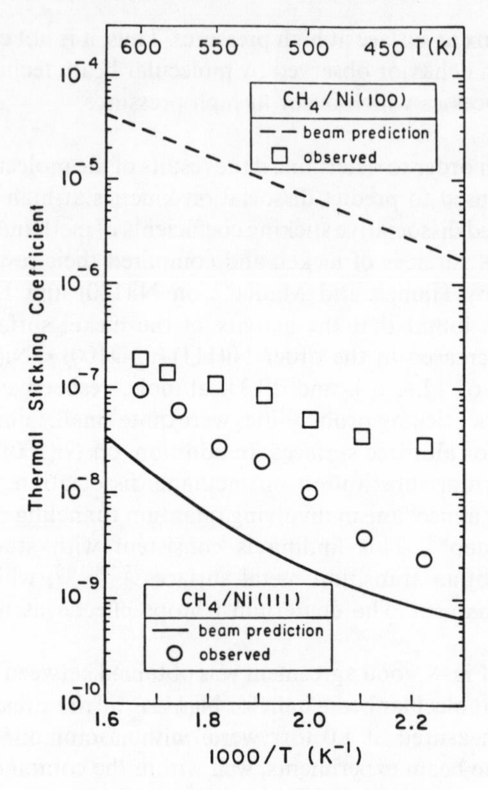

Fig. 8. Comparison of dissociative sticking coefficients of methane on Ni(100) and Ni(111) measured at 1.0 torr with sticking coefficients predicted from molecular beam data. (*From Ref. 99.*)

It should be noted that vibrational excitation of the incident methane molecules was also present in the molecular beam study of methane on Ni(111)[96]. In that case, however, the contributions of vibrational energy to the sticking coefficients were carefully subtracted out of the data, so that the reported sticking probabilities reflected only the effects of translational energy. Thus, the differences in vibrational excitation between the beam study[96] and the high pressure study[99] were negligible, and good agreement was obtained between the two studies.

(ii) Higher alkanes Sault and Goodman[103] have recently extended the study of Beebe *et al.*[99] to the higher alkanes. As was the case for methane, the object of this study was to compare sticking probabilities measured at high ambient pressures with the results of a molecular beam study by Hamza and Madix[93]. The work was confined to the Ni(100) surface as this is the only plane of nickel

on which dissociation of higher alkanes has been studied using molecular beam techniques.

Using alkane–helium mixtures with alkane partial pressures from 0.001 to 0.1 torr and total pressure of 1.0 torr to ensure thermal equilibrium between the surface and the ambient gas, Sault and Goodman[103] found that the sticking probabilities for alkanes on Ni(100) increased in the order methane < ethane < propane < n-butane, with activation energies of 6.4, 9.5, 3.8 and 3.1 kcal/mole, respectively. In contrast to methane, no measurable deuterium kinetic isotope effect was found for ethane dissociation. This result indicates that proton tunneling is not the dominant mechanism for dissociation of the higher alkanes. The possibility that proton tunneling is a minor pathway to ethane dissociation cannot be ruled out, however.

Comparison of the dissociative sticking probabilities measured at high pressures with the predictions of the molecular beam study are shown in Fig. 9. For all of the higher alkanes, the agreement is very poor. The absolute values of the sticking coefficients measured at high pressure are all greater than predicted by the molecular beam data. In addition, the activation energies for dissociation and the effect of increasing carbon chain length are also in disagreement. The discrepencies are believed to be due to the behavior of very low energy molecules, which dominate the flux at the surface at high pressures[103], but are difficult to examine using molecular beam techniques. Hamza and Madix[93] reported a linear increase in the dissociation probabilities of ethane, propane and n-butane with normal kinetic energy for energies above a threshold of 50 kJ/mole. The slope of this linear increase decreased in the order ethane > propane > n-butane. Below the threshold, no dissociation could be detected, but the smallest sticking probability which could be measured by Hamza and Madix was on the order of 10^{-4}[102]. From Fig. 9 it can be seen that all of the sticking probabilities measured at high pressures are well below 10^{-4}. Thus, if alkane molecules with energies below the threshold dissociate with nonzero probabilities which are less than 10^{-4}, then the discrepencies in the magnitude of the sticking probabilities and activation energies for dissociation shown in Fig. 9 can be accounted for[103].

To account for the fact that the dissociation probabilities measured in the beam experiments increased in the order n-butane < propane < ethane < methane, while the opposite trend was observed at high pressres, it must be assumed that different dissociation mechanisms operate for low and high energy molecules. At the high energies at which Hamza and Madix[93] were able to observe dissociation, a direct mechanism for dissociation occurs. Hamza and Madix[93] postulate that for the direct mechanism, dissipative transfer of translational enery out of the reaction channel into surface phonons and internal modes of the alkanes can explain the observed variation of the sticking probability with carbon chain length. As the chain length increases, the amount of dissipative transfer also increases, thereby decreasing the

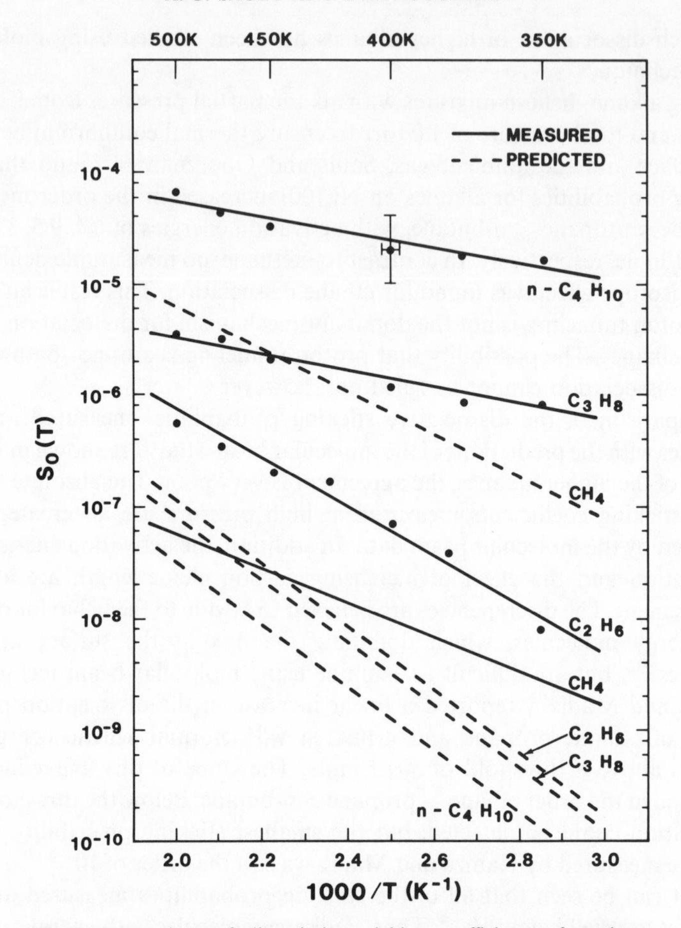

Fig. 9. Comparison of dissociative sticking coefficients of methane, ethane, propane, and n-butane on Ni(100) measured at high pressures with sticking coefficients predicted from molecular beam data. (*From Ref 103.*)

fraction of the incident translational energy which is transferred into the reaction channel and resulting in the observed decrease in sticking probability with increasing carbon chain length. The low energy molecules which dominate the dissociation kinetics at high pressures are believed to dissociate via a trapped molecular procursor[103]. Since both the trapping probability and the surface lifetime of the trapped alkanes increases in the order ethane < propane < n-butane, the sticking probabilities also increase in this order at high pressures.

Based on the studies of alkane dissociation at high pressures, two conclusions were reached regarding the applicability of molecular beam studies for predicting high pressure behavior[103]. First, the effects of vibrational excitation must be accounted for in order to make accurate predictions. Only in cases where the vibrational excitation produced during formation of the molecular beam is similar to that in the high pressure process being modeled will good agreement be obtained between high pressure and molecular beam studies. Second, the behavior of low energy molecules is critically important in determining the high pressure behavior, since the flux of molecules at a surface is dominated by low eneorgy molecules[103]. Only in cases where the behavior of low energy molecules is accurately measured will molecular beam techniques be able to reliably predict high pressure behavior.

Finally, it should be pointed out that the rates of alkane dissociation on the clean nickel surfaces are generally somewhat greater than the rates of hydrogenolysis and steam reforming of the alkanes[99,103]. This result is expected since both steam reforming and hydrogenolysis take place on partially carbon covered surfaces and carbon is known to inhibit alkane dissociation. Furthermore, it is not entirely clear that the dissociation step is rate limiting for either steam reforming or hydrogenolysis. Thus, the rates of alkane dissociation measured on the clean surfaces represent a theoretical upper limit for the rates of steam reforming and hydrogenolysis on unpromoted nickel surfaces.

(b) Hydrogenolysis of alkanes

(i) Reactions on nickel The activity of metal catalysts toward alkane hydrogenolysis (reaction (3)) generally depends markedly on the metal particle size and upon the nature of the metal[104,105]. These reactions have therefore been described as structure sensitive. Although there is a consensus in the literature regarding the relationship of particle size to activity, no such consensus exists in detailing the origins of the effect. Work by Martin[106] suggests that different activities of different crystal planes toward a given reaction could be responsible for the observed rate attenuation with increasing particle size. For example, if (111) orientations become dominant as the metal particles become larger and if these crystal planes exhibited a lower activity toward hydrogenolysis, then the activity of the catalyst would fall with particle growth. Martin further speculated that the lower activity of the (111) facet relative to other facets could be intrinsic or could arise from preferential poisoning of the (111) crystal plane by reactants or products. These studies emphasized the need for kinetic measurements over single metal catalysts. Such studies have been carried out for the hydrogenolysis of ethane[107], n-butane[108], and cyclopropane on Ni(111) and Ni(100)[109]. These results confirm that Ni(111) is indeed significantly less active toward hydrogenolysis than Ni(100). This reduced activity has been shown to be intrinsic and not due to selective poisoning of the (111) surface by carbon.

Figure 10 shows the specific reaction rate for methane formation from

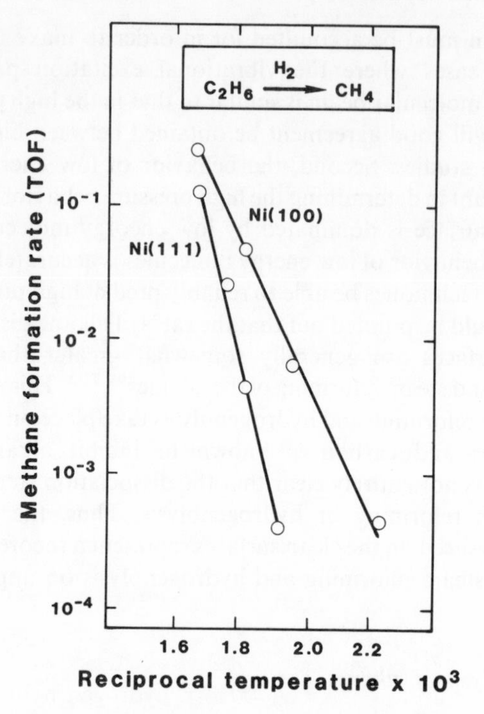

Fig. 10. Methane production from ethane hydrogenolysis over Ni(100) and Ni(111) catalysts at total reactant pressures of 100 torr. $H_2/C_2H_6 = 100$. (*From Ref. 107.*)

ethane over a Ni(100) catalyst. At a given temperature the rate of methane production over an initially clean crystal was constant with no apparent induction period. The carbon level during reaction remained constant at a submonolayer coverage. The methane turnover frequency during a fixed time (typically 1000 s) was determined using the Ni(100) atom density of 1.62×10^{15} atoms/cm^2.

Figure 10 also shows the kinetic data for ethane hydrogenolysis over a Ni(111) catalyst. As observed for the (100) catalyst, the carbon level during reaction remained constant at submonolayer coverages. The specific rate was determined using the Ni(111) atom density of 1.88×10^{15} atoms/cm^2. It is evident that the activity of the (111) surface toward ethane hydrogenolysis is considerably less than that observed for the (100) surface as suggested by Martin[106]. The kinetic results for the Ni(111) catalyst agree favorably with the data of Martin[106] for Ni/SiO_2 catalysts reduced at high temperature with a resulting larger particle size. Such severely sintered catalysts are expected to contain metal crystallites exposing predominantly (111) faces[110].

The origin of the lower activity associated with the (111) surface compared with the (100) is not presently understood. We can speculate that electronic differences, which most certainly exist, are responsible. An additional contribution could be the spatial distribution of the high coordination bonding sites[107], those on the (111) surface being more favorable for stabilization of C_2 fragments. The result of this stabilization is that the (111) surface is less effective in cleaving carbon–carbon bonds, an obvious crucial step in hydrogenolysis.

Although the chemistry of these observations is not well understood, it is clear that these kinds of studies are necessary in detailing the origins of this most intriguing aspect of catalysis by metals.

(ii) Reactions on iridium Both the Ir(110)–(1 × 2)[84–86] and Pt(110)–(1 × 2)[87] surfaces are unique in that they are the only known surfaces on which adsorbed alkanes dissociate during TPD in ultrahigh vacuum. This behavior is clearly related more to the structure of these reconstructed surfaces than to the specific nature of the metal as dissociation does not occur on the (111) surfaces of either Pt[83] or Ir[85]. In order to understand more fully the relationship between surface structure and catalytic activity and selectivity, Engstrom *et al.*[111–113] studied the hydrogenolysis of a number of alkanes on the (110)–(1 × 2) and (111) surfaces of iridium. These reactions were conducted at total pressures on the order of 100 torr, hydrogen:hydrocarbon ratios of 100:1 and temperatures in the range 425–625 K[111–113]. Total conversions were kept below 1% in order to avoid further hydrogenolysis of reaction products. Reaction orders in both the hydrocarbons and hydrogen were also investigated.

Before reviewing reactions of the individual hydrocarbons, some overall trends will be discussed. In general, post-reaction surface analysis by AES showed the presence of submonolayer coverages of carbon, with a fractional coverage fairly independent of reaction conditions for a given alkane. The amount of carbon left on the surface increased approximately linearly with the size of the parent hydrocarbon molecule. This carbon residue is not believed to be an important intermediate in the hydrogenolysis reactions[112,113], however, and will not be discussed further. For all of the linear hydrocarbons and both the Ir(111) and Ir(110)–(1 × 2) surfaces, Arrhenius plots of product formation rate *vs.* temperature were linear at low temperatures, but showed large deviations from linearity at higher temperature. Simultaneous with the deviations from linearity, the selectivity of the reaction moved toward more complete hydrogenolysis, i.e. the relative production of methane increased at the expense of higher alkanes. The high temperature deviations are related to a decrease in the steady-state coverage of hydrogen as the temperature approaches the desorption temperature of hydrogen[111,112]. Because the binding energy of hydrogen on Ir(110)–(1 × 2) is greater than on Ir(111)[114,115], the deviations from linearity occur at a higher temperature on Ir(110)–(1 × 2) than on Ir(111)[112].

Different kinetic parameters were observed for ethane hydrogenolysis on the two surfaces[112]. The activation energies and pre-exponential factors for hydrogenolysis are 35 kcal/mole and $1 \times 10^{13}\ s^{-1}$ on Ir(111) and 49 kcal/mole and $6 \times 10^{18}\ s^{-1}$ on Ir(110)–(1 × 2). The differences in these parameters indicate that a different mechanism operates on the two surfaces. Based on kinetic modeling, it is believed that hydrogenolysis proceeds through a C_2H_4 intermediate on Ir(111) and a C_2H_2 intermediate on Ir(110)–(1 × 2). The C_2H_2 species can be stabilized on the Ir(110)–(1 × 2) surface by adsorption in the trough sites, which allows greater coordination of the C_2H_2 species to the surface than is possible on the flat Ir(111) surface. Because of the differences in activation energy, the relative rates of hydrogenolysis on the two surfaces cross over near 550 K, with Ir(111) being more active below 550 K and Ir(110)–(1 × 2) being more active above 550 K. This crossover explains why no structure sensitivity was observed in a study of ethane hydrogenolysis on supported iridium catalysts[116] at temperatures of 525–550 K.

At low temperatures on both surfaces, propane hydrogenolysis involves cleavage of a single C—C bond to produce equal amounts of methane and

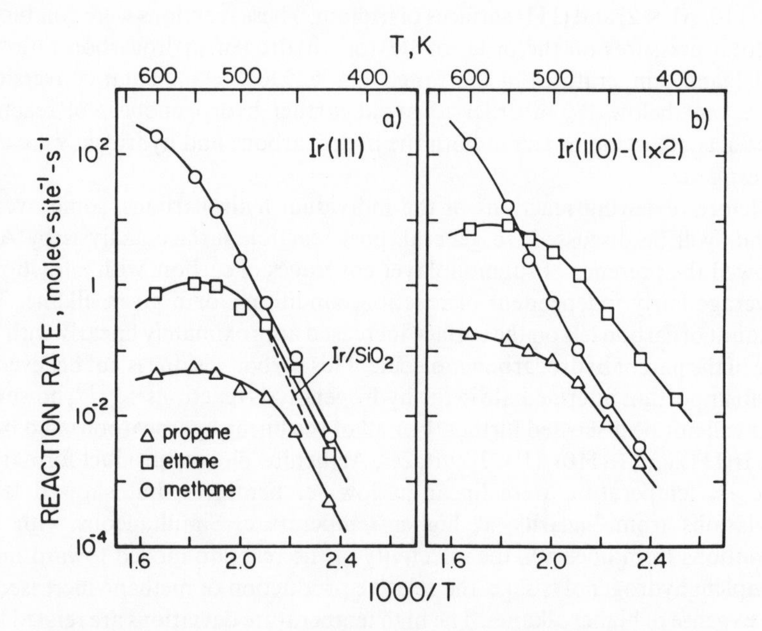

Fig. 11. Arrhenius plots for n-butane hydrogenolysis on (a) Ir(111) and (b) Ir(110)–(1 × 2). (*From Refs. 111, 112.*) The pressure of n-butane was 1 torr and that of hydrogen was 100 torr. The dashed line in (a) represents data for n-butane hydrogenolysis on a supported Ir/SiO_2 catalyst[116].

ethane[112]. At high temperatures the relative rate of methane production increases due to depletion of adsorbed hydrogen. Similarities in the reaction rates, kinetic parameters, and the stoichiometries of adsorbed hydrocarbon fragments (predicted from kinetic modeling) between the two surfaces suggest that the same reaction mechanism is operative on both surfaces. It was speculated that the reaction occurs through a C_3H_6 intermediate bound as either a metallacyclobutane or a binuclear metallacyclopentane. The similar kinetics observed on the two surfaces suggests that the binding site of the C_3H_6

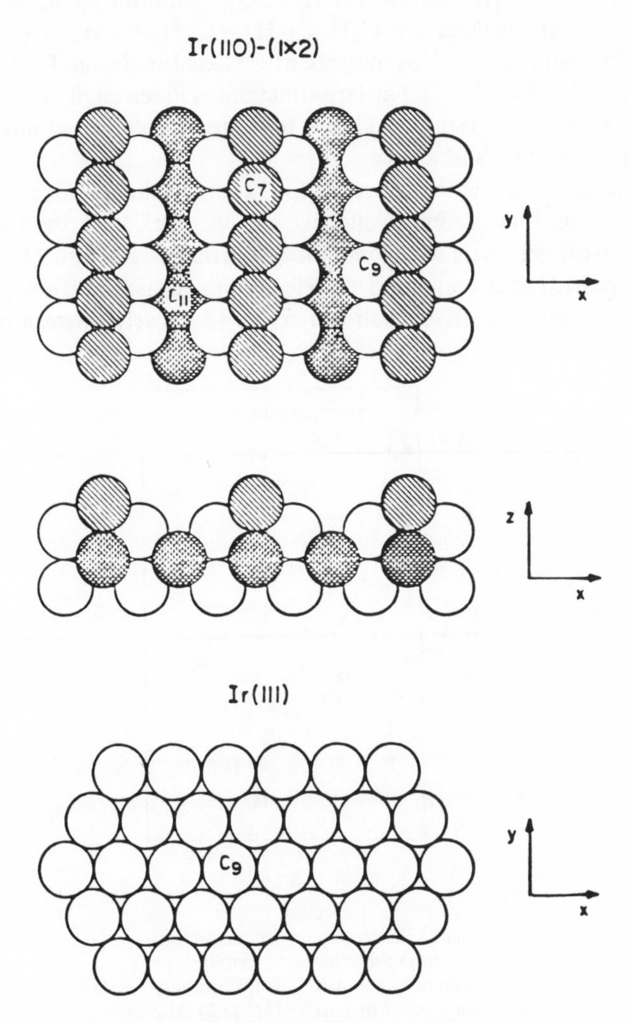

Fig. 12. Structural models for the Ir(110)–(1 × 2) and Ir(111) surfaces. (*From Ref. 112.*) C_n designates the coordination numbers of the metal atoms.

intermediate is the same on both Ir(111) and Ir(110)–(1 × 2). This situation could occur only if the intermediate were located on the (111) facets of the Ir(110)–(1 × 2) surface, such that the edge atoms of the Ir(110)–(1 × 2) surfaces were not involved in the metal–C_3H_6 bond.

For n-butane hydrogenolysis, the major reaction pathways on the two surfaces are clearly different in the linear ($T < 475$ K) region of the Arrhenius plots shown in Fig. 11[111,112]. The product distributions indicate that the major reaction pathways are n-$C_4H_{10} + H_2 \rightarrow 2CH_4 + C_2H_6$ on Ir(111) and n-$C_4H_{10} + H_2 \rightarrow 2C_2H_6$ on Ir(110)–(1 × 2). In addition, a minor reaction pathway on both surfaces is n-$C_4H_{10} + H_2 \rightarrow C_3H_8 + CH_4$. The activation energies for n-butane hydrogenolysis are 34 kcal/mole on Ir(111) and 22 kcal/mole on Ir(110)–(1 × 2). The large differences in selectivity and activation energy between the two surfaces clearly indicate that a different mechanism is operative on the two surfaces.

By using ratio the number of edge (C_7) atoms to the number of (111) face (C_9) atoms (Fig. 12) to define an effective particle size[111,112], the selectivity of n-butane hydrogenolysis as a function of particle size for the two surfaces could be plotted and compared to selectivities measured on supported Ir catalysts[116]. This comparison is shown in Fig. 13. Clearly, the results on single

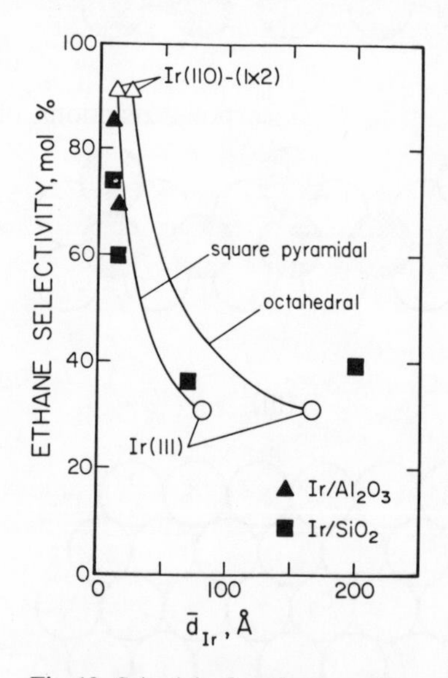

Fig. 13. Selectivity for ethane production from n-butane hydrogenolysis on iridium as a function of effective particle size. (*From Refs. 111, 112.*) Also shown are data for n-butane hydrogenolysis on supported Ir catalysts[116]. The temperature is 475 K in all cases.

crystal catalysts correlate well with the results of studies on supported catalysts. In addition, it is clear that high selectivities toward ethane formation are related to the presence of low coordination number surface atoms, i.e. edge atoms. It has been postulated[111,112] that the high selectivity toward ethane on the Ir(110)–(1 × 2) surface is the result of hydrogenolysis occurring via a metallacyclopentane intermediate. Thermal decomposition of transition metal metallacyclopentane complexes can result in cleavage of a C—C bond to form two C_2H_4 ligands[117,118]. A similar reaction the Ir(110)–(1 × 2) surface would likely result in rapid hydrogenation of the resulting adsorbed ethylene molecules due to the large excess of hydrogen present. The open nature of the Ir(110)–(1 × 2) surface makes it ideally suited for formation of a metallacyclopentane species on the edge (C_7) atoms, while the absence of any edge atoms on the Ir(111) surface makes metallacyclopentane formation sterically unfavorable[111,112]. As a result, ethane formation is favored on Ir(110)–(1 × 2), while on Ir(111) a different mechanism must operate, which apparently involves cleavage of the terminal C—C bonds to form both methane and ethane.

The hydrogenolysis of neopentane occurs primarily by breaking of a single C—C bond to produce methane and isobutane as the major reaction products[112]. Small amounts of ethane and propane were also produced. The reaction kinetics were similar on the two surfaces, indicating a similar mechanism for both surfaces.

In addition to saturated hydrocarbons, reactions of cyclopropane and methylcyclopropane with hydrogen on Ir(111) and Ir(110)–(1 × 2) have also been studied[113]. On both surfaces cyclopropane undergoes both hydrogenation to form propane and hydrogenolysis to form ethane and methane. Hydrogenation dominates below 500 K while hydrogenolysis dominates at higher temperatures. The Ir(110)–(1 × 2) surface is a factor of three to four times more active that the Ir(111) surface for both reactions. As was the case for n-butane hydrogenolysis, the greater activity for the Ir(110)–(1 × 2) surface is attributed to higher activity of the low coordination number sites present on this surface[113]. In addition to methane, ethane and propane, propylene production was also observed at high temperatures on Ir(III). Both propylene and propane appear to be formed via the same rate limiting step. The onset of propylene production is related to a decrease in the steady-state coverage of hydrogen at high temperatures or under hydrogen-lean conditions, which results in a shift in the selectivity toward the more dehydrogenated product. Propylene production was also observed on Ir(110)–(1 × 2) under very hydrogen-lean conditions.

In contrast to propane hydrogenolysis, the Arrhenius plots for cyclopropane hydrogenolysis on both surfaces exhibit no deviations from linearity over the entire temperature range studied, and the reaction orders in both hydrogen and cyclopropane are essentially zero[113]. These results have been interpreted as indicating that cyclopropane hydrogenolysis occurs by a facile, irreversible ring-opening step followed by rate limiting cleavage of a C—C bond in the resulting intermediate[113].

The reaction of methylcyclopropane with hydrogen produced n-C_4 hydrocarbons as the major reaction products on both surfaces[113]. At high temperatures or under hydrogen-lean conditions, n-butene production dominates, while at lower temperatures and hydrogen-rich conditions, n-butane production dominates. As for cyclopropane, this shift in selectivity is attributed to decreases in the steady-state coverage of hydrogen at high temperatures or under hydrogen-lean conditions. The shift toward n-butene production begins at a higher temperature on Ir(110)–(1 × 2) than on Ir(111). This difference is due to the higher binding energy of hydrogen on Ir(110)–(1 × 2) than on Ir(111)[114,115]. Overall, the Ir(110)–(1 × 2) surface is 3–4 times more active for reaction of cyclopropane with hydrogen than the Ir(111) surface.

3.2. Reactivity of chemically modified surfaces

It has long been recognized that the addition of impurities to metal catalysts can produce large effects on the activity, selectivity, and resistance to poisoning of the pure metal[119]. For example, the catalytic properties of metals can be altered greatly by the addition of a second transition or group 1B metal or by the addition of impurities such as potassium or sulfur. On the other hand, catalytic processing is often plagued by loss of activity and/or selectivity due to the inadvertent contamination of catalysts by undesirable impurities. Although these effects are well recognized in the catalytic industry, the mechanisms responsible for surface chemical changes induced by surface additives are poorly understood. However, the current interest and activity in this area of research promises a better understanding of the fundamentals by which impurities alter surface chemistry. A pivotal question concerns the underlying relative importance of ensemble (steric or local) versus electronic (nonlocal or extended) effects. A general answer to this question will critically influence the degree to which we will ultimately be able to tailor-make exceptionally efficient catalysts by fine tuning the electronic structure. If, indeed, low concentrations of surface impurity can profoundly alter the surface electronic structure and thus catalytic activity, then the possibility for the systematic manipulation of these properties via the selection of the appropriate additive would appear limitless. On the other hand, if steric effects dominate the mode by which surface additives alter the catalytic chemistry, then a different set of considerations for catalyst alteration come into play, a set which will most certainly be more constraining than the former. In the final analysis, a complete understanding will include components of both electronic and ensemble effects, the relative importance of each to be assessed for a given reaction and conditions. A major emphasis of our research has been in the area of addressing and partitioning the importance of these two effects in the influence of surface additives in catalysis.

Catalyst deactivation and promotion are extremely difficult questions to address experimentally[119]. For example, the interpretation of related data on dispersed catalysts is severely limited by the uncertainty concerning the structural characterization of the active surface. Specific surface areas cannot always be determined with adequate precision. In addition, a knowledge of the crystallographic orientation, the concentration and the distribution of impurity atoms, as well as their electronic states is generally poor. The degree of contamination may vary considerably along the catalytic bed and the impurity very well may alter the support as well as the metal. Moreover, the active surface may be altered in an uncontrolled manner as a result of sintering or faceting during the reaction itself.

The use of metal single crystals in catalytic reaction studies essentially eliminates the difficulties mentioned above and allows, to a large extent, the utilization of a homogeneous surface amenable to study using modern surface analytical techniques. Carefully prepared, single-crystal catalytic surfaces are particularly suited to the study of impurity effects on catalytic behavior because of the ease with which impurity atoms can be uniformly introduced to the surface. Although the studies to date are few, the results appear quite promising in addressing the fundamental aspects of catalytic poisoning and promotion.

3.2.1. Electronegative impurities

Impurities whose electronegativities are greater than those for transition metals generally poison a variety of catalytic reactions, particularly those involving H_2 and CO. Of these poisons sulfur is the best known and technologically the most important[119]. The first step in the systematic definition of the poisoning mechanism of this category of impuirties is the study of the influence of these impurities on the adsorption and desorption of the reactants.

The effect of preadsorbed electronegative atoms Cl, S, and P on the adsorption–desorption of CO and H_2 on Ni(100) has been extensively studied[120–127] using temperature programmed desorption, low energy electron diffraction and Auger spectroscopy. It has been found that the presence of the electronegative atoms Cl, S, and P causes a reduction of the sticking coefficient, the adsorption bond strength, and the adsorption capacity of the Ni(100) surface for CO and H_2. Furthermore, the poisoning effect becomes more prominent with increasing electronegativity of the preadsorbed atoms[128].

Figures 14 and 15 show the observed dependence of the saturation H_2 and

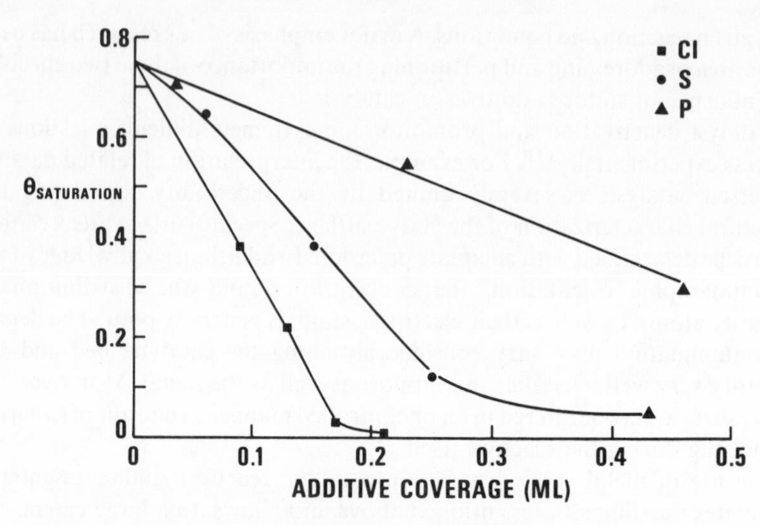

Fig. 14. Dependence of saturation H_2 coverage on Ni(100) on additive precoverage. (*From Refs. 120–127.*)

CO coverages on the coverage of Cl, S, and P. Coverages of impurities are expressed in terms of monolayers (ML) or the ratio of surface impurity atoms to the surface metal atoms. The data represent the total H_2 and CO desorption, as determined by TPD, for different impurity coverages after an exposure sufficient to reach the saturation adsorbate coverage. Both CO and H_2 adsorption decrease markedly in the presence of surface impurities. The effects of P, however, are much less pronounced than for Cl or S. As seen in Fig. 14, the reduction of H_2 coverage is most apparent in the presence of Cl atoms. The similarity in the atomic radii of Cl, S, and P (0.99, 1.04, and 1.10 Å, respectively[129]) suggests a relationship between electronegativity and the poisoning of chemisorptive properties by these surface impurities. Related studies[122] have been carried out in the presence of C and N. These impurities have the same electronegativities as S and Cl, 2.5 and 3.0, respectively. The comparison between the results for C and N and those for S and Cl are entirely consistent with the interpretation that electronegativity effects dominate the poisoning of chemisorption by surface impurities with similar atomic size, and which occupy the same adsorption sites. In the case of adsorbed impurities with the same electronegativity, but with different atomic radii (S and C, Cl and N), the effect becomes less pronounced with decreasing atomic radius.

Particularly noteworthy in the above studies is the general observation that those impurities that are strongly electronegative with respect to nickel, e.g. Cl, N, and S, modify the chemisorptive behavior far more strongly than would result from a simple site blocking model. The initial effects of these impurities as shown in Figs. 14 and 15 suggest that a single impurity atom can

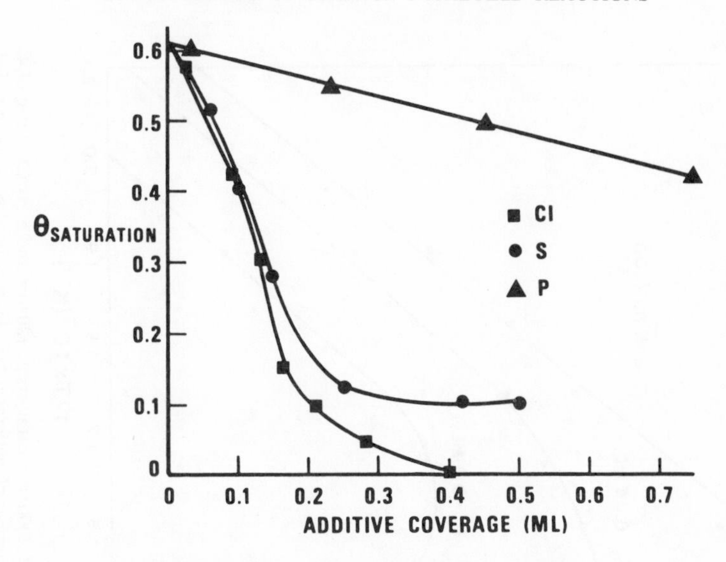

Fig. 15. Dependence of saturation CO coverage on Ni(100) on additive precoverage. (*From Refs. 120–127.*)

successfully poison more than just its four nearest neighbor nickel atoms This effect is especially apparent for hydrogen adsorption on the sulfur covered Ni(100) surface. Poisoning of next nearest neighbors supports an interaction that is primarily electronic in nature.

Kinetic studies have been carried out for several reactions as a function of sulfur coverage over single crystals of nickel[120] rhodium[130], and ruthenium[131]. For the methanation reaction over Ni(100)[120], the sulfided surface (Fig. 16a) shows behavior remarkably similar to results for the clean surface at a considerably reduced hydrogen partial pressure. For clean Ni(100)[3] a departure from Arrhenius linearity is observed at 700 K. Associated with the onset of this nonlinearity or 'rollover' is a rise in the surface carbon level. This rise in carbon level continues until the carbon level reaches 0.5 ML, i.e. the saturation level. This behavior of the Arrhenius plot has been interpreted[3] as reflecting the departure of atomically adsorbed hydrogen from a saturation or critical coverage. For a sulfur surface coverage of 4% the reaction rate at identical conditions departs similarly from linearity at 600 K, some 100 K lower in reaction temperature. Here too, an increase in surface carbon level is associated with this deviation from linearity. This behavior indicates that the sulfur is very effective in reducing the steady-state surface atomic hydrogen coverage which results in an attenuation of the rate of surface carbon hydrogenation. These results are consistent with the chemisorption results[121] discussed above for H_2 on sulfur poisoned Ni(100) surface. Similar results

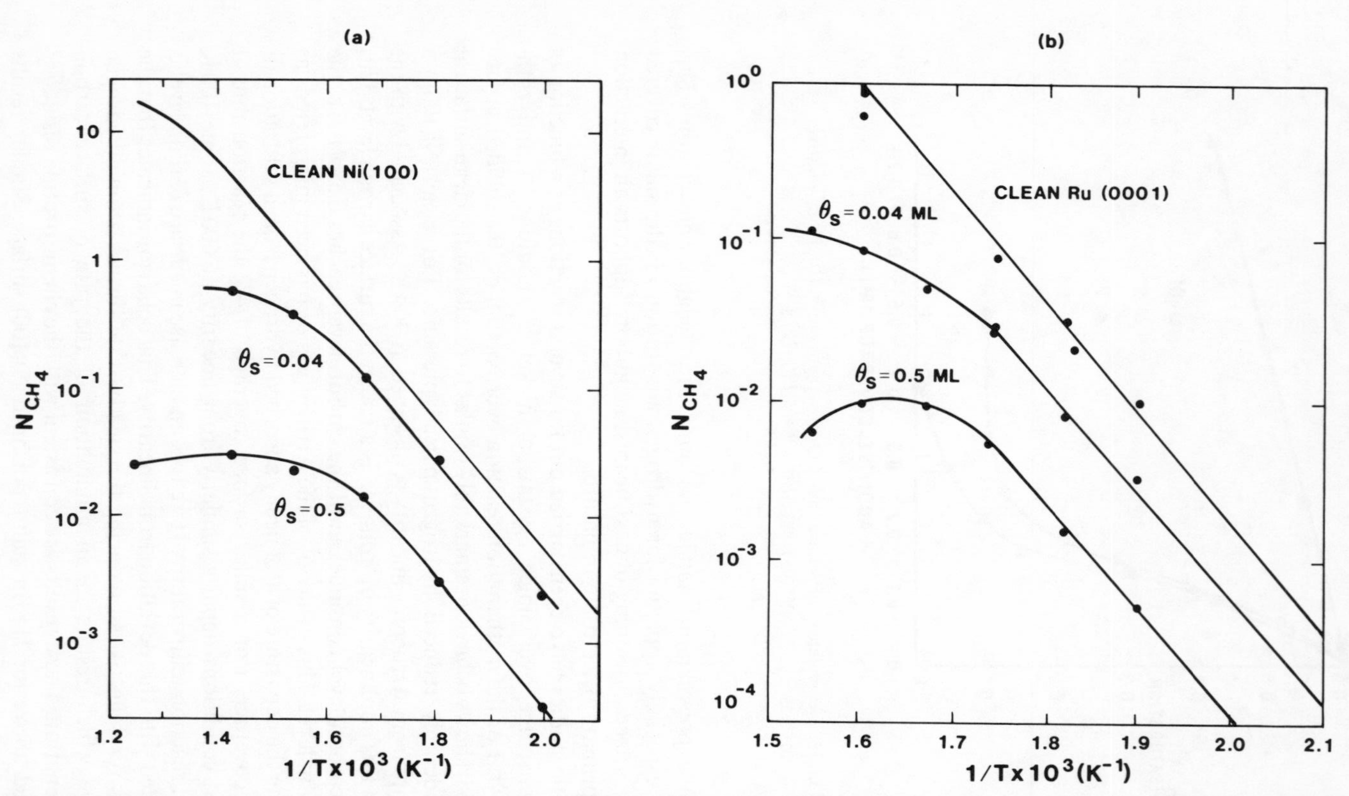

Fig. 16. An Arrhenius plot of the rate of methanation over sulfided (a) Ni(100)[120] and (b) Ru(0001)[131] catalysts at 120 torr and a H_2/CO ratio of 4. Coverages are expressed as fractions of a monolayer. N_{CH_4} is the turnover frequency or the number of methane molecules produced per surface nickel atom per second.

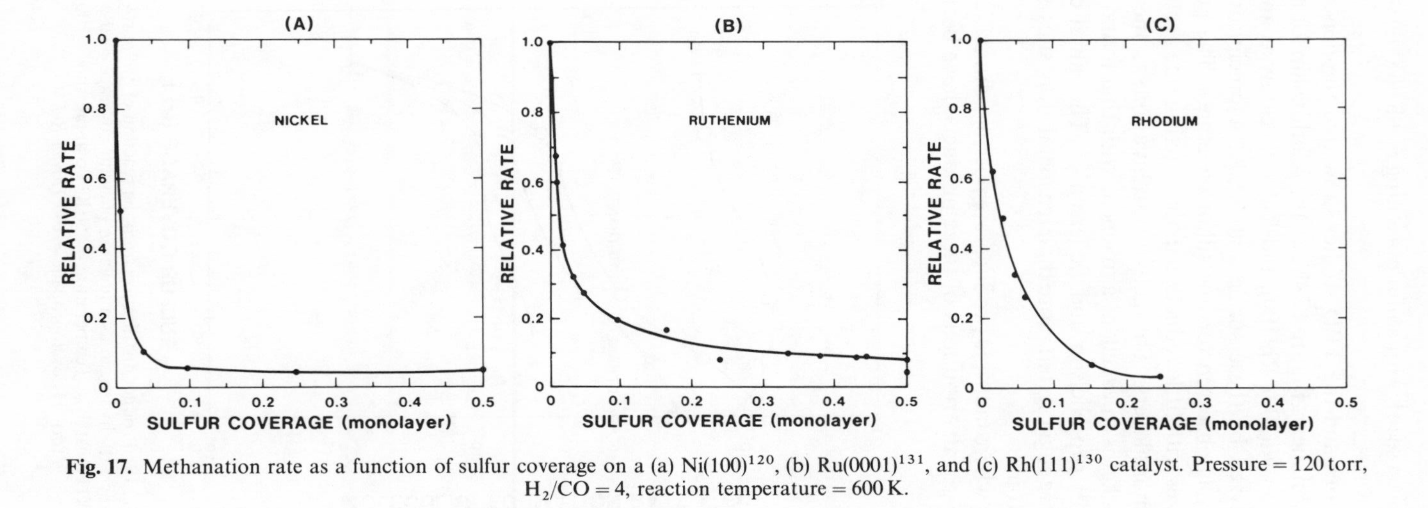

Fig. 17. Methanation rate as a function of sulfur coverage on a (a) Ni(100)[120], (b) Ru(0001)[131], and (c) Rh(111)[130] catalyst. Pressure = 120 torr, $H_2/CO = 4$, reaction temperature = 600 K.

(Fig. 16b) have been seen[131] for sulfur poisoning of a Ru(0001) surface towards CO hydrogenation.

Both the kinetics and the TPD studies show that the poisoning effects of sulfur are very nonlinear. Figure 17 shows the relationship found between the sulfur coverage on Ni(100), Rh(100), and Ru(0001) catalysts and the methanation rate catalyzed by these surfaces at 600 K. A precipitous drop in the catalytic activity is observed for low sulfur coverages. The poisoning effect quickly maximizes with little reduction in the reaction rate at sulfur coverages exceeding 0.2 monolayers. The activity attenuation at the higher sulfur coverages on nickel is in excellent agreement with that found for supported Ni/Al_2O_3 by Rostrup-Nielson and Pedersen[132]. The initial changes in the rates in Fig. 17a–c suggest that more than ten metal atom sites are deactivated by one sulfur atom.

In those cases where multiple products are possible, dramatic modifications in the selectivity, or distributions of these products, have been observed[109].

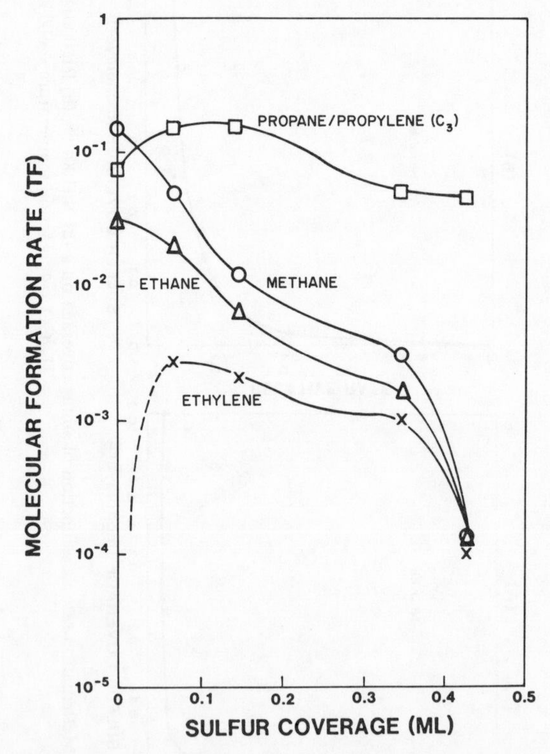

Fig. 18. Product distribution from the reaction of cyclopropane with hydrogen as a function of sulfur coverage over a Ni(111) catalyst. Temperature = 550 K. Total pressure = 100 torr. H_2/cyclopropane = 100. (*From Ref. 4.*)

Figure 18 shows the effect that progressive sulfiding of a Ni(111) catalyst has on the cyclopropane/hydrogen reaction. A small amount of sulfur (< 0.1 ML) exponentially lowers the rate of methane formation, the dominant product formed on the clean surface. Similarly, the rate of ethane formation falls in concert with the methane suggesting, as is expected, a close correlation between these hydrogenolysis products. In contrast to the methane and ethane products, the production of propane/propylene (C_3) product actually increases with the sulfur addition. Qualitatively, the increase in the C_3 product corresponds to the decrease in the methane rate. These results show rather directly that the initial sulfiding promotes the ring-opening reaction by reducing the tendency of the surface to break more than one carbon–carbon bond.

In contrast to the clean Ni(111) surface where no ethylene was observed as a reaction product, significant amounts of ethylene are found[109] for the sulfided surface. In addition to reducing the tendency of the surface to break carbon–carbon bonds, sulfur also lowers the hydrogenation activity. This tendency has been confirmed by measuring directly the attenuation of the hydrogenative character of Ni(111) versus sulfur coverage by monitoring the ethylene/hydrogenative reaction[133].

At first glance one might interpret these results as the simple poisoning of monitoring or defect sites on the surface and that these sites are crucial to the reactivity at steady-state reaction conditions and the TPD decomposition. However, this is a very unquickly explanation given the close correspondence between the steady-state rates measured for single crystal catalysts and those rates found for supported, small-particle catalysts seen in section 3.1. That the defect densities on these two very different materials would be precisely the same is highly unlikely. It is much more likely that these reactions are not defect controlled and that the surface atoms of the single crystals are uniformly active. There are two other possible explanations for this result: (1) an electronic or ligand effect, or (2) an ensemble effect, the requirement that a certain collection of surface atoms are necessary for the reaction to occur. Experimentally these two possibilities can be distinguished[123,134]. If an ensemble of more than ten nickel atoms is required for methanation, then altering the electronic character of the impurity should produce little change in the degree to which the impurity poisons the catalytic activity. That is, the impurity serves merely to block a single site in the reaction ensemble, nothing more. On the other hand, if electronic effects are playing a significant role in the poisoning mechansim, then the reaction rate should respond to a change in the electronic character of the impurity. Substituting phosphorus for sulfur (both atoms are approximately the same size) in a similar set of experiments results in a marked change in the magnitude of poisoning at low coverages as shown in Fig. 19. Phosphorus, because of its less electronegative character, effectively poisons only the four nearest neighbor metal atom sites.

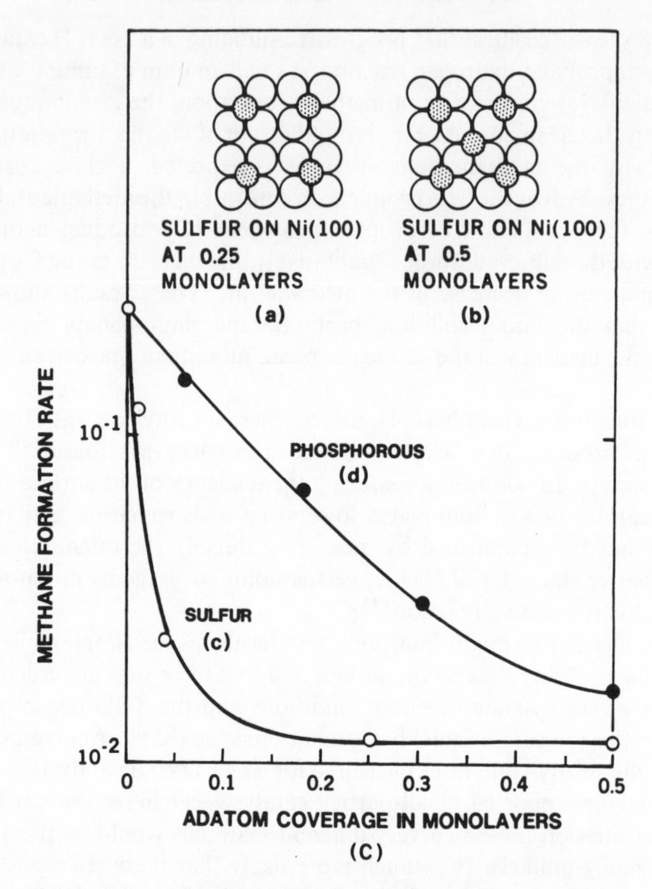

Fig. 19. A plot of the rate of CO methanation as a function of sulfur and phosphorous coverage over a Ni(100) catalyst at 120 torr and a H_2/CO ratio equal to 4. (*From Ref. 4.*)

Effective poisoning of catalytic activity at sulfur coverages less than 0.1 ML has been observed for other reactions including ethane and cyclopropane hydrogenolysis[109], ethylene hydrogenation[133], and CO_2 methanation[135]. The results of several studies on nickel are summarized in Fig. 20. These studies indicate that the sensitivity of the above reactions to sulfur poisoning are generally less than that for poisoning by sulfur of CO methanation. The rate attenuation is, nevertheless, strongly nonlinear at the lower sulfur levels. A direct consequence of the differing molecular sizes of the reactants (CO, ethylene, ethane, cyclopropane) involved in the reactions investigated is that electronic effects, rather than ensemble requirements, dominate the catalytic poisoning mechanism for these experimental conditions.

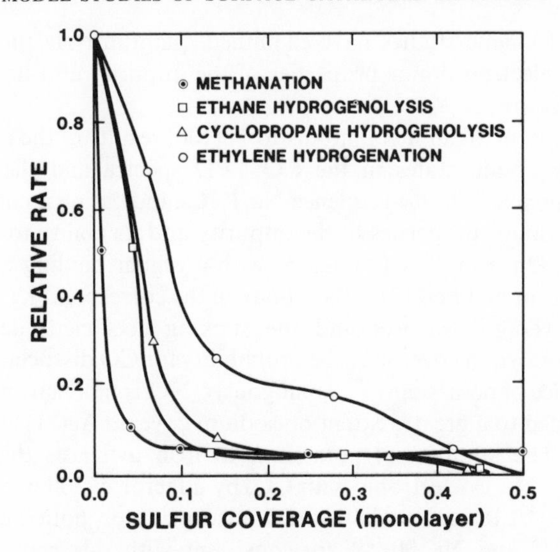

Fig. 20. Sulfur poisoning of various reactions over nickel. Hydrogen partial pressure = 100 torr. (*From Refs. 109, 133, 135.*)

Recent studies[136] using high resolution electron energy loss and photoelectron spectroscopy to investigate the effect of sulfur on the CO/Ni(100) system are consistent with an extended effect by the impurity on the adsorption and bonding of CO. Sulfur levels of a few percent of the surface nickel atom concentration were found sufficient to significantly alter the surface electronic structure as well as the CO bond strength.

3.2.2. *Electropositive impurities*

A direct consequence of interpreting the poisoning effects of electronegative impurities in terms of electronic surface modification is that additives with electronegativities less than that of the metal should promote a different chemistry reflecting the donor nature of the additive. For example, alkali atoms on a transition metal surface are known to exist in a partially ionic state, donating a large fraction of their valence electrons to the metal, resulting in a work function decrease. This additional electron density on the transition metal surface atoms is thought to be a major factor in alkali atoms altering the chemisorptive bonding of molecules such as N_2[137] or CO[138], and in promoting the catalytic activity in ammonia synthesis[139]. These results are consistent with the general picture that electron acceptors tend to inhibit CO hydrogenation reactions whereas electron donors typically produce desirable catalytic effects, including increased activity and selectivity. Recent chemisorption and kinetic studies have examined quantitatively the relationship between the electron donor properties of the impurity and its effect on the catalytic behavior.

The addition of alkali metal atoms to Ni(100) results in the appearance of more tightly bound states in the CO TPD spectra and the dissociation probability increases in the sequence Na, K, Cs, indicating a correspondence between the donor properties of the impurity and its ability to facilitate CO dissociation. On iron[140], CO adsorbs with a higher binding energy on the potassium promoted Fe(110) surface than on the corresponding clean surface. The CO coverage increases and the sticking coefficient decreases with increasing potassium coverage. The probability for CO dissociation increases in the presence of potassium[140]. Analogously, NO is more strongly adsorbed and dissociated to a greater extent on sodium covered Ag(111) than on clean Ag(111)[141]. The addition of potassium to iron increases the dissociative adsorption of N_2, isoelectronic with CO, by a factor of 300 over that for the clean surface[142]. Recent studies of CO adsorption on potassium-promoted Pt(111)[143–145] and Ni(100)[146] are consistent with this general picture of donor-enhanced metal–CO bonding. For H_2 chemisorption, Ertl and coworkers[147], using TPD techniques, have observed an increase in the adsorption energy of hydrogen on iron. They suggest that the empty state above the Fermi level created by the pronounced electron transfer from potassium to the d-band of iron may possibly be involved via interaction with the H 1s level.

Adsorbed potassium causes a marked increase in the rate of CO dissociation on a Ni(100) catalyst[148]. The increase of the initial formation rate of 'active' carbon or carbidic carbon via CO disproportionation is illustrated in Fig. 21. The relative rates of CO dissociation were determined for the clean and potassium covered surfaces by observing the growth in the carbon Auger signal with time in a CO reaction mixture, starting from a carbon-free surface. The rates shown in Fig. 21 are the observed rates of carbon formation extrapolated to zero carbon coverage. Of particular significance in these studies is the reduction of the activation energy of reactive carbon formation from 23 kcal mole^{-1} for the clean Ni(100) surface to 10 kcal mole^{-1} for a 10% potassium covered surface[148].

Kinetic measurements[148] over a Ni(100) catalyst containing well-controlled submonolayer quantities of potassium show a general decrease in the steady-state methanation rate with little apparent change in the activation energy associated with the kinetics (Fig. 22). However, the potassium did change the steady-state coverage of active carbon on the catalyst. This carbon level changed from 10% of a monolayer on the clean catalyst to 30% on the potassium covered catalyst.

Adsorbed potassium causes a marked increase in the steady-state rate and selectivity of nickel for higher hydrocarbon synthesis[148]. At all temperatures

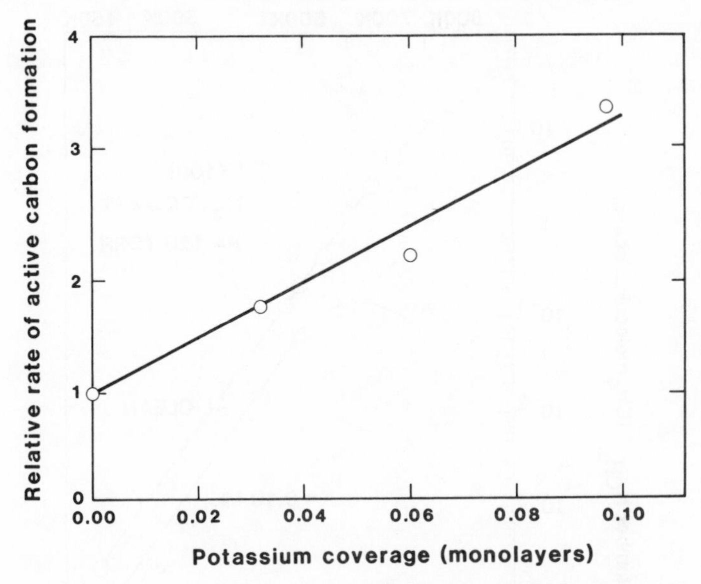

Fig. 21. The relative initial rate of reactive carbon formation from CO disproportionation as a function of potassium coverage. $P_{CO} = 24$ torr, $T = 500$ K. (*From Ref. 148.*)

studied, the overall rate of higher hydrocarbon production was faster on the potassium-dosed surface showing that potassium is a promoter with respect to Fischer–Tropsch synthesis. This increase in higher hydrocarbon production is attributed to the increase in the steady-state active carbon level during the reaction, a factor leading to increased carbon polymerization. Potassium impurities on a nickel catalyst, then, cause a significant increase in the CO dissociation rate and a decrease in the activation energy for CO dissociation at low carbon coverages. These effects can be explained in terms of an electronic effect, whereby the electropositive potassium donates extra electron density to the nickel surface atoms, which in turn donate electron density to the adsorbed CO molecule. This increases the extent of pi-backbonding in the metal–CO complex, resulting in an increased metal–CO bond strength and a decrease in C—O bond strength. This model satisfactorily explains the decrease in the activation energy for carbide build-up brought about by potassium.

Intrinsic to interpreting catalytic poisoning and promotion in terms of electronic effects is the inference that adsorption of an electropositive impurity should moderate or compensate for the effects of an electronegative impurity. Recent experiments have shown this to be true in the case of CO_2 methanation[135] where the adsorption of sulfur decreases the rate of methane formation significantly. The adsorption of potassium in the presence of sulfur indicates that the potassium can neutralize the effects of sulfur.

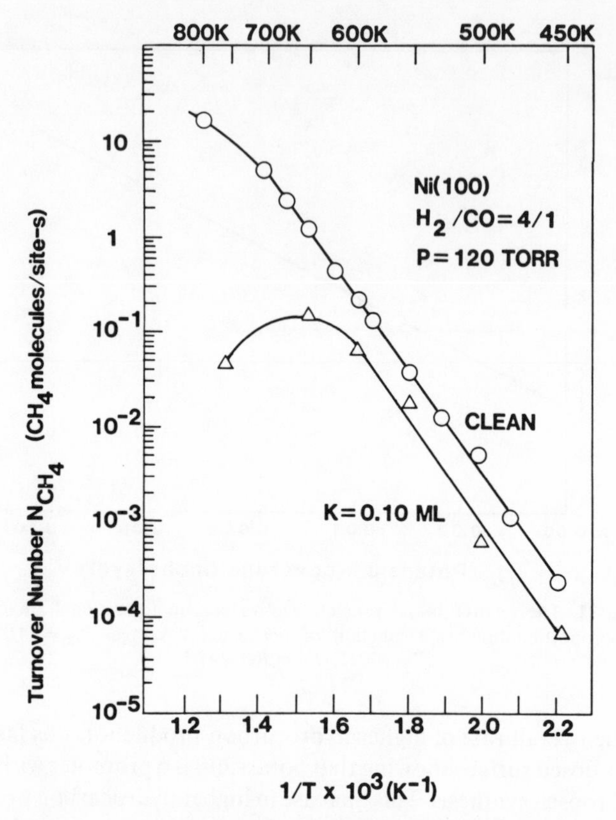

Fig. 22. A comparison of the rate of methane synthesis over a clean single crystal Ni(100) catalyst with the corresponding rate over a potassium-doped catalyst. Total reactant pressure is 120 torr, $H_2/CO = 4/1$. (*From Ref. 148.*)

3.2.3. Related theory

Theoretical work has been undertaken to address directly the predicted magnitude of the near surface electronic perturbations by impurity atoms. Early work by Grimley and coworkers[149,150] and Einstein and Schrieffer[151] concentrated on the indirect interactions between adsorbates which occur via the surface conduction electrons. These calculations suggested that atom–atom interactions through several lattice spacings can occur. More recently, Feibelman and Hamann[152] and Joyner *et al.*[153] have calculated the change in the surface one-electron density of states due to the adsorption of an electropositive or electronegative atom. Such changes are assumed to affect the interaction of an adsorbing atom or molecule with the surface. The lateral range of these changes is then a measure of the range of the interactions.

The calculations of Feibelman and Hamann have expressly addressed the surface electronic perturbation by sulfur[152] as well as by Cl and P[154]. The sulfur-induced total charge density vanishes beyond the immediately adjacent substrate atom site. However, the Fermi-level density of states, which is not screened, and which governs the ability of the surface to respond to the presence of other species, is substantially reduced by the sulfur even at nonadjacent sites. Finally, the results for several impurities indicate a correlation between the electronegativity of the impurity and its relative perturbation of the Fermi-level density of states, a result which could be very relevant to the poisoning of H_2 and CO chemisorption by S, Cl, and P[121] as discussed above.

Theoretical treatments taking into account the direct interaction between adsorbates due to an overlap between their orbitals have also been reported[155,156]. Considerations are given to the energy cost of orthogonalization of the orbitals to one another, a factor which dominates at very short adsorbate–adsorbate distances.

In addition, the direct electrostatic interaction between adsorbates has been treated[157–159]. At intermediates distances of the order of a surface lattice constant, Norskov, Holloway, and Lang[158] report that this interaction can give rise to substantial (> 0.1 eV) interaction energies, when both adsorbates in question induce electron transfer to or from the surface or have a large internal electron transfer.

Both sets of theories, that is, 'through bond' or 'through space', are consistent with adsorbate perturbations sufficiently large to effect chemically significant changes at next-nearest-neighbor metal sites. This perturbation length is sufficient to adequately explain the observed poisoning of chemisorptive and catalytic properties by surface impurities reported in the above discussion.

3.3. Surface properties of mixed-metal catalysts

In addition to modification of surfaces by non-metals, the catalytic properties of metals can also be altered greatly by the addition of a second transition metal[160]. Interest in bimetallic catalysts has arisen steadily over the years because of the commercial success of these systems. This success results from an enhanced ability to control the catalytic activity and selectivity by tailoring the catalyst composition[160,161]. A long-standing question regarding such bimetallic systems is the nature of the properties of the mixed-metal system which give rise to its enhanced catalytic performance relative to either of its individual metal components. These enhanced properties (improved stability, selectivity and/or activity) can be accounted for by one or more of several possibilities. First, the addition of one metal to a second may lead to an electronic modification of either or both of the metal constituents. This electronic perturbation can result from direct bonding (charge transfer) or from a structural modification induced by one metal upon the other. Second, a metal additive can promote a particular step in the reaction sequence and, thus, act in parallel with the host metal. Third, the additive metal can serve to block the availbility of certain active sites, or ensembles, required for a particular reaction step. If this 'poisoned' reaction step involves an undesirable reaction product, then the net effect is an enhanced overall selectivity. Further, the attenuation by this mechansim of a reaction step leading to undesirable surface contamination will promote catalyst activity and durability.

The studies reviewed here are part of a continuing effort[162-168] to identify those properties of bimetallic systems which can be related to their superior catalytic properties. A pivotal question to be addressed of bimetallic systems (and of surface impurities in general) is the relative importance of ensemble (steric or local) versus electronic (nonlocal or extended) effects in the modification of catalytic properties. In gathering information to address this question it has been advantageous to simplify the problem by utilizing models of a bimetallic catalyst such as the deposition of metals on single-crystal substrates in the clean environment familiar to surface science.

3.3.1. Copper–ruthenium

Many such model systems have been studied but a particularly appealing combination is that of Cu on Ru. Cu is immiscible in Ru which facilitates coverage determinations by TPD[162] and circumvents the complication of determining the 3-d composition. The adsorption and growth of Cu films on the Ru(0001) surface have been studied[162-177] by work function measurements, LEED, AES, and TPD. The results from recent studies[162-168] indicate that for submonolayer depositions at 100 K the Cu grows in a highly dispersed mode, subsequently forming 2-d islands pseudomorphic to the Ru(0001) substrate upon annealing to 300 K. Pseudomorphic growth of the copper indicates that the copper–copper bond distances are strained approximately 6% beyond the equilibrium bond distances found for bulk ckpper.

A comparison of CO desorption from Ru[165] from multilayer Cu (10 ML) on Ru and 1 ML Cu on Ru is shown in Fig. 23. The TPD features of the 1 ML Cu (peaks at 160 and 210 K) on Ru are at temperatures intermediate between Ru and bulk Cu. This suggests that the monolayer Cu is electronically perturbed and that this perturbation manifests itself in the bonding of CO. An increase in the desorption temperature relative to bulk Cu indicates a stabilization of the CO on the monolayer Cu suggesting a coupling of the CO through the Cu to the Ru. The magnitude of the CO stabilization implies that the electronic modification of the Cu by the Ru is significant and should be observable with a band structure probe. Recent angular resolved photoemission studies[167] indeed show a unique interface state which is likely related

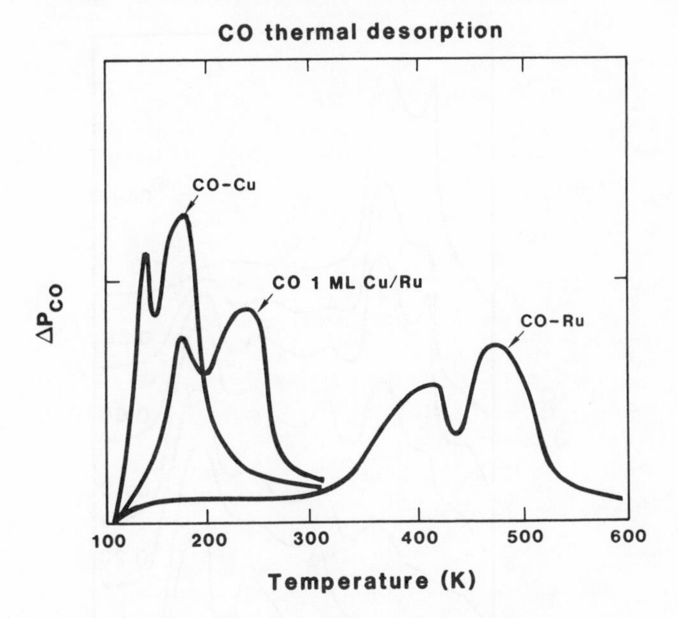

Fig. 23. TPD results for CO adsorbed to saturation levels on clean Ru(0001), on multilayer Cu, and on a 1 ML Cu covered Ru(0001). (*From Ref. 165.*)

to the altered CO bonding on Cu films intimate to Ru.

Figure 24 shows the results[165] of CO chemisorption on the Cu/Ru(0001) system as a function of the Cu coverage. In each case the exposure corresponds to a saturation coverage of CO. Most apparent in Fig. 24 is a monotonic decrease upon addition of Cu of the CO structure identified with Ru (peaks at 400 and 480 K) and in increase of the CO structure corresponding to Cu (peaks at 160 and 219 K). The build-up of a third feature at ~ 300 K (indicated by the dashed line) is assigned to correspond to CO desorbing from the edges of Cu islands. Integration of the 200, 275, and 300 K peaks provides information regarding island sizes, that is, perimeter-to-island area ratios, at various Cu coverages. For example, at $\theta_{Cu} = 0.66$ the average island size is estimated to be approximately 50 Å in diameter. This island size is consistent with an estimate of the 2-d island size corresponding to this coverage of 40–60 Å derived from the width of the LEED beam profiles[165].

Model studies of the Cu/Ru(0001) catalyst have been carried out[166] for methanation and hydrogenolysis reactions. These data suggest that copper merely serves as an inactive diluent, blocking sites on a one-to-one basis. Similar results have been found in analogus studies[130] introducing silver onto a Rh(111) methanation catalyst.

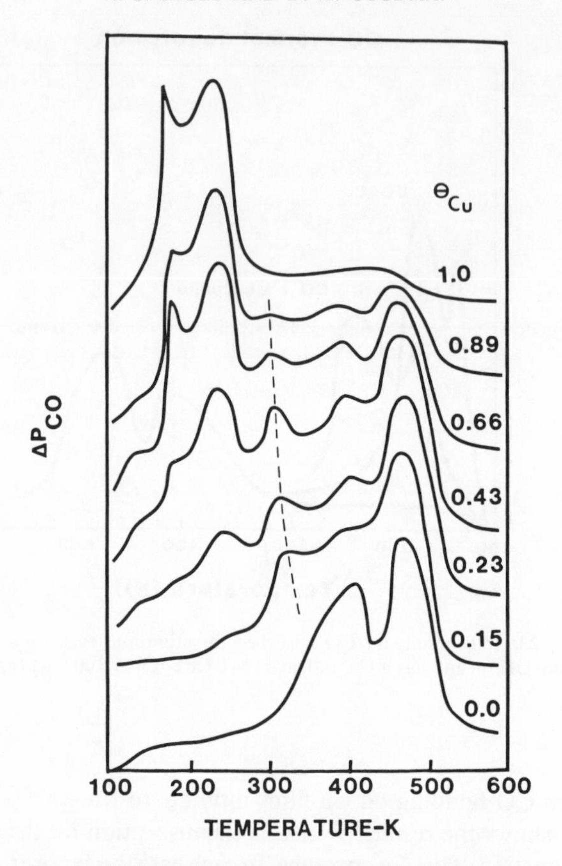

Fig. 24. TPD results corresponding to CO adsorbed to saturation levels on the clean Ru(0001) surface, and from this same surface containing various coverages of Cu. (*From Ref. 165.*)

Sinfelt *et al.*[178] have shown that copper in a Cu/Ru catalyst is confined to the surface of ruthenium. Results from the model catalysts discussed here then should be relevant to those on the corresponding supported, bimetallic catalysts. Several such studies have been carried out investigating the addition of copper or other group 1B metals on the rates of CO hydrogenation[179–181] and ethane hydrogenolysis[179] catalyzed by ruthenium. In general, these studies show a marked reduction in activity with addition of the group 1B metal suggesting a more profound effect of the group 1B metal on ruthenium than implied from the model studies. A critical parameter in the supported studies is the measurement of the active ruthenium surface using hydrogen chemisorption techniques. Haller and coworkers[182,183] have recently suggested that hydrogen spillover during chemisorption may occur from ruthenium to copper complicating the assessment of surface Ru atoms. Recent studies in our laboratory[163,164] have shown directly that spillover from Ru to Cu can take place and must be considered in the hydrogen chemisorption measurements. H_2 spillover would lead to a significant overestimation of the number of active ruthenium metal sites and thus to significant error in calculating ruthenium specific activity. If this is indeed the case, the results obtained on the supported catalysts, corrected for the overestimation of surface ruthenium, could become more comparable with the model data reported here. Finally, the activation energies observed on supported catalysts in various laboratories are generally unchanged by the addition of group 1B metal[182-184] in agreement with the model studies.

These arguments suggest that Ru specific rates for methanation and ethane hydrogenolysis on supported Cu/Ru catalysts approximate those values found for pure Ru. As a consequence, the rates for cyclohexane dehydrogenation reaction on supported Cu/Ru, similarly corrected, must *exceed* those specific rates found for pure Ru. The uncorrected specific rates for cyclohexane dehydrogenation on the supported Cu/Ru system remain essentially unchanged upon addition of Cu to Ru[184]. An activity enhancement for cyclohexane dehydrogenation in the mixed Cu/Ru system relative to pure Ru is most surprising given that Cu is *less active* for this reaction than Ru.

Figure 25 shows the effect of the addition of Cu to Ru on the rate of cyclohexane dehydrogenation to benzene. The overall rate of this reaction is seen to increase by approximately an order of magnitude at a copper coverage of 3/4 of a monolayer. This translates to a Ru specific rate enhancement of ~ 40. Above this coverage, the rate falls to an activity approximately equal to that of Cu-free Ru. The observation of nonzero rates at the higher Cu coverages is believed to be caused by three-dimensional clustering of the Cu overlayers[185]. Similar data have been obtained for this reaction on epitaxial and alloyed Au/Pt(111) surfaces[186].

The rate enhancement observed for submonolayer Cu deposits may relate to an enhanced activity of the strained Cu film for this reaction due to its altered geometric[185] and electronic[167] properties. Alternatively, a mechansim whereby the two metals cooperatively catalyze different steps of the reaction may account for the activity promotion. For example, dissociative H_2 adsorption on bulk Cu is unfavorable due to an activation barrier of approximately 5 kcal/mol[187]. In the combined Cu/Ru system, Ru may function as an atomic hydrogen source/sink via spillover to/from neighboring Cu. A kinetically controlled spillover of H_2 from Ru to Cu, discussed above, is consistent with an observed optimum reaction rate at an intermediate Cu coverage.

Finally, we note the differences between a Ru(0001) catalyst with or without added Cu with respect to attaining steady-state reaction rates. On the Cu-free

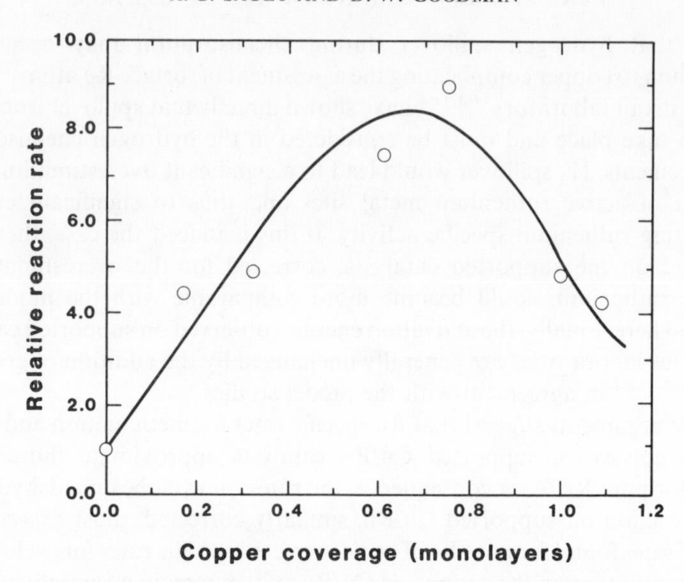

Fig. 25. Relative rate of reaction versus surface Cu coverage on Ru(0001) for cyclohexane dehydrogenation to benzene, $P_T = 101$ torr. H_2/cyclohexane = 100. $T = 650$ K. (*From Ref. 185.*)

surface, an induction time of approximately 10 minutes is required to achieve steady-state activity. During this time, production of benzene is quite low while the hydrogenolysis to lower alkanes, primarily methane, is significantly higher than at steady state. During this induction time the carbon level (as determined by Auger spectroscopy) rises to a saturation value coincidental with the onset of steady-state reaction. This behavior suggests that a carbonaceous layer on the metal surface effectively suppresses carbon–carbon bond scission, or hydrogenolysis, on the Ru surface.

Cu addition leads to an enhanced rate of benzene production with little or no production time. That is, the initial rate of cyclohexane hydrogenolysis, relative to the Cu-free surface, is suppressed. Further, Cu reduces the relative carbon build-up on the surface during reaction. Thus, Cu may play a similar role as the carbonaceous layer in suppressing cyclohexane hydrogenolysis while concurrently stabilizing those intermediates leading to the product benzene. In addition, copper may serve to weaken the chemisorption bond of benzene and thus limit self-poisoning by adsorbed product. This latter possibility has been proposed by Sachtler and Somorjai[186] to explain the role of Au in Au/Pt(111) catalysts for this reaction. A weakening of benzene chemisorption satisfactorily accounts for our observation that the reaction changes from zero order in cyclohexane on Ru(0001) to approximately first order upon the addition of Cu.

3.3.2. Nickel on tungsten

A second bimetallic system which has been thoroughly studied is nickel adsorbed onto tungsten[188,189]. Figures 26 and 27 show plots of the AES Ni(848 eV)/W(179 eV) peak height ratios as a function of the Ni (mass 58) TPD area following deposition of Ni onto W(110) and W(100) substrates, respectively. On both surfaces Ni is adsorbed layer by layer with clear breaks in the AES data of Figs. 26 and 27 at each successive monolayer. Annealing Ni layers with coverages less than 1.3 ML to 1200 K produced little change in the Ni(848 eV)/W(179 eV) AES ratio. However, for Ni coverages above 1.3 ML, a 1200 K anneal resulted in a very slow increase in this AES ratio with coverage, indicating either alloy or 3-dimensional island formation.

The growth of Ni layers on W(110) and (100) subsequent to the first monolayer, determined by the breaks in the AES vs. TPD area curves shown in Figs. 26 and 27, does not yield TPD areas that correspond to simple multiples of the TPD area found for the monolayer TPD feature. This result indicates that the second and successive Ni layers have significantly altered Ni atomic densities compared with the first Ni layer. On W(110) the ratio of the first to second monolayer TPD areas, 0.78, compares favorably with the ratio of the surface atomic densities of Ni(111) and W(110), 0.79, using the values 1.81×10^{15} and 1.43×10^{15} for the atomic densities of Ni(111) and W(110), respectively. On W(100) the ratio of Ni atoms in the first to the second layers,

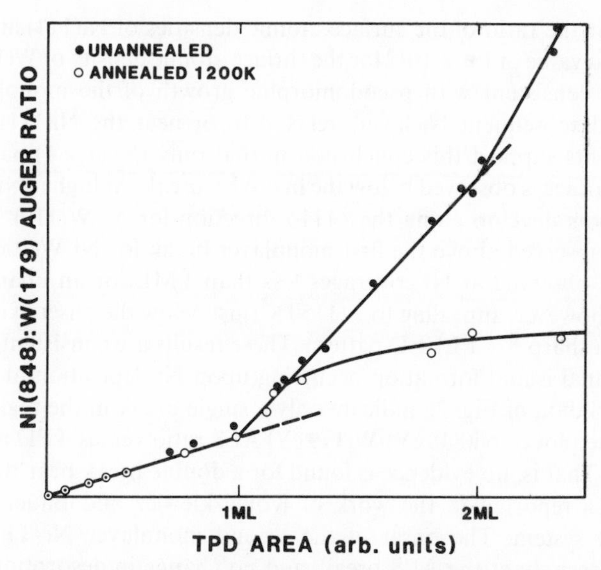

Fig. 26. Adsorption of Ni on W(110) plotted as the ratio Ni(848):W(179) Auger transitions versus Ni desorbed in TPD. Ni deposition was carried out at 100 K. (*From Ref. 188.*)

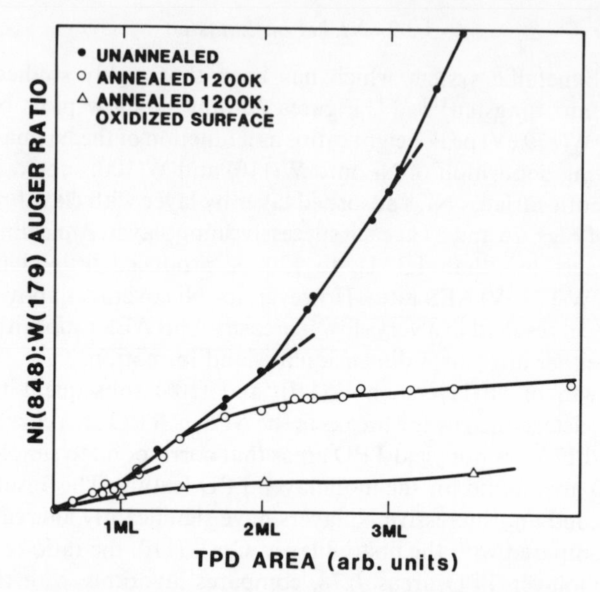

Fig. 27. Adsorption of Ni on W(100) plotted as the ratio of Ni(848):W(179) Auger transitions versus Ni desorbed in TPD. Ni deposition was carried out at 100 K. (*From Ref. 188.*)

0.52, is near the ratio of the surface atomic densities of Ni(111) and W(100), 0.55, using a value of 1.0×10^{15} for the surface atomic density of W(100). These results are consistent with pseudomorphic growth of the monolayer, with second and subsequent Ni layers relaxed to or near the Ni(111) structure. LEED results support this conclusion in that only the (1×1) pattern from either W surface is observed below the first AES break. At higher Ni coverages satellite spots develop along the $\langle 111 \rangle$ direction for Ni/W(110); a complex pattern is observed above the first monolayer break for Ni/W(100). Satellite spots were observed at Ni coverages less than 1 ML for an unannealed Ni overlayer; however, annealing to ~ 1150 K (just below the onset of desorption) produced a sharp 1×1 LEED pattern. These results are consistent with some 3-dimensional island formation occurring upon Ni deposition at 100 K.

The AES data of Fig. 26 indicate only a single break in the region around 1 ML in the plot of Ni(848 eV)/W(179 eV) AES ratio versus TPD area (i.e. Ni coverage). That is, no evidence is found for a double break near the 1 ML Ni coverage as reported in the work of Kolaczkiewicz and Bauer[190] for the Ni/W(110) system. The onset of the second monolayer Ni TPD peak is detected precisely at the AES break, and no change in desorption temperature for the first Ni monolayer is detected from 1.0 to 1.2 ML. Adsorption of Ni beyond the first monolayer results in the appearance of a second Ni TPD

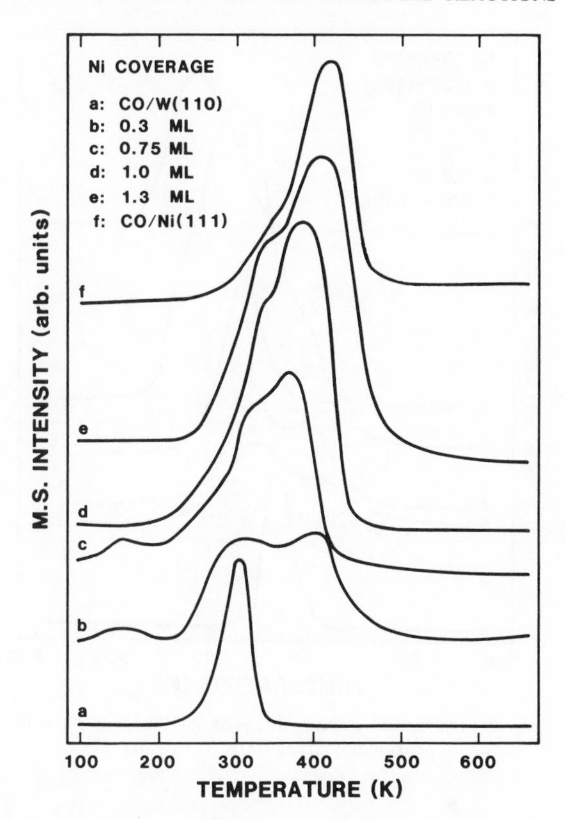

Fig. 28. TPD of CO vs. Ni coverage on W(110) for an exposure of approximately 2L at 100 K. (*From Ref. 188.*)

peak, and the introduction in the LEED of satellite spots, due to the presence of atoms in the second layer. Changes in the chemisorptive properties would be expected for a phase transition such as that reported by Kolaczkiewicz and Bauer[190], considering the structural and electronic modification that would accompany this transition. However, no apparent changes in the chemisorption properties of the first monolayer were observed near the atom density at which the phase transition was reported.

CO TPD data[188] as a function of Ni coverage on W(110) are shown in Fig. 28. As the Ni coverage is increased from 0.3 to 1.0 ML, adsorption on the W(110) substrate decreases, as evidenced by a reduction in the CO features between 225 and 350 K, while a feature at 380 K becomes more prominent. The 380 K CO TPD peak maximum for 1 ML Ni compares with 430 K for the CO TPD peak maximum for Ni(111)[191]. Increasing the Ni coverage above 1.0 ML results in a broadening of the 380 K CO TPD peak and in the

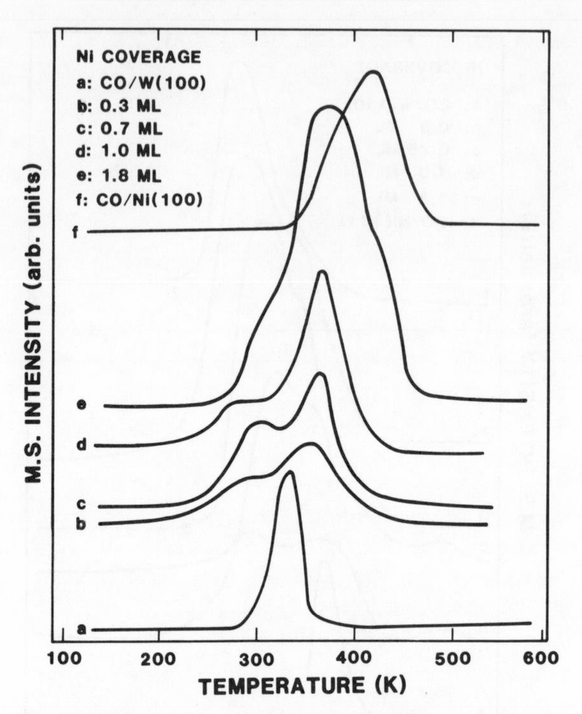

Fig. 29. TPD of CO vs. Ni coverage on W(100) for an exposure of approximately 2L at 100 K. (*From Ref. 188.*)

development of a shoulder feature, suggestive of bulk Ni CO desorption, at ~ 430 K.

CO chemisorption on the Ni/W(100) surface is similar to CO adsorption on Ni/W(110). Figure 29 shows CO desorption as a function of Ni coverage on W(100). As the Ni coverage is increased from 0.3 to 1.0 ML, decreasing intensity in the TPD features associated with W(100) are evident near 300 K. At a Ni coverage of 1 ML, the CO TPD peak maximum is reduced by approximately 50 K from the corresponding peak maximum on Ni(100)[192]. For coverages greater than 1 ML, a clear shoulder at 420–450 K is observed, indicating that second and successive Ni layers have chemisorptive properties very similar to bulk Ni. Thus the W substrates clearly alter the chemisorptive properties of the first Ni layer, but have only slight effects on the second and subsequent layers.

That CO chemisorption is perturbed on strained-layer Ni is not surprising in view of CO chemisorption behavior on other metal overlayer systems. For example, on Cu/Ru it has been proposed that charge transfer from Cu to Ru results in decreased occupancy of the Cu 4s level. This electronic modification makes Cu more 'nickel-like', and results in an increase in the binding energy for CO. Similarly Cu/W[193] also exhibits charge transfer to the substrate and an increase in CO binding strength to Cu. In another case where the CO binding energy increases, Ni/Ru[194], an increase in the density of states is observed close to the Fermi level. The increased electron density may result in increased metal–CO backbonding, which in turn would increase the binding energy of CO.

In contrast to the above examples, CO on Ni/W is less strongly bound to the Ni monolayer than to bulk Ni. One explanation for this effect is that the charge transfer observed from Ni to W[190] results in a shift of the Ni d levels, relevant to CO bonding, to higher binding energies (i.e. farther from the Fermi level). Indeed, such an effect has been observed in the case of Ni/Nb(110) and Pd/Ta(110)[195]. Similarly, results on other group VIII metal–W systems[196,197] have shown a decrease in the CO binding strength.

The catalytic activity of strained-layer Ni on W(110) for methanation and ethane hydrogenolysis has been studied as a function of Ni coverage[189]. The activity per Ni atom site for methanation, a structure-insensitive reaction, is independent of the Ni coverage and similar to the activity found for bulk Ni. The activation energy for this reaction is lower on the strained-metal overlayer, however, very likely reflecting the lower binding strength of CO on the bimetallic system.

In contrast, ethane hydrogenolysis, which is a structure-sensitive reaction over bulk Ni, displayed marked structural effects on the Ni/W system[189]. We have observed that the specific rate, or rate per surface metal atom, but not the activation energy, is a strong function of metal coverage on the Ni/W(110) surface, suggesting that the critical reaction step involves the need for a single, sterically unhindered Ni atom. On the Ni/W(100) surface the specific reaction rate was independent of Ni coverage. In addition, the rate on bulk Ni(100), Ni/W(110) in the limit of zero coverage, and Ni/W(100) were all equal, as were the activation energies. This implies that on Ni/W(100) the Ni atom geometry is sufficiently open to allow unhindered access to each Ni atom. Apparently on the Ni/W(110) surface only island edges and individual atoms display activity similar to the Ni(100) surface; the island interiors, in contrast, exhibit behavior similar to Ni(111) which has a much lower specific rate and higher activation energy. As the Ni coverage is reduced, the number of active, Ni(100)-like atoms increases, leading to an increase in the specific rate. The activation energy, however, remains unchanged.

3.3.3. Other metal overlayer systems

We have studied several other metal overlayers on W(110), W(100), and Ru(0001) substrates[198]. Table 1 lists properties of the metal overlayers, and the effect of the substrate on CO chemisorption. In general only the first monolayer grows pseudomorphically, though more than one monolayer may

TABLE 1
Comparison of strained-metal overlayer systems.

Adsorbate	Substrate	Atom density mismatch/ML (%)	Pseudomorphic/epitaxial layers	Change in CO desorption T(K)
Cu	Ru(0001)	6	1/2	50
Cu	W(110)	20	1/1	80
Ni	W(110)	21	1/1	– 50
Ni	W(100)	42	1/1	– 50
Ni	Ru(0001)	15	1/1	50
Pd	W(110)	10	1/1	– 200
Pd	W(100)	35	2/2	– 170
Pd	Ta(110)	18	1/1	– 230
Fe	W(110)	9	1/2	– 50
Fe	W(100)	35	2/2	– 60

be stable before three-dimensional islands are formed (e.g. Cu/Ru grows two stable layers). The binding strength of CO is always altered from the bulk metal, though the magnitude of the effect is seemingly more dependent on the metal overlayer than on the degree of strain induced by the substrate (represented as the atom density mismatch). As with Ni/W and Cu/Ru, the effect on CO binding energy extends primarily to only the first monolayer; subsequent layers exhibit behavior close to the bulk metal.

4. CONCLUSIONS

The use of a high pressure reactor combined with a UHV surface analytical chamber to study catalytic reactions on metal single crystal surfaces provides detailed information regarding the mechanisms of surface reactions. Using these techniques, the concepts of structure sensitivity and structure insensitivity have been investigated. For structure-insensitive reactions, excellent agreement is obtained between studies on single crystal surfaces and studies on high surface area supported catalysts, demonstrating the relevance of kinetics measured on well-ordered single crystal surfaces for modeling and behavior of practical catalysts. For structure-sensitive reactions, these model studies allow correlations to be drawn between surface structure and catalytic activity. Surface analysis following reaction allows the composition and structure of the surface under steady-state reaction conditions to be determined.

Model studies on single crystal surfaces are also helpful in developing an understanding of the effects of surface additives on catalyst performance. Electronegative, electroneutral (i.e. metals) and electropositive additives can all be studied. The influence of additives on the bond strengths and structure of adsorbed reactants, products and intermediates can be studied using UHV techniques, and this information can be related to the effects of the additives on catalytic activity. Of particular interest is the possibility that these types of studies will help to clarify the relative importance of electronic and geometric effects in determining additive effects. The types of studies reviewed here, in conjunction with studies on high surface area supported catalysts, hold great promise for contributing to the overall understanding of surface catalyzed reactions.

Acknowledgements

We acknowledge with pleasure the partial support of this work by the Department of Energy, Office of Basic Energy Sciences, Division of Chemical Sciences.

References

1. Albert, M. R., and Yates, J. T., Jr., *The Surface Scientist's Guide to Organometallic Chemistry*, American Chemical Society, Washington, 1987.
2. Blakely, D. W., Kozak, E., Sexton, B. A., and Somorjai, G. A., *J. Vac. Sci. Tech.*, **13**, 1091 (1976).
3. Goodman, D. W., Kelley, R. D., Madey, T. E., and Yates, J. T., Jr., *J. Catal.*, **63**, 226 (1980).
4. Goodman, D. W., *Ann. Rev. Phys. Chem.*, **37**, 425 (1986).
5. Krebs, H. J., Bonzel, H. P., and Gafner, G., *Surface Sci.*, **88**, 269 (1979).
6. Campbell, C. T., and Paffett, M. T., *Surface Sci.*, **139**, 396 (1984).
7. Sault, A. G., and Madix, R. J., *Surface Sci.*, **176**, 415 (1986).
8. Goodman, D. W., *J. Vac. Sci. Tech.*, **20**, 522 (1982).
9. Boudart, M., *Advan. Catalysis Related Subjects*, **20**, 153 (1968).
10. Vannice, M. A., *Catal. Rev.*, **14**, 153 (1976).
11. Kelley, R. D., and Goodman, D. W., *The Chemical Physics of Solid Surfaces and Heterogeneous Catalysts*, Vol. 4, Elsevier, 1982.
12. Kelley, R. D., and Goodman, D. W., *Surf. Sci.*, **123**, L743 (1982).
13. Vannice, M. A., *J. Catal.* **44**, 152 (1976).
14. Houston, J. E., Peebles, D. E., and Goodman, D. W., *J. Vac. Sci. and Technol.*, **A1**, 995 (1983).
15. Wentrek, R. R., Wood, B. J., and Wise, H., *J. Catal.*, **43**, 363 (1976).
16. Rabo, J. A., Risch, A. P., and Poutsma, M. L., *J. Catal.*, **53** 295 (1978).
17. Araki, M., and Ponec, V., *J. Catal.*, **44**, 439 (1979).
18. Goodman, D. W., Kelley, R. D., Madey, T. E., and Yates, J. T., Jr., *J. Catal.*, **64**, 479 (1980).
19. Goodman, D. W., and Yates, J. T., *J. Catal.*, **82**, 255 (1983).
20. Goodman, D. W., and White, J. M., *Surface Sci.*, **90**, 201 (1979).
21. Biloen, P., Helle, J. N., and Sachler, W. M. H., *J. Catal.*, **58**, 95 (1979).
22. (a) Ekerdt, J. G., and Bell, A. T., *J. Catal.*, **58**, 179 (1979).
    (b) Dalla Betta, P. A., and Shelef, M., *J. Catal.*, **48**, 111 (1977).
23. Sachtler, J. W. A., Kool, J. M., and Ponec, V., *J. Catal.*, **56**, 84 (1979).
24. Kummer, J. T., *J. Phys. Chem.*, **90**, 4747 (1986).

25. Cant, N. W., Hicks, P. C., and Lennon, B. S., *J. Catal.*, **54**, 372 (1978).
26. Kiss, J. T., and Gonzalez, R. D., *J. Phys. Chem.*, **88**, 892 (1984).
27. Kiss, J. T., and Gonzalez, R. D., *J. Phys. Chem.*, **88**, 898 (1984).
28. Oh, S. E., and Carpenter, J. E., *J. Catal.*, **80**, 472 (1983).
29. Yu, Y. Y., *J. Catal.*, **87**, 152 (1984).
30. Okamoto, H., Kawamura, G., and Kudo, T., *J. Catal.*, **87**, 1 (1984).
31. Cant, N. W., and Angove, D. E., *J. Catal.*, **97**, 36 (1986).
32. Shishu, R. C., and Kowalczyk, L. S., *Platinum Metals Rev.*, **18**, 58 (1974).
33. Voltz, S. E., Morgan, C. R., Liederman, D., and Jacob, S. M., *Ind. Eng. Chem.* Prod. Res. Dev., **12**, 294 (1973).
34. Nicholas, D. M., and Shah, Y. T., *Ind. Eng. Prod. Res. Dev.*, **15**, 35 (1976).
35. Taylor, J. L., Ibbotson, D. E., and Weinberg, W. H., *Surf. Sci.*, **90**, 37 (1979).
36. Matsushima, T., *J. Phys. Chem.*, **88**, 202 (1984).
37. Matsushima, T., and White, J. M., *J. Catal.*, **39**, 265 (1975).
38. Kawai, M., Onishi, T., and Tamaru, K., *Appl. Surf. Sci.*, **8**, 361 (1981).
39. Matsushima, T., Mussett, C. J., and White, J. M., *J. Catal.*, **41**, 397 (1976).
40. Matsushima, T., *J. Catal.*, **55**, 337 (1978).
41. Hori, G. K., and Schmidt, L. D., *J. Catal.*, **38**, 335 (1975).
42. White, J. M., and Golchet, A., *J. Chem. Phys.*, **66**, 5744 (1977).
43. McCarthy, E., Zahradnik, J., Kuczynski, G. C., and Carberry, J. J., *J. Catal.*, **39**, 29 (1975).
44. Matsushima, T., *Surf. Sci.*, **79**, 63 (1979).
45. Hagen, D. I., Nieuwenhuys, B. E., Rovida, G., and Somorjai, G. A., *Surf. Sci.*, **57**, 632 (1976).
46. Kuppers, J., and Plagge, A., *J. Vac. Sci. Technol.*, **13**, 259 (1986).
47. Zhdanov, V. P., *Surf. Sci.*, **137**, 515 (1984).
48. Zhdan, P. A., Boreskov, G. D., Boronin, A. I., Schepelin, A. P., Withrow, S. P., and Weinberg, W. H., *Appl. Surf. Sci.*, **3**, 145 (1979).
49. Zhdan, P. A., Boreskov, G. K., Egelhoff, W. F., Jr., and Weinberg, W. H., *Surf. Sci.*, **61**, 377 (1976).
50. Matolin, V., and Gillet, E., *Surf. Sci.*, **166**, L115 (1986).
51. Stuve, E. M., Madix, R. J., and Brundle, C. R., *Surf. Sci.*, **146**, 155 (1984).
52. Engel, T., and Ertl, G., *J. Phys. Chem.*, **69**, 1267 (1978).
53. Conrad, H. Ertl, G., and Kuppers, J., *Surf. Sci.*, **76**, 323 (1978).
54. Mantell, D. A., Ryali, S. B., and Haller, G. L., *Chem. Phys. Lett.*, **102**, 37 (1983).
55. Matolin, V., Gillet, E., and Gillet, M., *Surf. Sci.*, **162**, 354 (1985).
56. Engel, T., *J. Chem. Phys.*, **69**, 373 (1978).
57. Behm, R. J., Thiel, P. A., Norton, P. R., and Binder, P. E., *Surf. Sci.*, **147**, 143 (1984).
58. Akhter, S., and White, J. M., *Surf. Sci.*, **171**, 527 (1986).
59. Matsushima, T., and Asada, H., *J. Chem. Phys.*, **85**, 1658 (1986).
60. Palmer, R. L., and Smith, J. N., Jr., *J. Chem. Phys.*, **60**, 1453 (1974).
61. Campbell, C. T., Ertl, G., Kuipers, H., and Segner, J., *J. Chem. Phys.*, **73**, 5862 (1980).
62. Gland, J. L., and Kollin, E. B., *J. Chem. Phys.*, **78**, 963 (1983).
63. Engel, T., and Ertl, G., *Adv. Catal.*, **28**, 1 (1979).
64. Engel, T., and Ertl, G., *The Chemical Physics of Solid Surfaces and Heterogeneous Catalysts*, Vol. 4 (Ed. A. King and D. P. Woodruff), Elsevier, Holland, 1982.
65. Langmuir, I., *Trans. Farad. Soc.*, **17**, 621 (1921–22).
66. Oh, S. H., Fisher, G. B., Carpenter, J. E., and Goodman, D. W., *J. Catal.*, **100**, 360 (1986).

67. Schwartz, S. B., Schmidt, L. D., and Fisher, G. B., *J. Phys. Chem.*, **90**, 6194 (1984).
68. Klein, R. L., Schwartz, S., and Schmidt, L. D., *J. Phys. Chem.*, **89**, 4908 (1985).
69. Peden, C. H. F., and Goodman, D. W., *J. Phys. Chem.*, **90**, 1360 (1986).
70. Peden, C. H. F., Goodman, D. W., Blair, D. S., Berlowitz, P. J., Fisher, G. B., and Oh, S. H., *J. Phys. Chem.*, **92**, 1563 (1988).
71. Goodman, D. W., and Peden, C. H. F., *J. Phys. Chem.*, **90**, 4839 (1986).
72. Kellogg, G. L., *J. Catal.*, **92**, 167 (1985).
73. Berlowitz, P. J., Peden, C. H. F., and Goodman, D. W., *J. Phys. Chem.*, **92**, 5213 (1988).
74. Matsushima, T., *Surf. Sci.*, **87**, 665 (1979).
75. Savchenko, V. I., Boreskov, G. K., Kalinkin, A. V., and Salanov, A. N., *Kinetics and Catalysis*, **24**, 983 (1984).
76. Sales, B. C., Turner, J. E., and Maple, M. B., *Surf. Sci.*, **112**, 272 (1981).
77. Niehus, H., and Comsa, G., *Surf. Sci.*, **93**, L147 (1980).
78. Niehus, H., and Comsa, G., *Surf. Sci.*, **102**, L14 (1981).
79. Bonzel, H. P., Franken, A. M., and Pirug, G., *Surf. Sci.*, **104**, 625 (1981).
80. Anderson, J. R., and Baker, B. G., *Proc. Roy. Soc.*, **A271**, 402 (1963).
81. Wright, P. G., Asmore, P. G., and Kemball, C., *Trans. Far. Soc.*, **54**, 1692 (1958).
82. Trapnell, B. M. W., *Trans. Far. Soc.*, **52**, 1618 (1956).
83. Salmeron, M., and Somorjai, G. A., *J. Phys. Chem.*, **85**, 3835 (1981).
84. Wittrig, T. S., Szuromi, P. D., and Weinberg, W. H., *J. Chem. Phys.*, **76**, 3305 (1982).
85. Szuromi, P. D., Engstrom, J. R., and Weinberg, W. H., *J. Chem. Phys.*, **80**, 508 (1984).
86. Szuromi, D., Engstrom, J. R., and Weinberg, W. H., *Surface Sci.*, **149**, 226 (1985).
87. Szuromi, D., Engstrom, J. R., and Weinberg, W. H., *J. Phys. Chem.*, **89**, 2497 (1985).
88. Yates, J. T., Jr., and Madey, T. E., *Surface Sci.*, **28**, 437 (1971).
89. Madey, T. E., *Surface Sci.*, **29**, 571 (1972).
90. Madey, T. E., and Yates, J. T., Jr., *Surface Sci.*, **76**, 397 (1978).
91. Hamza, A. V., Steinruck, H.-P., and Madix, R. J., *J. Chem. Phys.*, **85**, 7494 (1986).
92. Hamza, A. V., Steinruck, H.-P., and Madix, R. J., *J. Chem. Phys.*, **86**, 6506 (1987).
93. Hamza, A. V., and Madix, R. J., *Surface Sci.*, **179**, 25 (1987).
94. Steinruck, H.-P., Hamza, A. V., and Madix, R. J., *Surface Sci.*, **173**, L571 (1986).
95. Lee, M. B., Yang, Q. Y., Tang, S. L., and Ceyer, S. T., *J. Chem. Phys.*, **85**, 1693 (1986).
96. Lee, M. B., Yang, Q. Y., and Ceyer, S. T., *J. Chem. Phys.*, **87**, 2724 (1987).
97. Rettner, C. T., Pfnur, H. E., and Auerbach, D. J., *Phys. Rev. Lett.*, **54**, 2716 (1985).
98. Rettner, C. T., Pfnur, H. E., and Auerbach, D. J., *J. Chem. Phys.*, **84**, 4163 (1986).
99. Beebe, T. P., Jr., Goodman, D. W., Kay, B. D., and Yates, J. T., Jr., *J. Chem. Phys.*, **87**, 2305 (1987).
100. Kay, B. D., and Coltrin, M. E., *Surface Sci.*, **198**, 2375 (1988).
101. Kay, B. D., in preparation.
102. Hamza, A. V., private communication.
103. Sault, A. G., and Goodman, D. W., *J. Chem. Phys.*, **88**, 7232 (1988).
104. Carter, J. T., Cusumano, J. A., and Sinfelt, J. H., *J. Phys. Chem.*, **70**, 2257 (1966).
105. Yates, D. J. C., and Sinfelt, J. H., *J. Catal.*, **8**, 348 (1967).
106. Martin, G. A., *J. Catal.*, **60**, 452 (1979).
107. Goodman, D. W., *Surface Sci.*, **123**, L679 (1982).
108. Goodman, D. W., *Proceedings of the 8th International Congress on Catalysis*, (1984).

109. Goodman, D. W., *J. Vac. Sci. Technol.*, **A2**, 873 (1984).
110. Clark, J. E. A., and Rooney, J. J., *Advan. Catalysis*, **25**, 125 (1976).
111. Engstrom, J. R., Goodman, D. W., and Weinberg, W. H., *J. Amer. Chem. Soc.*, **108**, 4653 (1986).
112. Engstrom, J. R., Goodman, D. W., and Weinberg, W. H., *J. Amer. Chem. Soc.*, **110**, 8305 (1988).
113. Engstrom, J. R., Goodman, D. W., and Weinberg, W. H., Submitted to *J. Amer. Chem. Soc.*
114. Ibbotson, D. E., Wittrig, T. S., and Weinberg, W. H., *J. Chem. Phys.*, **72**, 4885 (1980).
115. Engstrom, J. R., Tsai, T. S., and Weinberg, W. H., *J. Chem. Phys.*, **87**, 3104 (1987).
116. Foger, K., and Anderson, J. R., *J. Catal.*, **59**, 325 (1979).
117. Grubbs, R. H., and Miyoshita, A., *J. Amer. Chem. Soc.*, **100**, 1300 (1978).
118. Grubbs, R. H., Miyoshita, A., Liu, M., and Burk, P., *J. Amer. Chem. Soc.*, **100**, 2418 (1978).
119. Imelik, B., Naccache, C., Coudurier, G., Praliaud, H., Meriaudeau, P., Gallezot, P., Martin, G. A., and Vedrine, J. C., (Eds.), *Metal-Support and Metal-Additive Effects in Catalysts*, Elsevier, 1982.
120. Goodman, D. W., and Kiskinova, M., *Surf. Sci.*, **105**, L265 (1981).
121. Kiskinova, M., and Goodman, D. W., *Surf. Sci.*, **108**, 64 (1981).
122. Kiskinova, M., and Goodman, D. W., *Surf. Sci.*, **109**, L555 (1981).
123. Goodman, D. W., *Appl. Surf. Sci.*, **19**, 1 (1984).
124. Johnson, S., and Madix, R. J., *Surf. Sci.*, **108**, 77 (1981).
125. Madix, R. J., Thornburg, M., and Lee, S.-B., *Surf. Sci.*, **133**, L447 (1983).
126. Madix, R. J., Lee, S.-B., and Thornburg, M., *J. Vac., Sci. Technol.*, **A1**, 1254 (1983).
127. Hardegree, E. L., Ho., P., and White, J. M., *Surf. Sci.*, **165**, 488 (1988).
128. Kiskinova, M., and Goodman, D. W., *Surf. Sci.*, **108**, 64 (1981).
129. Kittel, C., *Introduction to Solid State Physics*, John Wiley & Sons, Inc., New York, 1971.
130. Goodman, D. W., in 'Heterogeneous catalysis' (*Proceedings of IUCCP Conference*), Texas A & M University (1984).
131. Peden, C. H. F., and Goodman, D. W., ACS Symposium Series, *Proceedings of Sym. on the Surface Science of Catalysis* (Eds. M. L. Devinny and J. L. Gland), Philadelphia, 1984.
132. Rostrup-Nielsen, J. R., and Pedersen, K., *J. Catal.*, **59**, 395 (1979).
133. Goodman, D. W., unpublished results.
134. Goodman, D. W., *Accts. Chem. Res.*, **17**, 194 (1984).
135. Peebles, D. E., Goodman, D. W., and White, J. M., *J. Phys. Chem.*, **87**, 4378 (1983).
136. Houston, J. E., Rogers, J. R., Goodman, D. W., and Belton, D. N., *J. Vac. Sci. Technol.*, **A2**, 882 (1984).
137. Anderson, S., and Jostell, U., *Surf. Sci.*, **46**, 625 (1974).
138. Kiskinova, M., *Surf. Sci.*, **111**, 584 (1981).
139. Ertl, G., *Catalysis Rev.-Sci. Eng.*, **21**, 201 (1980).
140. Broden, G., Gafner, G., and Bonzel, H. P., *Surf. Sci.*, **84**, 295 (1979).
141. Goddard, P. J., West, J., and Lambert, R. M., *Surf. Sci.*, **71**, 447 (1978).
142. Ertl, G., Weiss, M., and Lee, S. B., *Chem. Phys. Lett.*, **60**, 391 (1979).
143. Crowell, J. E., Garfunkel, E. L., and Somorjai, G. A., *Surf. Sci.*, **121**, 303 (1982).
144. Garfunkel, E. L., Crowell, J. E., and Somorjai, G. A., *J. Phys. Chem.*, **86**, 310 (1982).
145. Kiskinova, M. P., Pirug, G., and Bonzel, H. P., *Surf. Sci.*, **140**, 1 (1984).
146. Luftman, H. S., Sun, Y.-M., and White, J. M., *Appl. Surf. Sci.*, **19**, 59 (1984).
147. Ertl, G., Lee, S. B., and Weiss, M., *Surf. Sci.*, **111**, L711 (1981).

148. Campbell, C. T., and Goodman, D. W., *Surf. Sci.*, **123**, 413 (1982).
149. Grimley, T. B., and Walker, S. M., *Surf. Sci.*, **14**, 395 (1969).
150. Grimley, T. B., and Torrini, M., *J. Phys. Chem.*, **87**, 4378 (1973).
151. Einstein, T. E., and Schrieffer, J. R., *Phys. Rev.* B, **7**, 3629 (1973).
152. Feibelman, P., and Hamann, D., *Phys. Rev. Lett.*, **52**, 61 (1984).
153. Joyner, R. W., Pendry, J. B., Saldin, D. K., and Tennison, S. R., *Surf. Sci.*, **138**, 84 (1984).
154. Feibelman, P. J., and Hamann, D., *Surf. Sci.*, **149**, 48 (1985).
155. Muda, Y., and Hanawa, T., *Jap. J. Appl. Phys.*, **17**, 930 (1974).
156. Benziger, J., and Madix, R. J., *Surf. Sci.*, **94**, 119 (1980).
157. Lau, K. H., and Kohn, W., *Surf. Sci.*, **65**, 607 (1977).
158. Norskov, J. K., Holloway, S., and Lang, N. D., *Surf. Sci.*, **137**, 65 (1984).
159. Luftman, H. S., and White, J. M., *Surf. Sci.*, **139**, 362 (1984).
160. Sinfelt, J. H., *Bimetallic Catalysts: Discoveries, Concept, and Applications*, John Wiley & Sons, New York, 1983.
161. Schwab, G. M., *Disc. Faraday Soc.*, **8**, 166 (1950).
162. Yates, J. T., Jr., Peden, C. H. F., and Goodman, D. W., *J. Catal.*, **94**, 576 (1985).
163. Goodman, D. W., Yates, J. T., Jr., and Peden, C. H. F., *Surf. Sci.*, **164**, 417 (1985).
164. Goodman, D. W., and Peden, C. H. F., *J. Catal.*, **95**, 321 (1985).
165. Houston, J. E., Peden, C. H. F., Blair, D. S., and Goodman, D. W., *Surf. Sci.* **167**, 427 (1986).
166. Houston, J. E., Peden, C. H. F., Feibelman, P. J., and Goodman, D. W., *I & EC Fundamentals*, **25**, 58 (1986).
167. Houston, J. E., Peden, C. H. F., Feibelman, P. J., and Hamann, D. R., *Phys. Rev. Lett.*, **56**, 375 (1986).
168. Peden, C. H. F., and Goodman, D. W., *J. Catal.*, **100**, 5209 (1988).
169. Christmann, K., Ertl, G., and Shimizu, H., *J. Catal.*, **61**, 397 (1980).
170. Shimizu, H., Christmann, K., and Ertl, G., *J. Catal.*, **61**, 412 (1980).
171. Vickerman, J. C., Christmann, K., and Ertl, G., *J. Catal.*, **71**, 175 (1981).
172. Shi, S. K., Lee, H. I., and White, J. M., *Surf. Sci.*, **102**, 56 (1981).
173. Richter, L., Bader, S. D., and Brodsky, M. B., *J. Vac. Sci. Techn.*, **18**, 578 (1981).
174. Vickerman, J. C., and Christmann, K., *Surf. Sci.*, **120**, 1 (1982).
175. Vickerman, J. C., Christman, K., Ertl, G., Heiman, P., Himpsel, F. J., and Eastman, D. E., *Surf. Sci.*, **134**, 367 (1983).
176. Bader, S. D., and Richter, L., *J. Vac. Sci. Technol.*, **A1**, 1185 (1983).
177. Park, C., Bauer, E., and Poppa, H., *Surf. Sci.*, **187**, 86 (1987).
178. Sinfelt, J. H., Via, G. H., and Lytle, F. W., *Catal. Rev.-Sci. Eng.*, **26**, 81 (1984).
179. Datye, A. K., and Schwank, J., *J. Catal.*, **93**, 256 (1985).
180. Bond, G. C., and Turnham, B. D., *J. Catal.*, **45**, 128 (1976).
181. Luyten, L. J. M., Eck, M. v., Grondelle, J. v., and Hooff, J. H. C. v., *J. Phys. Chem.*, **82**, 2000 (1978).
182. Rouco, A. J., Haller, G. L., Oliver, J. A., and Kemball, C., *J. Catal.*, **84**, 297 (1983).
183. Haller, G. L., Resasco, D. E., and Wang, J., *J. Catal.*, **84**, 477 (1983).
184. Sinfelt, J. H., *J. Catal.*, **29**, 308 (1973).
185. Houston, J. E., Peden, C. H. F., Blair, D. S., and Goodman, D. W., *Surface Sci.*, **167**, 427 (1986).
186. Sachtler, J. W. A., and Somorjai, G. A.., *J. Catal.*, **89**, 35 (1984).
187. Balooch, M., Cardillo, M. J., Miller, D. R., and Stickney, R. E., *Surf. Sci.*, **50**, 263 (1975).
188. Berlowitz, P. J., and Goodman, D. W., *Surf. Sci.*, **187**, 463 (1988).
189. Greenlief, C. A., Berlowitz, P. J., Goodman, D. W. and White, J. M., *J. Phys. Chem.*, **91**, 6669 (1987).

190. Kolaczkiewicz, J., and Bauer, E., *Surf. Sci.*, **144**, 495 (1984).
191. Christmann, K., Schober, O., and Ertl, G., *J. Chem. Phys.*, **60**, 4719 (1974).
192. Goodman, D. W., Yates, J. T., Jr., and Madey, T. E., *Surf. Sci.*, **93**, L135 (1980).
193. Hamedeh, I., and Gomer, R., *Surf. Sci.*, **154**, 168 (1985).
194. Berlowitz, P. J., Houston, J. E., White, J. M., and Goodman, D. W., **205**, 1 (1988).
195. Ruckman; M. W., Strongin, M., and Pan, X., *J. Vac. Sci. Tech. A*, **5**, 805 (1987).
196. Prigge, D., Schlenk, W., Bauer, E., *Surf. Sci.*, **123**, L698 (1982).
197. Judd, R. W., Reichelt, M. A., Scott, E. G., and Lambert, R. M., *Surf. Sci.*, **185**, 515 (1987).
198. Berlowitz, P. J., Peden, C. H. F., and Goodman, D. W., *Mat. Res. Soc. Symp. Proc.*, **83**, 161 (1987).

Advances in Chemical Physics
Edited by K. P. Lawley

HELIUM-SCATTERING STUDIES OF THE DYNAMICS AND PHASE TRANSITIONS OF SURFACES

KLAUS KERN and GEORGE COMSA

Institut für Grenzflächenforschung und Vakuumphysik, KFR Jülich, P.O. Box 1913, D-5170 Jülich, Federal Republic of Germany

CONTENTS

1. INTRODUCTION

The access to the bulk of a solid is obviously through the surface. In spite of this, we know much more about the properties of the three-dimensional (3D) bulk than about the two-dimensional (2D) surfaces. One of the main reasons – from the experimentalist's point of view – is the relatively small number of surface atoms. As a consequence, the interaction cross-section of the probe particles with the surface atoms has to be large enough but, at the same time, the interaction should not disturb too much the sampled object. These conditions can be certainly contradictory. The most widespread surface probe is the electron. The overwhelming part of our present knowledge about surfaces is due to the use of electron beams as probe, as information carrier and as both. Even in domains for which they do not seem to be particularly appropriate, like phase transitions of physisorbed layers and low energy phonons, electron based methods have supplied valuable information. A remarkable effort had to be made for these achievements: the energy resolution had to be pushed to its limits and the disturbing effect of the electrons on the investigated phases – which are by definition particularly delicate near transitions – has always to be accounted for. In contrast, at least for these domains of investigation, the He atom seems to be the natural probe. Due to its large mass, the He atom has the right wavelength (~ 1 Å) for structural studies of surface phases at thermal energies. He atoms at these energies have two outstanding properties: they do not disturb the surface in any way and they are absolutely surface sensitive. In addition, they are extremely sensitive with respect to surface disorder; this allows the detailed characterization of surface disorder, which is always present even on the best prepared surfaces and which plays an important role in phase formation and transition. Due again to their large mass, the He atoms have, like the neutrons, a favorable relationship between momentum and energy: the momentum of He atoms is large enough to create a phonon anywhere in the Brillouin zone already at kinetic energies of the same order with the energy transferred. Accordingly, the energy resolution can be high enough for a detailed monitoring of dispersion curves and even for lifetime broadenings of characteristic energy exchange peaks. The analogy with the capabilities of thermal neutrons, which have allowed an almost complete understanding of the dynamics of the 3D bulk of solids, suggests that the use of thermal He-scattering will lead to comparable results for the 2D surfaces.

After discussing specific, basic and experimental aspects of the use of thermal He as a surface probe we will review recent results obtained in the study of surface dynamics and of 2D phase transitions.

2. THERMAL HE-BEAM SCATTERING AT SURFACES

The structural and dynamical properties of solid surfaces at atomic level are conveniently derived from scattering experiments. Thermal neutrons, X-ray photons, low energy electrons or thermal He atoms can be chosen as probes. Neutrons and X-rays interact weakly with the sample atoms. This is advantageous because the structural information is obtained within the straightforward kinematical approximation; it has, however, the disadvantage that these probe particles are basically surface insensitive. Surface sensitivity can only be obtained by using grazing angles of incidence (X-rays) or by using substrates with large surface-to-bulk ratio like powder samples (neutrons). On the other hand, electrons and He atoms have stronger scattering interactions which complicate in general the structural analysis, but which also ensure an

TABLE 1
Characteristics of the various surface probe particles.

Properties	Probe particles			
	He atoms	Electrons	Neutrons	X-ray photons
Energy (meV)	$5–10^2$	$10^4–3 \times 10^5$	$1–10^2$	$4 \times 10^6–3 \times 10^7$
Wavelength–energy relation (λ in Å, E in meV)	$\lambda = \frac{4.54}{\sqrt{E}}$	$\lambda = \frac{389}{\sqrt{E}}$	$\lambda = \frac{9.04}{\sqrt{E}}$	$\lambda = \frac{12.39 \times 10^8}{\sqrt{E}}$
Information depth (number of layers)	1	3–5	$\gg 1$	$\gg 1$
Energy resolution (meV)	0.2	3	0.001	> 8
Momentum resolution (Å^{-1})	10^{-2}	10^{-2}	$10^{-2}–10^{-3}$	10^{-4}
Sample	single crystal	single crystal	powder	single crystal powder
Maximum ambient pressure (Pa)	10^{-3}	10^{-3}	$> 10^5$	$> 10^5$

information depth confined to a few layers (electrons) or to the outermost surface layer only (He atoms).

In Table 1 we give the main parameters which characterize the surface analytical properties of the various probe particles.

2.1. He atoms as probe particles

It is primarily the exclusive surface sensitivity of thermal atom scattering which makes this method to an outstanding surface probe. This was recognized more than sixty years ago by T. H. Johnson[1]:

'These experiments are of interest not only because of their confirmation of the predictions of quantum mechanics, but also because they introduce the possibility of applying atom diffraction to investigations of the atomic constitution of surfaces. A beam of atomic hydrogen, for example, with ordinary thermal velocities, has a range of wavelengths of the right magnitude for this purpose, centering aroung 1 Å, and the complete absence of penetration of these waves will insure that the effects observed arise entirely from the outermost atomic layer'.

In spite of this early recognition of the potential provided by thermal atoms to study surfaces, the lack of an appropriate He-beam source was the main hurdle in the development of this now very powerful analytical tool. For a long time the Knudsen (effusion) cell was the only means for producing molecular beams. The Maxwellian effusive beams have low intensity ($I_0 \sim 10^{14}$ particles $s^{-1}sr^{-1}$) and low monochromaticity ($\Delta v/v = \Delta\lambda/\lambda = 0.95$). Monochromaticity improvement by means of mechanical velocity selectors reduces the already low intensity to a level which in view of the inefficient He detection is unacceptable for a decent analysis. The major breakthrough has been the development of high pressure nozzle sources. The effect achieved by the invention of these sources is only comparable to that of laser technology: simultaneous increase of intensity and monochromaticity by several orders of magnitude. Indeed, intensities of 10^{19} particles $s^{-1}sr^{-1}$ and monochromaticities of $\Delta v/v = \Delta\lambda/\lambda \approx 0.01$ are obtained routinely today.

The nautre of the He–surface interaction potential determines the major characteristics of the He beam as surface analytical tool. At larger distances the He atom is weakly attracted due to dispersion forces. At a closer approach, the electronic densities of the He atom and of the surface atoms overlap, giving rise to a steep repulsion. The classical turning point for thermal He is a few angstroms in front of the outermost surface layer. This makes the He atom sensitive exclusively to the outermost layer. The low energy of the He atoms and their inert nature ensures that He scattering is a completely non-destructive surface probe. This is particularly important when delicate phases, like physisorbed layers, are investigated.

The de Broglie wavelength of thermal He atoms is comparable with the interatomic distances of surfaces and adsorbed layers. Thus, from measurements of the angular positions of the diffraction peaks the size and orientation of the 2D unit cell, i.e. the structure of the outermost layer, can be straightforwardly determined. Analysis of the peak intensities yields the potential corrugation, which usually reflects the geometrical arrangement of the atoms within the 2D unit cell[2].

The energy of thermal He atoms is comparable with the energy of collective excitations (phonons) of surfaces and overlayers. Thus, in a scattering experiment the He atom may exchange an appreciable part of its energy with the surface. This energy can be measured in time-of-flight experiments with a resolution ~ 0.2 meV; the resolution can be brought even to ~ 0.1 meV (when using very low beam energies, $\simeq 8$ meV). Thus, surface phonon dispersion curves can be mapped out by measuring energy loss spectra at various momentum transfers in definite crystallographic directions. This is a substantial advantange of inelastic He scattering over inelastic neutron scattering. (In view of the random orientation of powdered samples, which have to be used in neutron scattering, only average phonon density of states, but not dispersion curves, can be obtained.) The range of energy transfer that can be covered by thermal He atoms is limited at the low end by the present maximum resolution of ~ 0.1 meV and at the high end by the nature of the scattering mechanism. The interaction time of thermal He atoms with the surface being larger than 10^{-13} s, the upper limit for the observable phonon modes is about 40 meV. From this point of view high resolution electron scattering is more advantageous. So far only modes with a large component perpendicular to the surface have been clearly detected; this seems to be less a fundamental, but rather a technical problem.

Besides the inelastic component, always a certain number of He atoms are elastically scattered in directions lying between the coherent diffraction peaks. We will refer to this scattering as diffuse elastic scattering. This diffuse intensity is attributed to scattering from defects and impurities. Accordingly, it provides information on the degree and nature of surface disorder. It can be used for example to study the growth of thin films[3] or to deduce information on the size, nature and orientation of surface defects[4]. Very recently from the analysis of the diffuse elastic peak width, information on the diffusive motion of surface atoms has been obtained[5].

Another remarkable way to use He scattering for the study of adsorbed layers is based on the large total cross-section Σ for diffuse He scattering of isolated adsorbates (e.g. $\Sigma > 100\,\text{Å}^2$ for a number of adsorbates like Xe, CO, NO; and even for H, $\Sigma > 20\,\text{Å}^2$). This large cross-section is attributed to the long range attractive interaction between adatom and the incident He atom, which causes the He atoms to be scattered out of the coherent beams. The remarkable size of the cross-section allows the extraction of important information concerning the lateral distribution of adsorbates, mutual interactions between adsorbates, dilute–condensed phase transitions in 2D, adatom mobilities, etc.[6], simply by monitoring the attenuation of one of the coherently scattered beams, in particular of the specular beam. This technique also allows the detection of impurities (including hydrogen!) in the permill range, a level hardly attainable with almost all other methods. For a detailed discussion of the application of this He scattering mode we refer to Ref. 6.

2.2. Experimental aspects

Like all analytical methods, the thermal He-scattering device can be represented schematically as in Fig. 1. It consists mainly of a source and a detector. The high pressure nozzle source provides a highly monochromatized flux of thermal He atoms which is then collimated to a narrow beam of thermal He atoms. This well-defined flux of probe particles impinges on the sample at a polar angle ϑ_i and azimuthal angle φ_i. The detector measures the properties of the He atoms which, after interacting with the sample, serve as information carriers. By determination of the whole set of spatial and energetic distributions of the scattered He atoms, the double differential scattering cross-section, containing the whole set of structural and dynamical information, can be determined. In contrast to electron spectrometers, where either the elastic or the inelastic scattering cross-section are measured in order to obtain structural information (LEED) or dynamical information (EELS), respectively, the currently operated high resolution He scattering spectrometers are hybrids. They are designed to measure structural and dynamical surface properties in one experiment, i.e. designed to measure the double differential scattering cross-section. This duality is achieved at the expense of very high diffraction capabilities. The instrumental transfer width, i.e. the momentum resolution of a He diffractometer, is governed by the angular spread of the He beam and of the detector and by the monochromaticity of the He wave. The now available primary He beam intensities are large enough to allow the

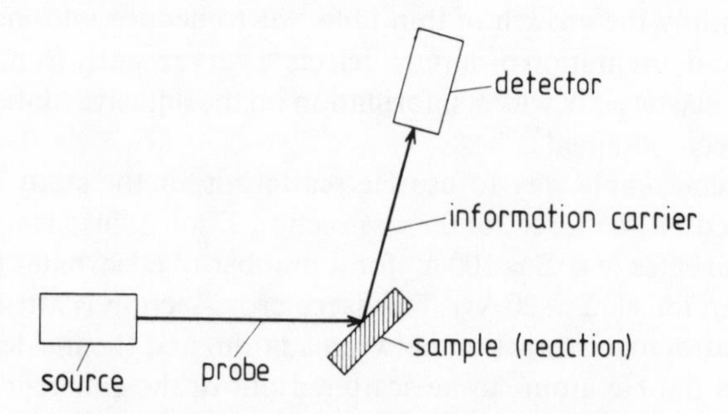

Fig. 1. Schematic of a surface scattering experiment.

construction of He diffractometers with transfer widths of several thousand angstroms, by monochromatizing the primary He beam (e.g. with Fizeau selectors) and by reducing the limiting apertures. This transfer width size is comparable to that of synchrotron X-ray diffractometers. However, in view of the low probability for the excitation of surface short wavelength phonons a compromise has to be found when building a hybrid spectrometer. This results in transfer widths of a few hundred angstroms, which are still larger than in conventional LEED systems.

The highly monochromatic He beam source is a nozzle beam generator which has been described in detail in Refs. 7 and 8. In brief, the monochromatic He beam is produced by expanding extra-high-purity helium from a high pressure reservoir (150 bar) through a narrow nozzle orifice ($5\,\mu$m) into a chamber which is evacuated by a $1500\,\mathrm{l\,s^{-1}}$ turbomolecular pump boostered by a $150\,\mathrm{m^3\,h^{-1}}$ roots blower. During expansion the translational energy distribution of the gas sharpens by about two orders of magnitude; the beam flux energy becomes $E_{\mathrm{He}} = \frac{5}{2}kT_0$, i.e. 20% larger than the flux energy of an effusion beam from a reservoir at the same temperature, T_0. By varying the nozzle temperature, T_0, beam energies in the range 100 meV ($\lambda = 0.45$ Å) to 5 meV ($\lambda = 2.03$ Å) are easily obtained. Cooling the source with liquid nitrogen (which leads here to an effective nozzle temperature of $T_0 = 85$ K) results in a 18.3 meV beam with an energy width of 0.25 meV (FWHM) and an intensity of $\sim 2 \times 10^{19}$ He atoms $\mathrm{s^{-1}\,sr^{-1}}$. Upon expansion, the collimation is obtained by a skimmer and one or two subsequent apertures separating differentially pumped chambers.

The major problems in high resolution He scattering are connected with the detection: large background and low detection efficiency. While with some technical effort the background can be efficiently reduced, no real breakthrough has been so far achieved in the attempt to increase the detection efficiency of He beams without impairing the time resolution. The method of choice is still electron bombardment ionization, followed by ion mass analysis. The real detection efficiency hardly surpasses 10^{-6}, while the time resolution is fully sufficient for time-of-flight He velocity analysis. Recent attempts to monitor the He metastables (generated also by electron bombardment) instead of He ions may be advantageous, when the He background pressure in the detector can not be made low enough. In this type of detection, the relatively lower efficiency is compensated by a favorable directionality effect: the He atoms in the beam are more efficiently detected than those in the isotropic background.

A relatively high He background is obviously inherent in all He-beam experiments. Even a highly collimated beam puts a continuous, heavy He load on the pumps evacuating the sample chamber. In the case of the apparatus shown in Fig. 2, the beam collimated to 0.2° (1.5×10^{-6} sr) supplies the sample chamber with about 1.5×10^{13} He/sec (5.7×10^{-7} mbar l/s). In view of some

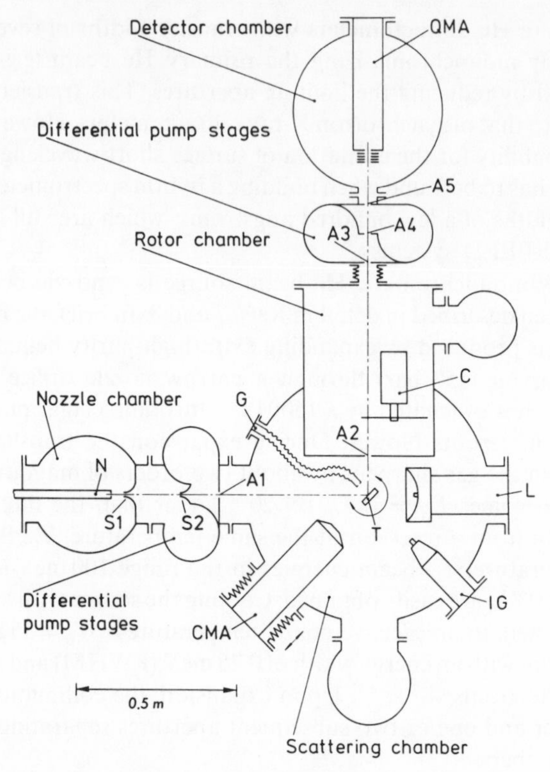

Fig. 2. Schematic diagram of a high resolution He time-of-flight spectrometer. N – nozzle beam source, S1, 2 – skimmers, A1–5 – apertures, T – sample, G – gas doser, CMA – Auger Spectrometer, IG – ion gun, L – LEED, C – magnetically suspended pseudorandom chopper, QMA – detector, quadrupole mass analyzer with channeltron.

additional He load from the beam generator system and with a reasonably sized pumping system, the He background level in the sample chamber is rarely below 10^{-9} mbar. A detector located in this chamber would have a dynamical range of $\sim 10^2$–10^3. This is acceptable when the information is inferred from the intense specular beam, but out of question when diffraction patterns of adlayers or inelastic scattering spectra are sought. The only practical way for a radical background reduction is the differential pumping. In the example in Fig. 2, the scattering (sample) chamber is separated by three differential pumping chambers (one of them contains the chopper) which, together with the very efficient (high compression ratio) pumping of the detector chamber, leads to a He background pressure of $\sim 10^{-15}$ mbar in the latter. This results in an effective dynamical range of $\sim 10^6$. Very low intensity diffraction peaks (like second-order diffraction peaks from clean packed metal surfaces) or weak inelastic resonances can be clearly resolved out of the background.

The total of eight individually pumped chambers separated by small orifices keeps a pressure ratio of 20 orders of magnitude between the He source reservoir and detector. This is done at the expense of the flexibility of the scattering geometry. In the case of the apparatus shown in Fig. 2, the angle between incident and outgoing beam is fixed at $\vartheta_i + \vartheta_f = 90°$.

The UHV sample chamber has a base pressure in the low 10^{-11} mbar range. The chamber contains conventional crystal cleaning (ion sputter gun) and analyzing devices (LEED, CMA–Auger). Special emphasis has been put on the design of the sample holder, because the fixed scattering geometry requires the diffraction scans to be made by rotations of the crystal. In Ref. 8 we have described in detail a sample holder which allows independent polar and azimuthal rotation as well as tilt with the sample temperature continuously adjustable between 25 K and 1800 K. Recently we have improved the cooling device of the sample holder, allowing now for temperatures as low as 15 K. Both the polar and azimuthal angles are varied by step motors. This ensures a continuous and accurate scanning, which is essential when monitoring diffraction scans. The step motors and the transmission used here allow an angular resolution of the polar and azimuthal sample rotation of 0.018° and 0.009°, respectively.

The velocity distribution of atoms scattered from a surface is usually determined by measuring the time they need to cover a given distance (the so-called time-of-flight (TOF) method). The flight path is defined by the position of the chopper blade, which is placed between sample and detector, and the center of the electron beam bombardment ionizer. For a maximal resolution the chopper opening function (gate function) should be in the ideal case a δ-function. In Fig. 3 the gate function and the TOF distribution of a 18 meV He nozzle beam are shown schematically. The time interval between gate

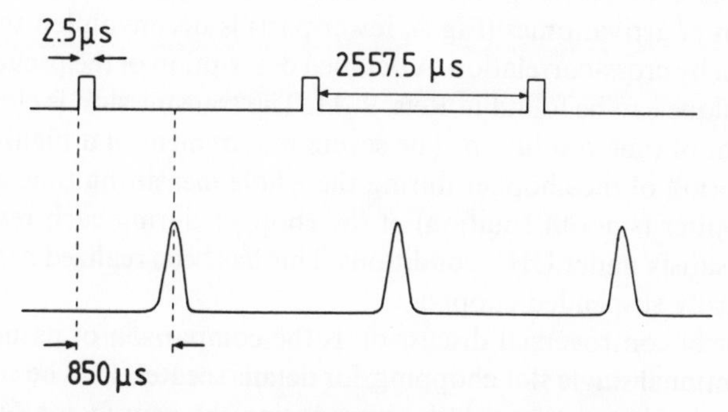

Fig. 3. Gate function and the TOF distribution of He atoms from an 18 meV nozzle beam (flight path 790 mm).

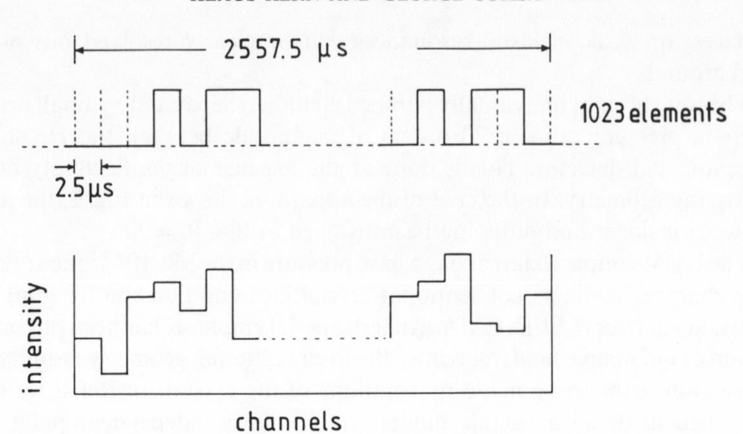

Fig. 4. Gate function of a pseudorandom sequence of slots and bars and a schematic intensity distribution.

openings has to be large enough so that superpositions of slow and rapid atoms originating from successive openings can be neglected. The transmission is equal to the ratio between gate opening time and the interval between openings. Thus the width of the gate function, i.e. the time resolution, is a compromise between the ideal δ-function and an acceptable transmission. The compromise shown in Fig. 3 results in a resolution of about 0.3% with respect to the average TOF and a rather low transmission of 0.1%. In view of the signal-to-noise problems due to the background this transmission is unsatisfactory.

An elegant and efficient alternative is the use of pseudorandom chopping. This method was first introduced in inelastic neutron scattering studies in the late 1960s. The gate function is a 'pseudorandom' sequence of slots and bars (Fig. 4, upper part). We use here a binary shift register sequence of 512 slots and 511 bars, of 2.5 μs width each (at a rotor speed of 391 Hz). The distribution of arrival times (Fig. 4, lower part) is deconvoluted to the TOF distribution by cross-correlation. A detailed description of the pseudorandom TOF technique can be found in Refs. 9–11. The transmission is close to 50%, independent of time resolution. The severe requirement of a highly constant rotation period of the chopper during the whole measuring time and a very low phase jitter (smooth rotation) of the chopper during each revolution is difficult to satisfy under UHV conditions. This has been realized here by using a magnetically suspended chopper.

A matter of controversial discussion is the comparison of pseudorandom and conventional single slot chopping; for details see Ref. 11. The comparison can be expressed quantitatively by introducing the gain factor $G(f_k)$, which represents the ratio of the variances resulting in channel k from the two kinds of chopping. For binary shift register sequences, the gain factor is given by

$$G(f_k) = \frac{\sigma^2 \text{ single slot}}{\sigma^2 \text{ pseudorandom}} \simeq \frac{1}{2} \frac{f_k + \bar{u}}{\bar{f} + \bar{u}/n}, \tag{1}$$

where f_k is the number of counts of the time-dependent signal in channel k, $\bar{f}$ is its mean value, $\bar{u}$ is the mean number of counts per channel of the time-independent signal (the real background), and n the number of slots of the pseudorandom sequence. The pseudorandom chopping has to be preferred when $G(f_k) > 1$. This is always the case when $\bar{u} > 2\bar{f}$ because $n \gg 1$. The general interest, so far, is focussed on the position, height, and shape of the peaks in the TOF spectra and much less on the shape of a possibly present energy-dependent background. In this case a much less restrictive condition can be deduced: the pseudorandom chopping has to be preferred when in the 'interesting' channels $f_k > 2\bar{f}$, irrespective of the value of $\bar{u}$. This condition is fulfilled for all significant peaks and thus the pseudorandom chopping is of advantage, even if technically more demanding.

The effective resolution of the He scattering spectrometer as a whole is in fact the figure of primary interest. For the spectrometer used in the authors' laboratory the effective resolution function of the chopper, taking into account the finite width of the beam, of the chopper slots and of the channels, has a full width at half-maximum (FWHM) of ~ 1.27 times the width of one chopper slot, i.e. 3.2 μs. Together with the velocity spread of the beam generated by the source ($\sim 0.7\%$ FWHM) and the finite length of the ionization region (estimated as ~ 5 mm FWHM) the overall instrumental resolution amounts to $\sim 8\,\mu$s FWHM for a typical flight time of 800 μs corresponding to 0.40 meV FWHM at 20 meV beam energy. These values are in agreement with the experimental width of the elastic peaks (specular and diffuse elastic) in the energy spectra taken from a clean Pt(111) surface.

3. SURFACE DYNAMICS

3.1. Some basic lattice dynamics

3.1.1. Surface phonons

This section deals with the dynamics of collective surface vibrational excitations, i.e. with surface phonons. A surface phonon is defined as a localized vibrational excitation of a semi-infinite crystal, with an amplitude which has wavelike characteristics parallel to the surface and decays exponentially into the bulk, perpendicular to the surface. This behavior is directly linked to the broken translational invariance at a surface, the translational symmetry being confined here to the directions parallel to the surface.

The phonon spectrum of a crystal surface consists of two parts. The bulk bands, which are due to the projection of bulk phonons onto the two-dimensional Brillouin zone of the particular surface, and the specific surface phonon branches which are due to the broken translational invariance. This is exemplified in Fig. 5 for a close-packed (111) surface of a face-centered cubic crystal. The theoretical dispersion curves have been calculated by Allen *et al.*[12] Already in 1971, they analyzed the vibrational spectra of low index surfaces of f.c.c. crystals in the quasiharmonic approximation using Lennard–Jones pair

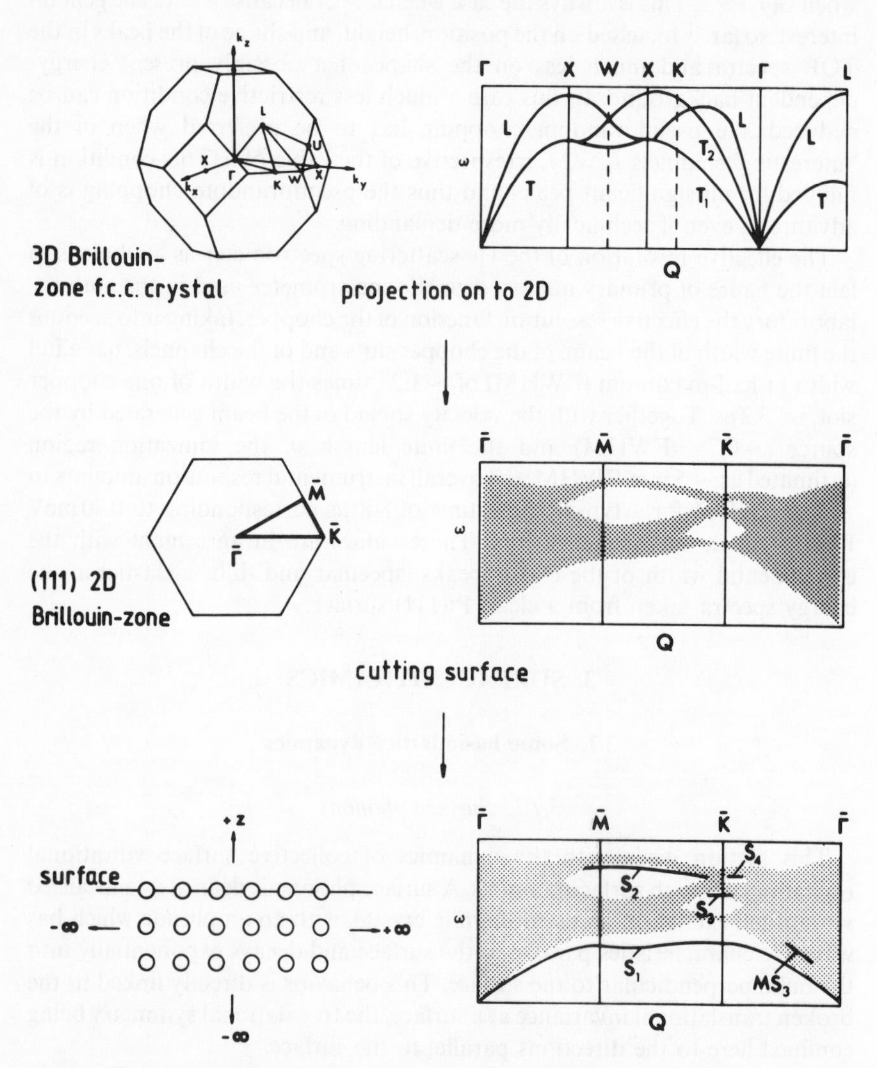

Fig. 5. Schematics of the formation of the surface phonon dispersion of a (111) f.c.c. crystal.

potentials. The solid lines, termed S_1–S_4, are specific surface phonon branches. These true surface modes only exist outside or in gaps of the projected bulk phonon bands of equivalent symmetry. However, inside the bulk bands there may exist well-defined modes with maximum amplitudes in the surface layer, but non-vanishing amplitudes in the bulk. Mode MS_3 in Fig. 5 is an example of such a 'surface resonance'. Of particular interest is the lowest frequency mode below the transverse bulk band edge. In this mode, the atoms are preferentially vibrating in the plane defined by the surface normal and the propagation direction, i.e. in the sagittal plane. This wave is called the Rayleigh wave.

Surface waves are usually discussed in two limits. Phonons whose wavelength is much larger than the atomic spacing of the crystal can be treated in the continuum limit, i.e. the elastic response of the crystal surface can be described in terms of average interatomic forces. When the ratio of wavelength and atom spacing is smaller than, say ~ 100, this approximation is no longer justified. The forces between the atoms have now to be taken into account in detail. This is the reason why the short wavelength phonons are of primary interest to surface physicists. The surface phonon dispersion directly reflects the interatomic force field at the surface.

3.1.2. *Continuum limit*

It was Lord Rayleigh[13] who was first to derive the dispersion relations for surface acoustic waves propagating along the surface of an isotropic elastic continuum with a planar, stress-free surface. Phonons are called acoustic when the frequency goes to zero for diverging wavelengths, otherwise they are called optic. The frequency ω_R of the Rayleigh wave varies linearly with wave vector $Q_\parallel$, the proportionality constant being the speed of sound c_R at the surface:

$$\omega_R = c_R Q_\parallel \tag{2}$$

The Rayleigh wave speed of sound c_R is determined by the equation:

$$[(K+2\mu)/\mu]\left[\left(\frac{c_R^2}{c_t^2}\right)^3 - 8\left(\frac{c_R^2}{c_t^2}\right)^2 + 8\left(\frac{3c_t^2}{c_l^2} - 2\right)\right] + 16\left(1 - \frac{c_t^2}{c_l^2}\right) = 0 \tag{3}$$

with $c_R/c_t < 1$ and K, μ being the Lamé coefficients of an elastic body. Here, c_t and c_l are the transverse and longitudinal sound velocities of the bulk, respectively. Resonable values for the ratio c_R/c_t range from 0.96 (non-compressible solids) to 0.69 (limit of stability for an elastic body). The displacement pattern of the Rayleigh wave is described by ellipsoids in the sagittal plane. The decay length of the amplitudes into the elastic bulk is of the order of the wavelength $\lambda = (2\pi/Q_\parallel)$. In Fig. 6 we show the displacement pattern and the decay length of a square lattice given by Farnell[14].

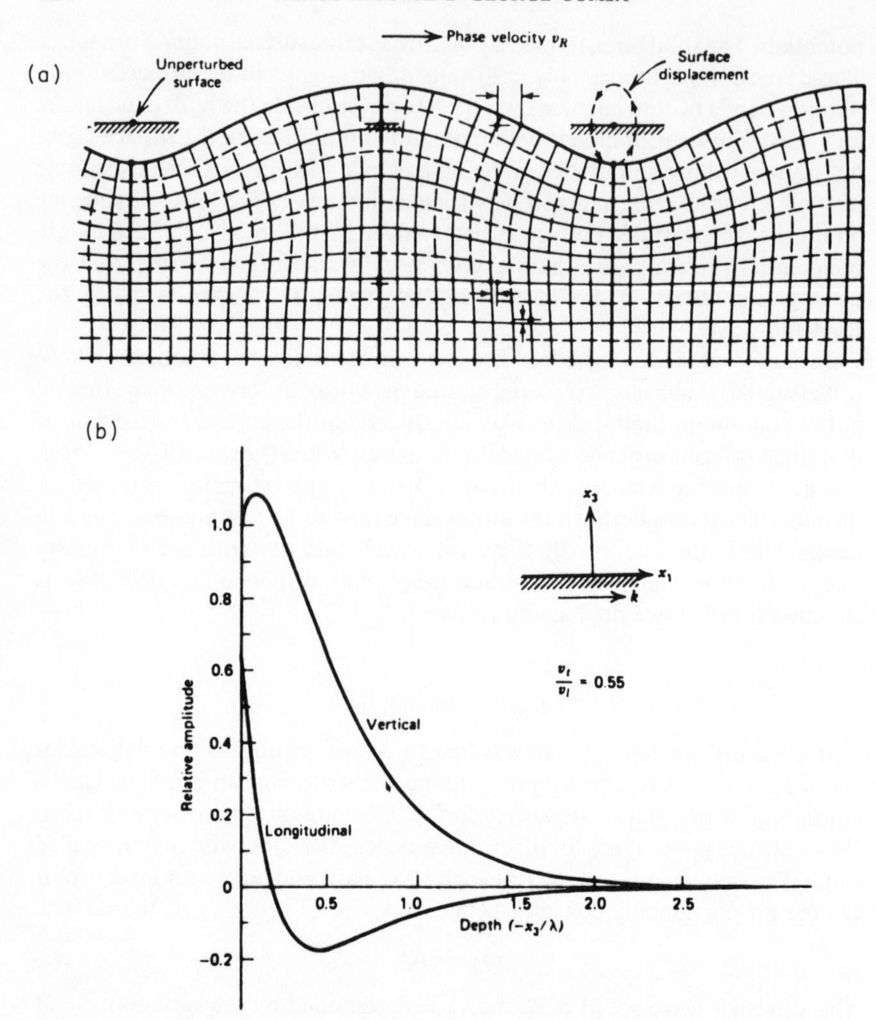

Fig. 6. (a) Distortion of a square lattice in a plane containing the surface normal and the propagation direction of the Rayleigh wave. (*After Ref. 14.*). (b) Relative amplitudes of the vertical and longitudinal displacements of the Rayleigh wave as a function of penetration depth into the bulk. (*After Ref. 14.*)

3.1.3. Lattice dynamics

The dispersion curves of surface phonons of short wavelength are calculated by lattice dynamical methods. First, the equations of motion of the lattice atoms are set up in terms of the potential energy of the lattice. We assume that the potential energy φ can be expressed as a function of the atomic positions $\vec{R}(\vec{l}_{\parallel}, l_z)$ in the semi-infinite crystal. The location of the nth atom can be written:

$$\vec{R}(\vec{l}_{\parallel}, l_z) = \vec{R}_0(\vec{l}_{\parallel}, l_z) + \vec{u}(\vec{l}_{\parallel}, l_z) \tag{4}$$

The vector $\vec{l}_{\parallel}$ specifies the position of the atom in a two-dimensional lattice in a plane parallel to the surface and the value of l_z identifies the lattice plane with respect to the surface ($l_z = 1$ for the surface layer. $\vec{u}$ is the displacement of the vibrating atom from its equilibrium position $\vec{R}_0$. The total kinetic energy of the vibrating lattice is

$$T = \frac{1}{2} \sum_{\vec{l}_{\parallel}, l_z, \alpha} m(\vec{l}_{\parallel}, l_z) \dot{u}_{\alpha}^2(\vec{l}_{\parallel}, l_z) \tag{5}$$

with m being the mass of the particular atom and α indexing the Cartesian coordinates of $\vec{u}$. For small displacements, we may expand the potential energy in a Taylor series:

$$\begin{aligned} \varphi = \varphi_0 &+ \sum_{\vec{l}_{\parallel}, l_z, \alpha} \Phi_{\alpha}(\vec{l}_{\parallel}, l_z) u_{\alpha}(\vec{l}_{\parallel}, l_z) + \frac{1}{2} \sum_{\vec{l}_{\parallel}, l_z, \alpha} \sum_{\vec{l}_{\parallel}, l_z, \beta} \\ &\times \Phi_{\alpha\beta}(\vec{l}_{\parallel}, l_z; \vec{l}'_{\parallel}, l'_z) u_{\alpha}(\vec{l}_{\parallel}, l_z) u_{\beta}(\vec{l}'_{\parallel}, l'_z) \end{aligned} \tag{6}$$

with

$$\Phi_{\alpha}(\vec{l}_{\parallel}, l_z) = \left. \frac{\partial \phi}{\partial u_{\alpha}(\vec{l}_{\parallel}, l_z)} \right|_{u=0}$$

$$\Phi_{\alpha\beta}(\vec{l}_{\parallel}, l_z; \vec{l}'_{\parallel}, l'_z) = \left. \frac{\partial^2 \phi}{\partial u_{\alpha}(\vec{l}_{\parallel}, l_z) \partial u_{\beta}(\vec{l}'_{\parallel}, l'_z)} \right|_{u=0}$$

being the first and second derivatives of the potential evaluated at the equilibrium positions. In the harmonic approximation only terms up to the second order are retained. The negative gradient $-\partial\phi/\partial u_{\alpha}(\vec{l}_{\parallel}, l_z)$ is the force, acting on the lattice atom $(\vec{l}_{\parallel}, l_z)$ in direction α. When all atoms are at their equilibrium sites they experience no force by definition and we have:

$$\Phi_{\alpha}(\vec{l}_{\parallel}, l_z) = 0, \qquad \sum_{\vec{l}_{\parallel}, \vec{l}'_{z}} \Phi_{\alpha\beta}(\vec{l}_{\parallel}, l_z; \vec{l}'_{\parallel}, l'_z) = 0 \tag{7}$$

In the framework of the Euler–Lagrange formalism, we write the equation of motion for the displacements of the atoms as:

$$M(\vec{l}_{\parallel}, l_z) \ddot{u}_{\alpha}(\vec{l}_{\parallel}, l_z) = - \sum_{\vec{l}_{\parallel}, l_z, \beta} \Phi_{\alpha\beta}(\vec{l}_{\parallel}, l_z; \vec{l}'_{\parallel}, l'_z) u_{\beta}(l'_{\parallel}, l'_z) \tag{8}$$

We take advantage of the translational symmetry parallel to the surface by seeking solutions to Eq. (8) of the form

$$u_{\alpha}(l_{\parallel}, l_z) = \sqrt{M(\vec{l}_{\parallel}, l_z)}\, u_{\alpha}(\vec{Q}_{\parallel}; l_z) \exp(\mathrm{i}\vec{Q}_{\parallel} \vec{R}_0(\vec{l}_{\parallel}, l_z) - \mathrm{i}\omega t) \tag{9}$$

with $\vec{Q}_{\parallel}$ being a wave vector parallel to the surface. If Eq. (9) is substituted into Eq. (8) we obtain a set of linear equations in the amplitudes

$w_\alpha(\vec{Q}_\parallel; l_z) = \sqrt{M(\vec{l}_\parallel, l_z)}\, u_\alpha(\vec{Q}_\parallel; l_z)$:

$$\omega^2 w_\alpha(\vec{Q}_\parallel; l_z) = \sum_{l'_z, \beta} D_{\alpha\beta}(\vec{Q}_\parallel; l_z, l'_z) w_\beta(\vec{Q}_\parallel; l'_z) \tag{10}$$

with the dynamical matrix:

$$D_{\alpha\beta}(\vec{Q}_\parallel, l_z, l'_z) = \sum_{\vec{l}'_\parallel} [\Phi_{\alpha\beta}(\vec{l}_\parallel, l_z; l'_\parallel, l_z)/\sqrt{M(\vec{l}_\parallel, l_z) M(\vec{l}'_\parallel, l'_z)}]$$
$$\times \exp[iQ_\parallel(\vec{R}_0(\vec{l}'_\parallel, l'_z) - \vec{R}_0(\vec{l}_\parallel, l_z))] \tag{11}$$

The sum $D_{\alpha\beta}$ is called the partially Fourier transformed dynamical matrix, which depends only on $\vec{Q}_\parallel$, l'_z and l_z. For each wave vector $\vec{Q}_\parallel$ the normal mode frequencies of the crystal can be found by setting the secular determinant equal to zero:

$$|D_{\alpha\beta}(\vec{Q}_\parallel; l_z, l'_z) - \omega^2 \delta_{\alpha\beta}\delta_{l_z, l'_z}| = 0 \tag{12}$$

Due to the hermitian character of the dynamical matrix, the eigenvalues are real and the eigenvector satisfies the orthonormality and closure conditions.

The coupling coefficients $\Phi_{\alpha\beta}$ are given by

$$\Phi(ij) = \delta_{ij} \sum_{j' \neq i} K_{\alpha\beta}(ij') - (1 - \delta_{ij}) K_{\alpha\beta}(ij) \tag{13a}$$

with

$$K_{\alpha\beta}(ij) = \frac{\Phi'_{ij}}{|\vec{r}_i - \vec{r}_j|}\delta_{\alpha\beta} + \left(\Phi''_{ij} - \frac{\Phi'_{ij}}{|\vec{r}_i - \vec{r}_j|}\right) n_\alpha(ij) n_\beta(ij) \tag{13b}$$

where Φ'_{ij} and Φ''_{ij} are the first and second derivatives of the two-body potential connecting atom i and j and $n_\alpha(ij)$ the αth Cartesian component of a unit vector pointing from atom i to atom j.

The solution of Eq. (12) provides the dispersion curves of the surface phonons. However, this solution requires the knowledge of the coupling constants Φ''_{ij} in the bulk and at the surface. The determination of the coupling constants can be done in several ways. The most direct way, but also the most difficult way, is the *ab initio* calculation of the electronic ground state energy as a function of the displacement of the atom cores; the second derivative of the energy with respect to the displacements provides the force constants. So far, this method has only been used to evaluate the surface phonon dispersion of low index Al surfaces[15]. The most common way to obtain coupling parameters, however, is to fit the data with calculated dispersion curves, with the force constants of surface atoms (intra- and interlayer force constants) as fitting parameters. The surface constants can obviously differ from the bulk ones due to the broken translational invariance: surface atoms have less neighboring atoms than bulk atoms. The changed force field at the surface can lead to relaxation phenomena in the selvedge. Various attempts have been undertaken to elucidate the forces responsible for the surface relaxation. An intuitive model is the point ion model of Finnis and Heine[16] based on the electron smoothing concept of Smoluchowsky[17]. This is exemplified in Fig. 7. The electrons have the tendency to spill over the surface in order to create a geometrically smooth surface and thus lower their kinetic energy. The new electron distribution causes electrostatic forces on the ion cores of the surface atoms, resulting in an inward relaxation of the first layer. The Finnis and Heine model is certainly too simple to be more than qualitatively correct. However, there is still interest in improving the model by more sophisticated electron density distributions and accounting for electron screening of the ion motion[18].

It would be most desirable to have a direct link between the geometrical relaxations and the force constant changes. From molecular vibrational spectroscopy it is well known that the intramolecular force constants show a scaling behavior, Badger's rule[19]:

$$\Phi''_{ij} = \Phi''_b(r_b/r_{ij})^\alpha \tag{14}$$

with r_{ij} being the distance between atom i and j and r_b nearest neighbor distance in equilibrium.

Recently, Baddorf *et al.* have studied the removal of the relaxation of the Cu(110) surface at 100 K upon hydrogen adsorption[20]. It is well established that the (110) surfaces of face-centered cubic metals exhibit large oscillatory relaxations. For Cu(110) a contraction of about 7.5% of the first interlayer spacing d_{12} and an expansion of about 2.5% of the second interlayer spacing d_{23} have been measured[21]. Baddorf *et al.*[20] observed that both lattice spacings return linearly to the bulk spacing with hydrogen coverage. The H atoms appear to eliminate the charge redistribution which originally caused the relaxation. The same authors have also monitored, by means of specular

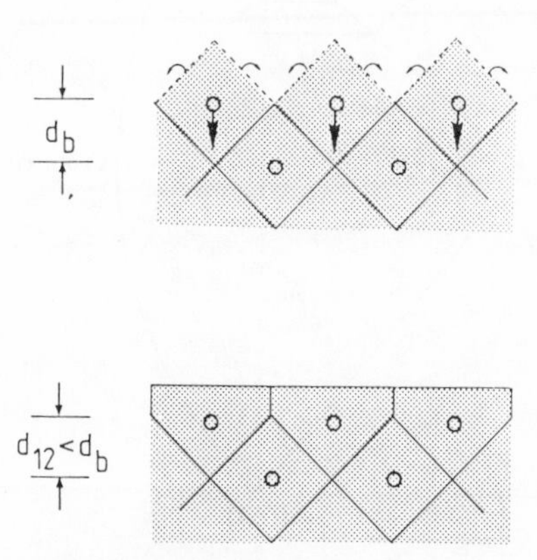

Fig. 7. Surface relaxation in the charge smoothing model of Smoluchowski[17].

EELS, the frequency of the A_1-phonon at the $\bar{\Gamma}$-point (see section 3.3) upon hydrogen adsorption. A decrease in phonon energy has been observed, which could be related to the increasing lattice spacing d_{12} through $\hbar\omega \sim d_{12}^{1.8\pm0.7}$. In the framework of a simple central force constant model, including only nearest neighbor interactions, this dependency can be use to link surface force constants with the interlayer spacing of Cu(110). Baddorf and Plummer[22] obtained:

$$\Phi''_{12}/\Phi''_b \sim (d_b/d_{12})^{7.3\pm3} \tag{15}$$

3.2. He atoms and surface phonons

He atoms, which are scattered at a surface, are characterized by their incoming and outgoing wavevectors $\vec{k}_i$ and $\vec{k}_f$ and energies $E_i = \hbar k_i^2/2m$ and $E_f = \hbar k_f^2/2m$, with m being the mass of the He atom. If we denote the incident and outgoing angle with respect to the surface normal by ϑ_i and ϑ_f, the projected wavevectors parallel to the surface are given by $\vec{K}_i = \vec{k}_i \sin\vartheta_i$ and $\vec{K}_f = \vec{k}_f \sin\vartheta_f$. The kinematic conditions for scattering are derived from energy and momentum conservation. Because of the loss of vertical translational invariance at the surface, only the momentum parallel to the surface is

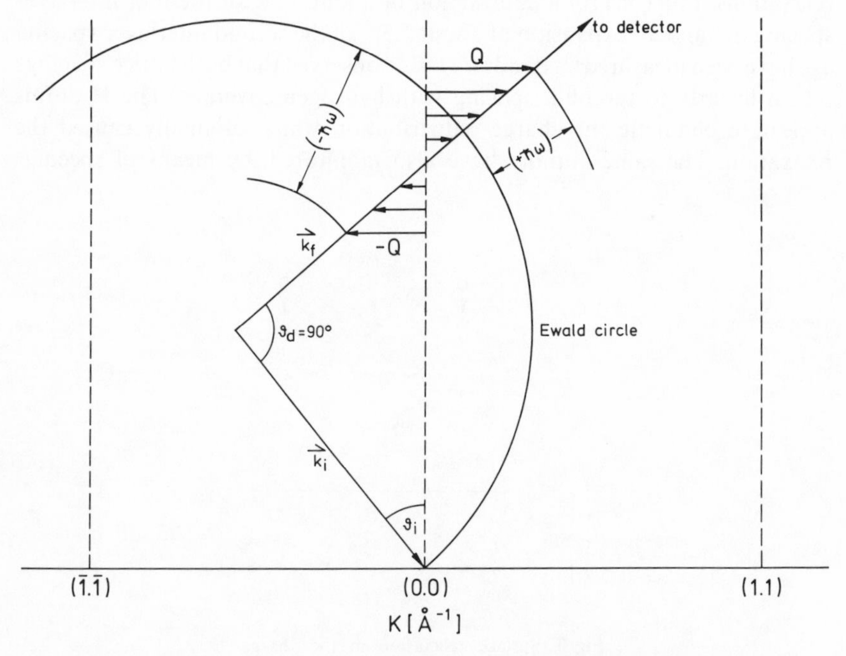

Fig. 8. Ewald diagram of inelastic He scattering from a clean metal surface, for $\vec{G} = 0$.

conserved in the scattering process:

$$E_i = E_f - \hbar\omega(\vec{Q}_\parallel) \tag{16}$$

$$\vec{K}_i = \vec{K}_f - (\vec{Q}_\parallel + \vec{G}) = \vec{K}_f + \Delta\vec{K} \tag{17}$$

with $\hbar\omega$ being the phonon energy, $\vec{Q}_\parallel$ the wavevector of the phonon parallel to the surface and $\vec{G}$ a reciprocal lattice vector of the crystal surface. According to the sign of ω a phonon is annihilated ($\omega > 0$) or created ($\omega < 0$) during the scattering process. In Fig. 8 the inelastic scattering process is described in terms of an Ewald diagram. For planar scattering with fixed scattering angle $\vartheta_d = \vartheta_i + \vartheta_f = 90°$, the combination of energy and momentum conservation leads to the following relation between ω and ΔK, known as a 'scan curve':

$$\omega = \frac{\hbar k_i^2}{2m}\left[\frac{\sin\vartheta_i + (\Delta K/k_i)}{\cos\vartheta_i}\right]^2 - \frac{\hbar k_i^2}{2m} \tag{18}$$

In Fig. 9 we show a family of scan curves, with the incident angle ϑ_i as parameter, for He atoms of incident energy 18 meV (thick lines). The

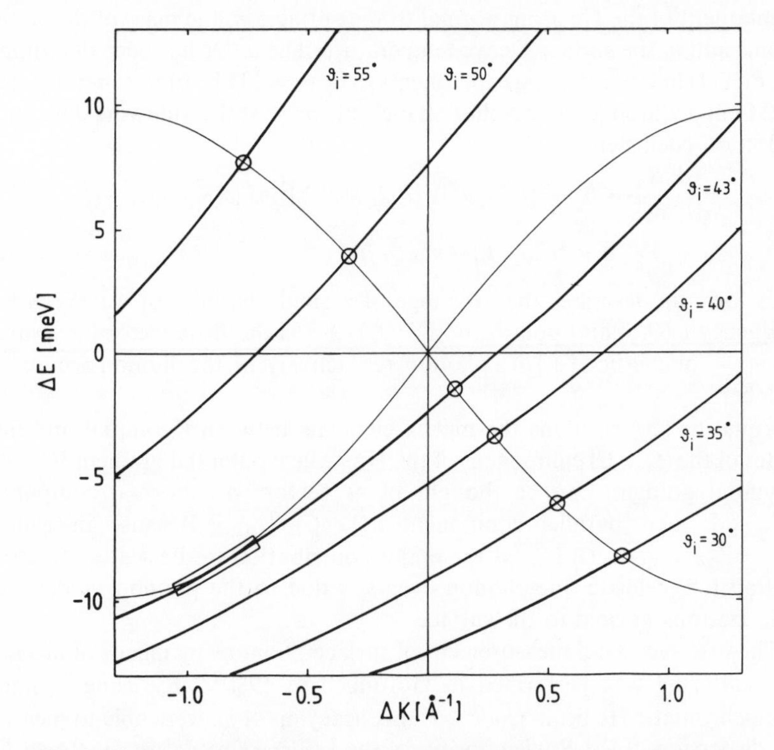

Fig. 9. Kinematics of surface phonon He spectroscopy. The thick lines correspond to scan curves of a 18 meV He beam. The thin lines display the Rayleigh phonon dispersion curve of Pt(111) along the $\bar{\Gamma}\,\bar{M}$ azimuth.

theoretical dispersion curve of the Pt(111) Rayleigh wave along the $\bar{\Gamma}\bar{M}$ azimuth is also plotted (thin lines). Phonon excitation can only occur when the scan and dispersion curves intersect. Figure 9 illustrates in a very simple way the experimental procedure to obtain dispersion curves: TOF spectra are measured for various incident angles at a fixed incident beam energy.

The fulfillment of the kinematical relation is only a necessary but not a sufficient condition for detecting a particular surface phonon in an inelastic He scattering experiment; the dynamical scattering cross-section for this process has to be large enough. Manson and Celli have calculated the dynamical reflection coefficients for He scattering in the distorted wave Born approximation[23]. This approximation is valid as long as single phonon scattering only is considered. Weare[24] gives an estimate of the importance of multiphonon events. For surface temperatures

$$T_s < \frac{m_s}{m_{He}} \frac{k_B \theta_s^2}{100 E_{iz}} \tag{19}$$

he estimates one-phonon processes to be dominant. Here, E_{iz} is the energy component of the He atom normal to the surface, m_s the mass of the surface atoms and θ_s the surface Debye temperature. The latter has been determined for Pt(111) in He scattering experiments to be $\theta_s = 231$ K. In this limit Manson and Celli deduced for a one-phonon inelastic process the following differential reflection coefficient:

$$\frac{d^2 R}{dE_f \, d\Omega_f} \sim \sum_{\vec{Q},j} |\vec{e}(\vec{Q},j) \langle \psi_G^{f*} | \nabla U(\vec{Q},z) | \psi_G^{i} \rangle|^2 |n^{\pm}| \delta(\vec{K}_f - \vec{K}_i - \vec{Q}) \\ \times \delta(E_f - E_i - h\omega(\vec{Q},j)) \tag{20}$$

This formula describes the exchange of a single phonon of wavevector $\vec{Q}$, frequency $\omega(\vec{Q},j)$ and polarization $\vec{e}(\vec{Q},j)$. $n^{\pm}$ is the Bose factor for annihilation $(-)$ or creation $(+)$ of a phonon, respectively, i.e. the phonon occupation number.

Equation (20) contains the matrix elements between the initial and final states of the $(\vec{Q} - \vec{G})$ component of the He surface potential gradient ∇U. The potential gradient can be thought of as vector with normal component $dU_{\vec{Q}-\vec{G}}/dz$ and parallel component $i(\vec{Q} - \vec{G})\ U_{\vec{Q}-\vec{G}}$. Because in general $dU_{\vec{Q}-\vec{G}}/dz > |i(\vec{Q} - \vec{G})\ U_{\vec{Q}-\vec{G}}|$, the major contribution to the scattering cross-section for inelastic one-phonon events is due to the phonon modes with polarizations normal to the surface.

The first successful measurement of surface phonons by means of inelastic He scattering was performed in Göttingen in 1980[25]. By using a highly monochromatic He beam ($\Delta v/v \approx 1\%$) Brusdeylins *et al.* were able to measure the dispersion of the Rayleigh wave of the LiF(001) crystal surfae. In earlier attempts[26,27] the inelastic events could not be resolved satisfactorily due to the low beam monochromaticity. In Fig. 10a we show a typical TOF spectrum,

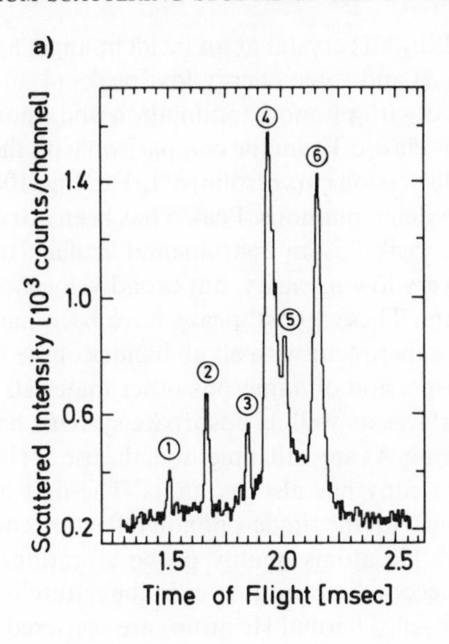

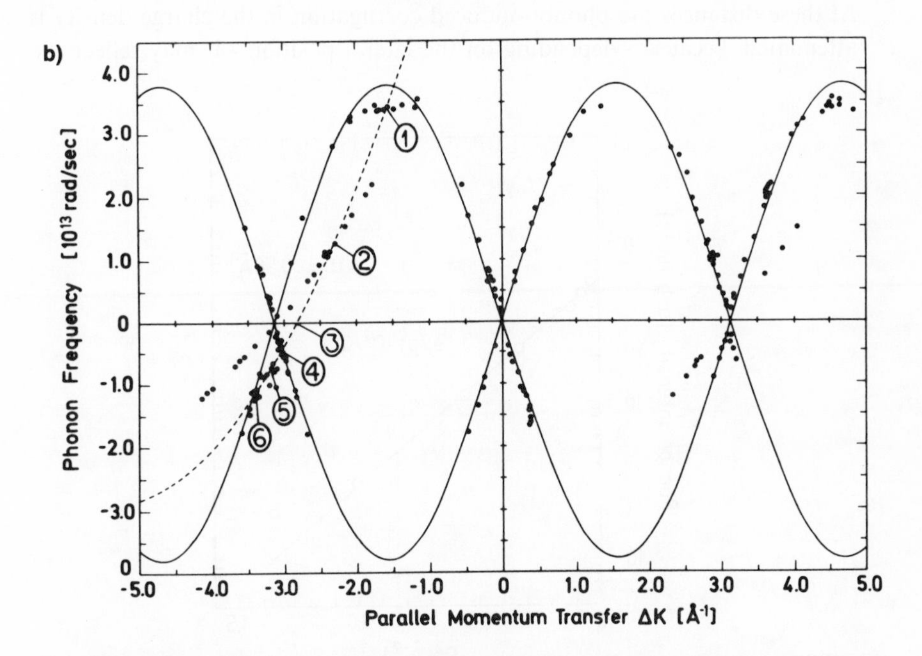

Fig. 10. (a) He time-of-flight spectrum taken from a LiF(001) surface along the ⟨100⟩ azimuth at an incident angle $\vartheta_i = 64.2°$. The primary beam energy was 19.2 meV. (*After Ref. 25.*). (b) Measured Rayleigh phonon dispersion curve of LiF(001) ⟨100⟩, including a scan curve (dashed) for the kinematical conditions in (a). (*After Ref. 25.*)

measured from a LiF(001) crystal at an incident angle $\vartheta_i = 64.2°$. Two energy gain peaks (1 and 2) and three energy loss peaks (4–6) are clearly resolved. They are associated with phonon annihilation and phonon creation, respectively. The peak 3 is elastic. From the comparison with the scan curve (dashed) and the Rayleigh dispersion curve (solid) of LiF in Fig. 10b, peaks 1, 4 and 6 are associated with Rayleigh phonons. Peak 5 has been attributed by the authors to a bulk phonon. Peak 2 is an instrumental artifact. It is due to the elastic diffraction of the very low intensity, but broad velocity distribution wings of the He nozzle beam. These 'ghost' peaks have been named 'deceptons'.

Since the initial experiment with alkali halides of the Göttingen group, the surface phonon dispersion of numerous other materials including metal and semiconductor surfaces as well as adsorbate systems has been measured by inelastic He scattering. As any other method, the use of He scattering as surface vibrational spectroscopy has also its limits. The first has been emphasized above in connection with the single-phonon reflection coefficient evaluated by Manson and Celli: He atoms mainly probe vibrations with perpendicular polarization. The second limit originates in the nature of the gas surface scattering interaction itself. Thermal He atoms are scattered by the repulsive part of the surface potential at appreciable distances above the surface ($\sim$3–4 Å). At these distances the phonon-induced corrugation in the charge density is attenuated, because – depending on the lateral position – it may reflect the

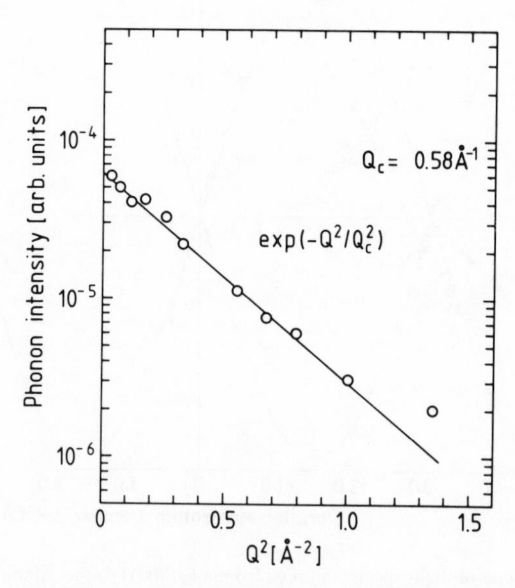

Fig. 11. Measured decrease of inelastic scattered He intensity (Rayleigh wave) from Pt(111) with increasing momentum transfer.

vibration of more than one surface atom. Taking into account, in addition, that the incoming He atom has a size comparable to the lattice spacing, its interaction with the charge density can be viewed as a collision with more than one surface atom. As a result, surface vibrations may be averaged out. The effect is usually cited as the 'Armand effect'[28]. Obviously, the effect dominates at the zone boundary, where neighboring atoms are vibrating in antiphase.

In Fig. 11 we have plotted the intensity of the inelastic Rayleigh phonon peak in He energy loss spectra taken from a Pt(111) surface along the $\bar{\Gamma}\bar{M}$ azimuth. The intensities have been extracted from TOF spectra taken with a primary He beam energy of 27 meV at a surface temperature of 105 K. All data have been corrected for the Bose factor and plotted on a semilogarithmic scale as a function of the square of the momentum transfer. The intensity rapidly decreases with increasing wavevector. The data quantitatively follow a Gaussian fall-off [$\sim \exp(-Q^2/Q_c^2)$], with a cut-off wavevector $Q_c = 0.58\,\text{Å}^{-1}$. Such a Gaussian fall-off of the Rayleigh phonon intensity has been predicted

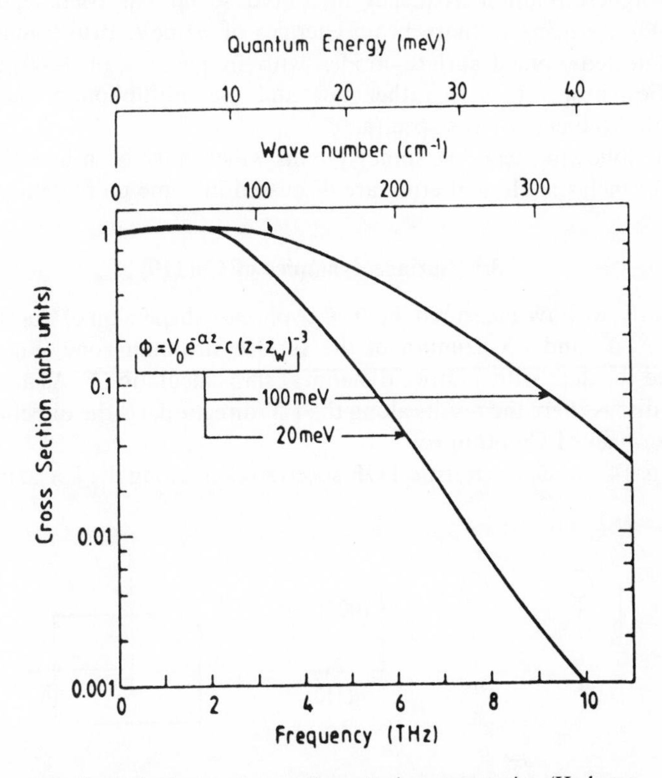

Fig. 12. Inelastic one-phonon He scattering cross-section (He beam energy 20 meV and 100 meV) as a function of phonon frequency. (*After Ibach*[32].)

by Bortolani *et al.*[29] in the framework of the distorted wave Born approximation. For Ag(111) they have calculated a cut-off value $Q_c = 0.74\,\text{A}^{-1}$.

In addition to the Armand correction, the substrate vibrations can be averaged out due to the finite interaction time between the He atom and the surface (the 'Levi effect'[30]). This is the reason why it is impossible to map out complete dispersion curves with inelastic scattering of the even slower Ne atoms[31]. This effect is also important for He scattering, especially when looking for high frequency modes. Ibach[32] has quantified this 'slow collision efect' by evaluating the single-phonon cross-section for He scattering from Cu(110), using the Harris–Liebsch[33] potential, as a function of phonon frequency. Ibach's results for 20 meV and 100 meV beams are shown in Fig. 12. The strong decrease of the cross-section with increasing phonon frequency is obvious. The effect can be partially counterbalanced by using higher beam energies. According to Eq. (19), however, the multiphonon background increases strongly with the beam energy; thus this alternative appears not that desirable.

The highest phonon frequency measured so far has been reported for NaF(100). By using primary beam energies of 90 meV, Brusdeylins *et al.*[34] have detected optical surface modes with frequencies of ~ 40 meV. The intensities, however, were rather low and the multiphonon background dominated the energy loss spectra.

In the following sections, some systems which have been investigated by means of inelastic He scattering are discussed in some more detail.

3.3. Surface dynamics of Cu(110)

Recently, we have measured the surface phonon dispersion of Cu(110) along the $\bar{\Gamma}\bar{X}$, $\bar{\Gamma}\bar{Y}$, and $\bar{\Gamma}\bar{S}$ azimuth of the surface Brillouin zone (Fig. 13) and analyzed the data with a lattice dynamical slab calculation[35]. As an example we will discuss here the results along the $\bar{\Gamma}\bar{X}$-direction, i.e. the direction along the close-packed Cu atom rows.

In Fig. 14 two characteristic TOF spectra taken along the $\bar{\Gamma}\bar{X}$ azimuth are

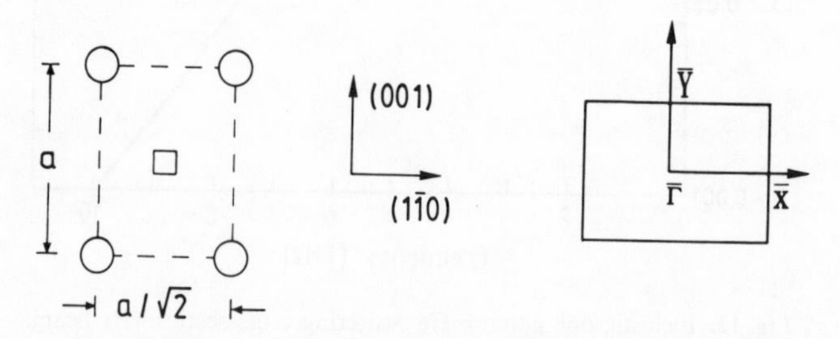

Fig. 13. Real and reciprocal lattice of the Cu(110) surface.

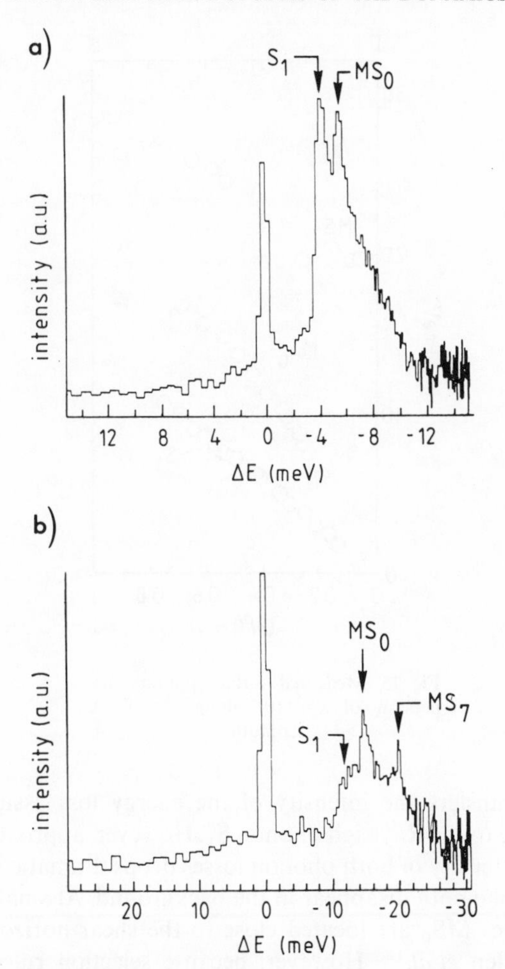

Fig. 14. He energy loss spectra taken from Cu(110) along the $\bar{\Gamma}\,\bar{X}$ azimuth at $\vartheta_i = 38°$ (a) and 35° (b), respectively. The primary He beam energy was 39.6 meV.

shown. The primary He beam energy was 39.6 meV and the angles of incidence 38° and 35°, respectively. The Rayleigh phonon S_1 and two other surface phonon losses, indicated by MS_0 and MS_7 are detected. In Fig. 15 the surface phonon dispersions are plotted, as obtained from numerous TOF spectra like those in Fig. 14 taken at different beam energies and incident angles. According to the early analysis of Allen *et al.*[12], the lowest energy phonon branch is the Rayleigh wave (S_1). The Rayleigh phonon loss S_1 is the dominant feature in the energy loss spectra for small energy transfers. Upon sampling

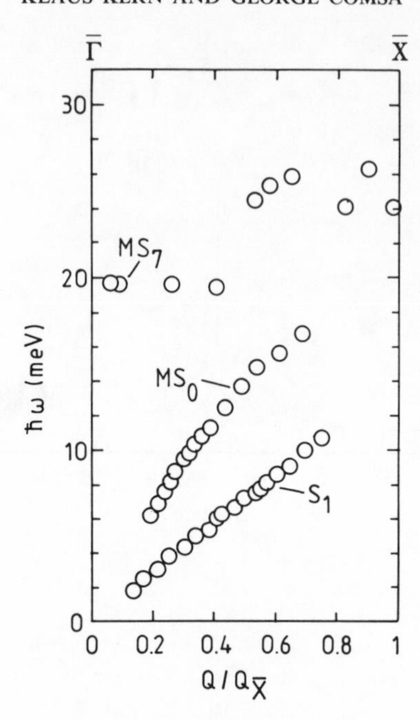

Fig. 15. Measured surface phonon dispersion of Cu(110) along the $\bar{\Gamma}$ $\bar{X}$ azimuth.

larger energy transfers the intensity of the energy loss assigned by MS_0 increases relative to the Rayleigh phonon S_1. However, approaching the zone boundary, the intensity of both phonon losses drops dramatically; in the last quarter of the zone both disappear in the background. At small wavevectors the phonon losses MS_0 are located close to the shear horizontal mode S_2 predicted by Allen *et al.*[12]. However, because selection rules exclude the coupling of He atoms to odd modes, the MS_0 energy losses observed correspond to a resonance not predicted by Allen *et al.*[12].

In addition to the acoustical modes S_1 and MS_0, we observe in the first half of the Brillouin zone a weak optical mode MS_7 at 19–20 meV. This particular mode has also been observed by Stroscio *et al.*[36] with electron energy loss spectrocopy. According to Persson *et al.*[37] the surface phonon density of states along the $\bar{\Gamma}\bar{X}$-direction is a region of depleted density of states, which they call pseudo band-gap, inside which the resonance mode MS_7 peals of. This behavior is explained in Fig. 16: (a) top view of a (110) surface; (b) and (c) schematic plot of the structure of the layers in a plane normal to the (110) surface and containing the (110) and (100) directions, respectively. Along the (110) direction each bulk atom has six nearest neighbors in a lattice plane, while in the (100) direction it has only four. As exemplified in Fig. 17, where inelastic

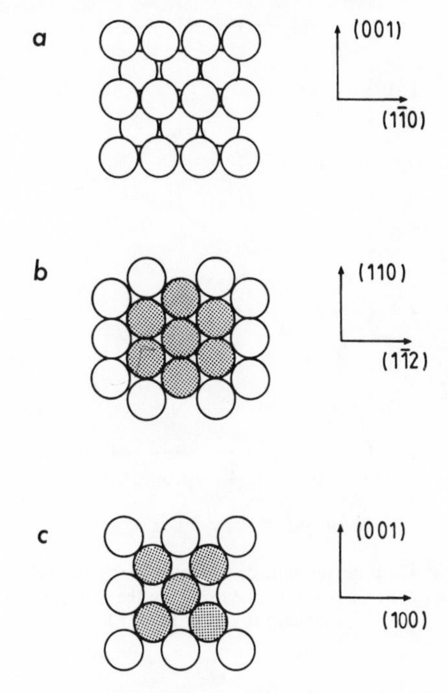

Fig. 16. Structure of f.c.c crystals; (a) top view of a Cu(110)-surface; (b) and (c) view of bulk planes normal to the (110)-surface plane and containing the (110) and (100) directions, respectively.

neutron scattering data for bulk Cu along the (110) and (100) directions are shown, this causes the restoring force along the (110) direction to be stronger (and thus the frequency to be larger) in the middle of the Brillouin zone than at the zone boundary, whereas in the (100) direction the frequency increases monotonically. Due to the non-monotonic behavior of the dispersion curve along the (110) direction, the projected phonons on a bulk plane have a large density of states in the range 21 meV $\lesssim$ hω $\lesssim$ 26.5 meV and a very low density below 21 meV (see Fig. 18, upper part). This allows a localized mode to peal off at the surface (surface density of states shown in the lower part of Fig. 18). Due to the low, but still not vanishing density of states, this mode turns to be a surface resonance.

The frequency of the MS_7 mode is well suited to give some fix points for the lattice dynamical calculation. This is obvious by inspection of Fig. 19 which shows the displacement pattern of this mode at the $\bar{\Gamma}$ point (at $\bar{\Gamma}$ the MS_7 phonon corresponds to the A_1 symmetry group). The motions of the atoms being shear vertical, the lattice layers remain rigid planes; i.e. the frequency of

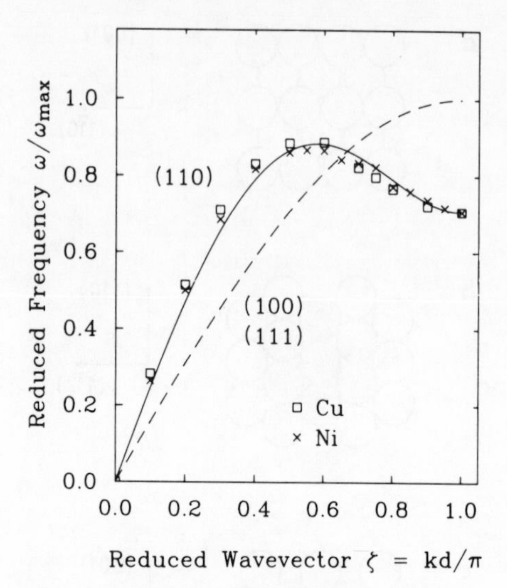

Fig. 17. Bulk phonon dispersion of longitudinal modes in Cu and Ni crystals, along the (110) and (100) directions. (*After Ref. 37.*)

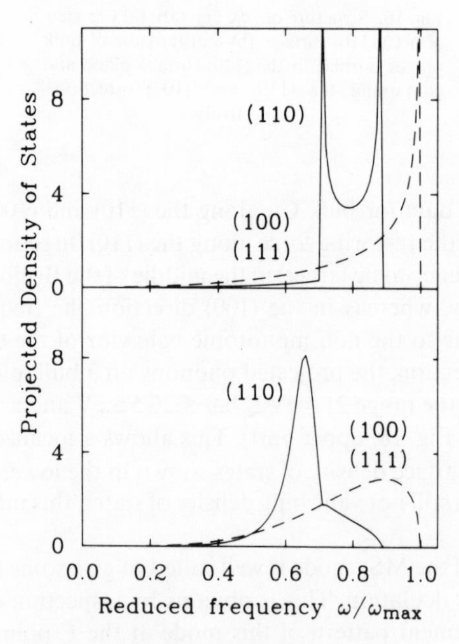

Fig. 18. Longitudinal bulk phonon density of states projected on bulk (top) and surface layers (bottom). (*After Ref. 36.*)

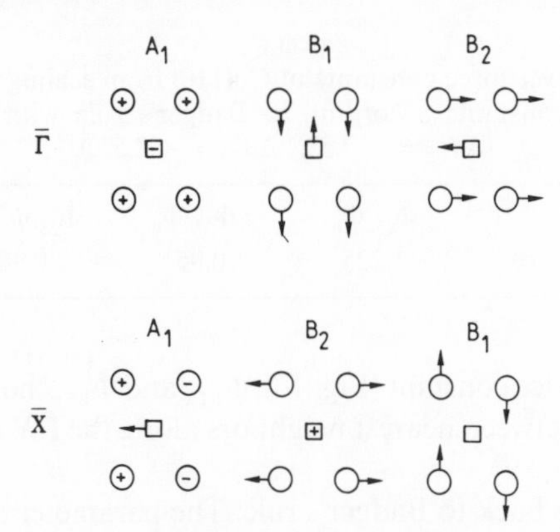

Fig. 19. Displacement patterns of the Cu(110) surface eigenmodes at $\bar{\Gamma}$ and $\bar{X}$.

the mode, in a simple nearest-neighbor central force model, is only affected by the interlayer force constant. Thus, we can determine the coefficient α in Badger's rule (Eq. (14)) by fitting the MS_7 frequency[38].

Before doing so, let us say a few words to the lattice dynamical model we have used to analyze the data. The lattice dynamics of the Cu atoms are treated in the harmonic approximation using pair potentials. The bulk phonon dispersion curves of Cu are well described by a single nearest-neighbor force constant model[39]. The potential energy term in the model Hamiltonian is based on nearest-neighbor interactions and central forces. Equilibrium conditions then require that the first derivative of the pair potential vanishes, i.e. each atom has it equilibrium position at the minimum of the pair potential connecting the atom to its next neighbor. However, in section 3.1 we have seen that upon creation of a surface, the electron charge density between the first few surface layers and also in between the atoms of a particular layer is rearranged. This charge rearrangement leads to the already noted interlayer relaxation, but also to an intralayer surface stress by modifying the pair potential in that its first derivative Φ'_{ij} becomes nonzero at equilibrium. The surface relaxation is actually the response to the interlayer stress; a relaxation of the atom positions in the surface plane which would be induced by the intralayer stress would be in fact a reconstruction. It is also obvious that surface stress affects the frequency of phonon modes with polarizations perpendicular to the stress (in order to increase the pitch of a violin string we have to tighten it). Lehwald *et al.*[38] were the first to account for this lateral surface stress in a lattice dynamical study of Ni(110). In this model, the stress along the high symmetry directions is given by the stress tensors:

$$\tau_{\bar{\Gamma}\bar{X}} = \Phi'_{11x}/a \tag{21}$$

$$\tau_{\bar{\Gamma}\bar{Y}} = \Phi'_{11y}\sqrt{2}/a \tag{22}$$

TABLE 2

Intralayer force constants of Cu(110) from scaling of the force constants according to Badger's rule with $\alpha = 8$ ($\Delta d_{12} = -7.5\%$, $\Delta d_{23} = +2.5\%$).

$\Phi^{\parallel}_{12}/\Phi^{\parallel}_{b}$	$\Phi^{\parallel}_{13}/\Phi^{\parallel}_{b}$	$\Phi^{\parallel}_{23}/\Phi^{\parallel}_{b}$	$\Phi^{\parallel}_{24}/\Phi^{\parallel}_{24}/\Phi^{\parallel}_{b}$
1.158	1.225	0.951	0.905

with a being the lattice constant (Fig. 13), Φ'_{11x} and Φ'_{11y} the first derivatives of the pair potential between nearest neighbors along the $\bar{\Gamma}\bar{X}$ and $\bar{\Gamma}\bar{Y}$ directions, respectively.

Let us now come back to Badger's rule. The parameter α is expected to lie somewhere between 5 and 10. By fitting the experimental MS_7 frequency of $\hbar\omega = 19.4$ meV with the lattice dynamical model described above, we find

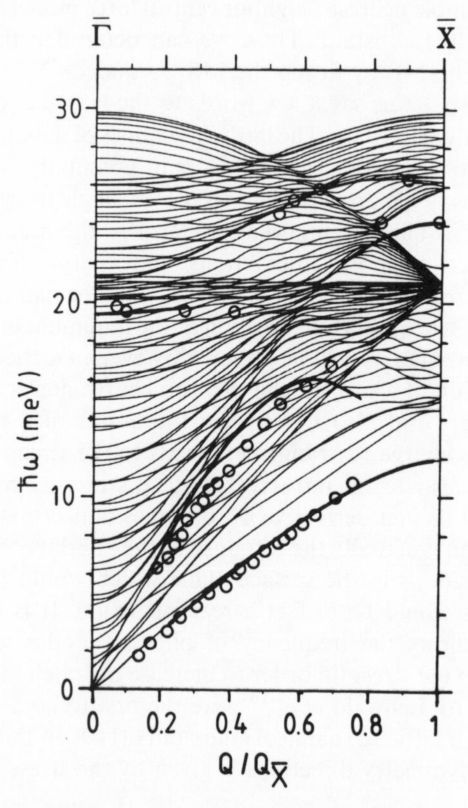

Fig. 20. Surface phonon dispersion curves of Cu(110) along the $\bar{\Gamma}\bar{X}$ azimuth calculated for a 30-layer slab with $\alpha = 8$; the data points are from Fig. 15.

a best-fit exponent $\alpha = 8$. This value is in good agreement with the relaxation analysis of Baddorf *et al.*[20,22] (Eq. (15)). In Table 2 we summarize the various force constants obtained with this value of the exponent α: The bulk force constant $\Phi_b'' = 2.73 \times 20^4$ dyn/cm is matched to the maximum bulk frequency $\hbar\omega_b = 29.7$ meV.

In Fig. 20 we show a theoretical dispersion plot using these parameters and a tensile stress $\tau_{\bar{\Gamma}\bar{X}} = 2.7 \times 10^3$ dyn/cm. Due to the symmetry of the modes at $\bar{X}$ the stress tensor $\tau_{\bar{\Gamma}\bar{Y}}$ does not affect the surface eigenmodes at this symmetry point. In addition, we have softened the intralayer force constant Φ_{11}'' in the first layer by about 10%. With these parameters, we find good agreement between experimental data and theoretical dispersion curves.

As already noted, the lowest mode S_1 is the Rayleigh wave. Throughout the Brillouin zone this mode exhibits significant vertical polarization. The resonance MS_0 along the edge of the longitudinal acoustic bulk modes has a large parallel component in the first third of the Brillouin zone. With increasing wavevector the vertical component increases and becomes dominant in the middle of the Brilliuin zone. In the last third, polarization changes again and at the $\bar{X}$ point the mode is a pure longitudinal mode. This is the reason why the MS_7 mode is observed in experiment only in the first part of the Brillouin zone. The MS_7 resonance is nearly dispersionless and mainly vertically polarized. At reduced wavevectors 0.5–0.6 the resonance interacts with the MS_0 resonance producing an 'avoided crossing' with an interchange of character. In the bulk gap near the $\bar{X}$ point the resonance develops in a well-defined surface mode S_7 which is mainly longitudinally polarized.

3.4. Au(111), the influence of reconstruction on surface phonon dispersion

The (111) surfaces of the noble metals Cu, Ag, Au and Pt were the first metal surfaces to be studied systematically by means of inelastic He scattering[40–44]. Figures 21 and 22 show a series of He energy transfer spectra taken from Cu(111) and Au(111) along the $\bar{\Gamma}\bar{M}$ azimuth[41]. In the case of Cu(111) in all spectra at least two energy losses are resolved. The surface phonon dispersion curves obtained from numerous energy loss spectra, are shown in Fig. 23 for the (111) surface of Cu, Ag and Au by the black dots. The theoretical dispersion curves, also shown, have been calculated by Jayanthi *et al.* using a new approach[45] and will be discussed below. The low energy branch is the surface Rayleigh wave (S_1). The high energy phonon branch is located within the projected bulk phonon band. This feature is not consistent with a lattice dynamical model assuming a simple geometrical termination of the bulk crystal[12]. The feature is much too soft to be ascribed to the longitudinally polarized acoustical mixed mode[40]. Note that the same discrepancy is observed for the Ag and Pt(111) surface along this azimuth. All three surfaces (Cu, Ag, Pt) exhibit a similar anomalously soft acoustic phonon branch along the $\bar{\Gamma}\bar{M}$-azimuth.

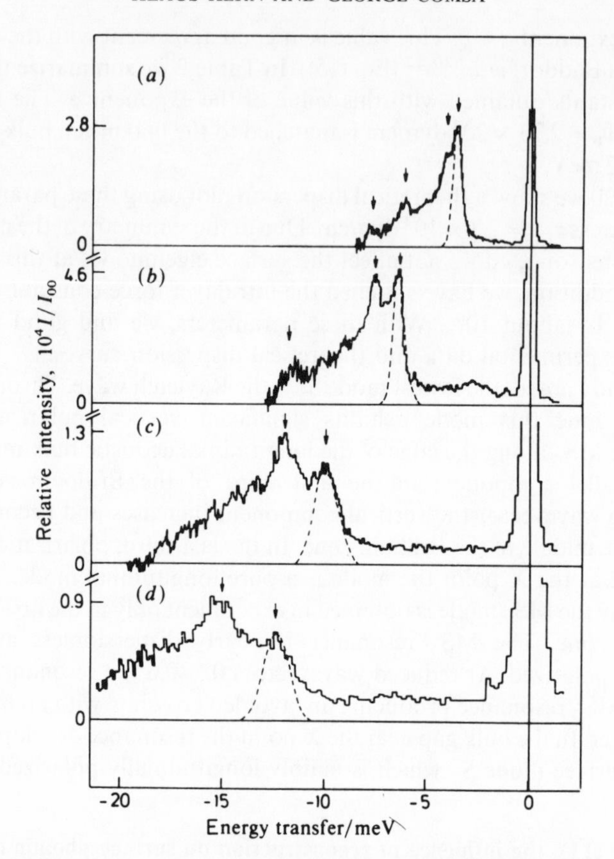

Fig. 21. He energy loss spectra taken from Cu(111) along the $\bar{\Gamma}\bar{M}$ azimuth. (a) $E_i = 12.7$ meV, $\vartheta_i = 38°$, (b) $E_i = 18.2$ meV, $\vartheta_i = 35°$, (c) $E_i = 31.5$ meV, $\vartheta_i = 35°$, (d) $E_i = 41.4$ meV, $\vartheta_i = 35°$. (*After Ref. 41.*)

It has been shown[46] that in the framework of a phenomenological force-constant model, it is possible to generate the anomalous phonon branch by assuming a substantial reduction of the in-plane surface force constant with respect to the bulk value. This force constant reduction causes a significant frequency decrease of the longitudinal phonon modes at the zone boundary. Bortolani *et al.*[46] have attributed this force constant reduction to the spilling of charge outside the surface and the resulting modification of the screening of the surface atoms. The force constant reduction necessary to fit the data amounted to about 50% (Cu, Ag) and 60% (Pt).

In the case of Au(111), the He energy loss spectra exhibit significantly broader inelastic peaks. Only in a few cases the existence of two components can be inferred. This behavior is only partly due to the lower Debye

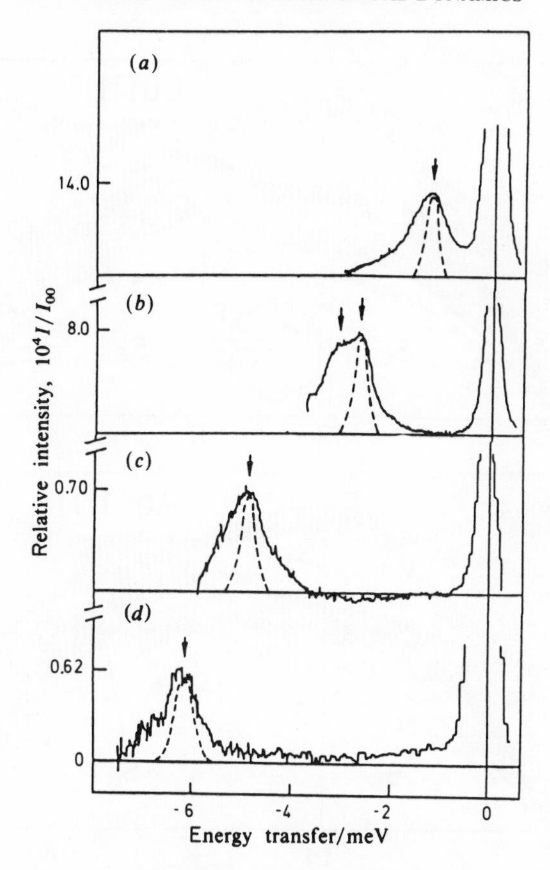

Fig. 22. He energy loss spectra taken from Au(111) along the $\bar{\Gamma}$ $\bar{M}$ azimuth. (a) $E_i = 12.2\,\text{meV}$, $\vartheta_i = 41°$, (b) $E_i = 12.2\,\text{meV}$, $\vartheta_i = 36°$, (c) $E_i = 12.2\,\text{meV}$, $\vartheta_i = 28°$, (d) $E_1 = 14.6\,\text{meV}$, $\vartheta_i = 24°$. (*After Ref. 41.*)

temperature (multiphonons). As will be shown in section 4.2.3, the clean Au(111)-surface is reconstructed with a $(23 \times \sqrt{3})$ unit cell. This reconstruction has a marked influence on the surface phonon displacement field, which becomes obvious by inspection of Fig. 23. Two extremely soft acoustic phonon branches, one of which is the Rayleigh wave, exist below the transverse bulk band edge. No acoustic resonance mode within the projected bulk bands is observed.

The model of Jayanthi *et al.*[45] overcomes the phenomenological force-constant models and thus avoids the large number of hypothetical force constants, sometimes used in these calculations. Jayanthi *et al.* calculate the charge density in each unit cell by an expansion over many-body interactions, which arise from the coupling of the electronic deformations to

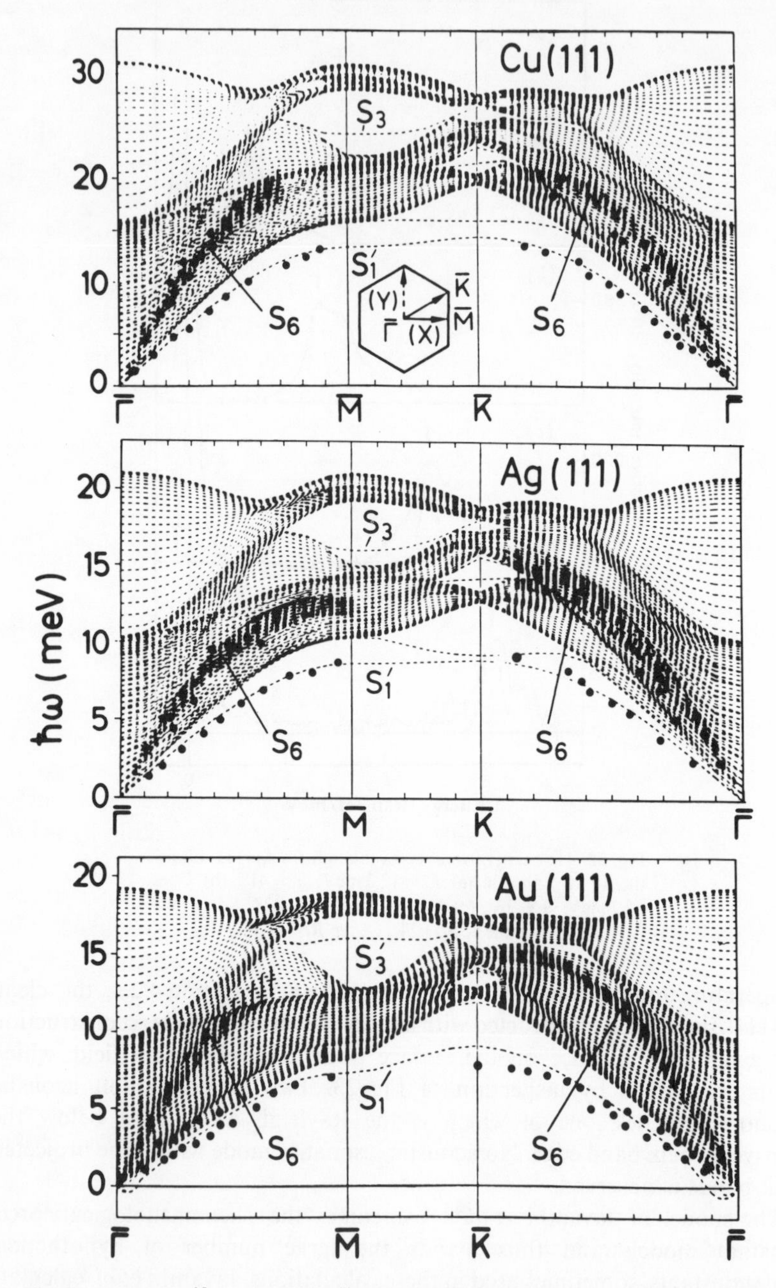

Fig. 23. Calculated[45] and measured[41] surface phonon dispersion curves of the (111) surfaces of the noble metals Cu, Ag and Au. (*After Ref. 45.*)

the displacement field. This model is appealing because of the small number of parameters and because all parameters have a direct microscopic meaning and may be obtained from *ab initio* calculations.

The calculated Rayleigh mode (S_1), the lowest lying phonon branch, is in good agreement with the experimental data of Harten *et al.*[41] for all three metals. Due to symmetry selection rules the shear horizontal mode just below the transverse bulk band edge can not be observed by scattering methods. The mode denoted by S_6 is the anomalous acoustic phonon branch discussed above. Jayanthi *et al.* ascribed this anomalous soft resonance to an increased Coulomb attraction at the surface, reducing the effective ion–ion repulsion of surface atoms. The Coulomb attraction term is similar for all three metals ($\sim -4 \times 10^4$ erg/cm^2); the repulsive forces are, however, larger in Au than in Cu and Ag (~ 3.5 and $\sim 2.5 \times 10^4$ erg/cm^2, respectively), preventing the development of a *soft* S_6 resonance on the Au(111) surface. In addition, Jayanthi *et al.* deduced for the Au(111) soft surface phonons with imaginary frequencies (in the first tenth of the Brillouin zone), indicative for surface reconstruction (see section 4.2.3). The S_3 mode in the band gaps were not detected by Harten *et al.*[41] for the metals Cu, Ag, Au, but have been detected by Kern *et al.*[43] for Pt.

3.5. Dynamical coupling between adsorbate and substrate

Despite the high energy resolution and extreme surface sensitivity only few studies on the dynamics of adsorbate covered surfaces have been performed so far by He scattering[47–51]. However, the vibrational states of adsorbates are relevant for most dynamical surface processes like scattering, accommodation, desorption, or diffusion, and therefore, their nature and relaxation dynamics deserve more attention.

In chemisorbed systems, the molecular orbitals of the adsorbate are mixed with the electronic states of the substrate, producing strong adsorption bonds, i.e. the frequency of the adsorbate mode is well above the highest phonon frequency of the substrate. The relaxation of these vibrational excited states via emission of substrate phonons has only a low probability, because many phonons have to be emitted during the decay. Non-radiative damping by electron–hole pair excitation appears to be the dominant relaxation path in these systems.

In physisorbed systems, the electronic ground state of the adsorbate is only weakly perturbed upon adsorption. The physisorption potential is rather flat and shallow, i.e. the restoring force of the vertical motion of adatoms is weak, and thus, the corresponding adsorbate–substrate vibrations are low-frequency modes[50,51]. Radiative phonon processes are expected to dominate the relaxation and coupling processes.

Hall, Mills and Black[52] have explored the phonon-mediated coupling between physisorbed rare gas atoms and Ag substrate atoms in the framework of a simple lattice dynamical model. The observed coupling effects can best be illustrated with the normal mode spectrum of a physisorption system, shown in Fig. 24. At the zone boundary of the Brillouin zone the adlayer mode which is assumed to be dispersionless in accordance with experimental observations lies well below the substrate modes. Approaching the zone center, the adlayer mode inevitably intersects the substrate Rayleigh wave and finally, close to the zone center, the adlayer mode and the projected substrate bulk bands overlap. The results of the calculations show that near the zone boundary $\bar{M}$ (the $\bar{\Gamma}\bar{M}$ direction has been explored), where the substrate phonon frequencies are well above those of the adlayer, the influence of the substrate–adlayer coupling is small.

The predicted anomalies introduced by the coupling near the zone center $\bar{\Gamma}$ are twofold:

1. A dramatic hybridization splitting around the crossing between the dispersionless adlayer mode and the substrate Rayleigh wave (and a less dramatic one around the crossing with the $\omega = c_1 Q_\parallel$ line – due to the Van Hove singularity in the projected bulk phonon density of states).
2. A substantial linewidth broadening of the adlayer modes in the whole region near $\bar{\Gamma}$ where they overlap the bulk phonon bands of the substrate: the excited adlayer modes may decay by emitting phonons into the substrate; they become leaky modes. These anomalies were expected to extend up to trilayers even if more pronounced for bi- and in particular for monolayers.

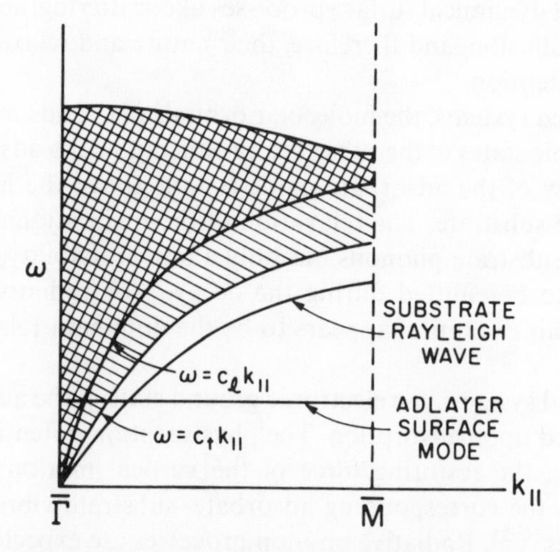

Fig. 24. Schematic normal mode spectrum of a rare gas monolayer physisorbed on a single crystal surface. (*After Ref. 52.*)

Experimental data of Gibson and Sibener[53] appears to confirm qualitatively these predictions at least for monolayers. The phonon linewidths were broadened around $\bar{\Gamma}$ up to half of the Brillouin zone. The hybridization splitting could not be resolved, but an increase of the inelastic transition probability centered around the crossing with the Rayleigh wave and extending up to 3/4 of the zone has been observed and attributed to a resonance between the adatom and substrate modes.

Recent measurements performed on Ar, Kr and Xe layers on Pt(111)[54] with a substantially higher energy resolution ($\Delta E \leqslant 0.4$ meV) have now confirmed quantitatively the theoretical predictions on the coupling effects within almost every detail (the hybridization around the van Hove singularity has only been seen in the case of Ar). The sequence of He TOF spectra in Fig. 25 taken along the $\bar{\Gamma}\bar{K}$-direction of a full Kr monolayer at 25 K gives a vivid picture of the coupling effects. The last spectrum $\vartheta_i = 37°$, taken near the zone boundary $\bar{M}$ exhibits a unique, sharp loss $\Delta E \approx -3.7$ meV resulting from the creation of an Einstein Kr monolayer phonon (perpendicular Kr–Pt vibration); its width corresponds to the instrumental width of $\Delta E \simeq 0.38$ meV, there is no linewidth broadening near the zone boundary. On the other hand, the Kr phonon peak in the first spectrum $\vartheta_i = 40°$ taken near the $\bar{\Gamma}$ point and located at $\Delta E \simeq -3.9$ meV is broadened by more than 0.5 meV. Of particular interest is also the small peak at $\Delta E \simeq -3.1$ meV, close to the position of the Pt substrate Rayleigh wave. The next two spectra, $\vartheta_i = 39.5°$ and 39°, taken always closer to the crossing between the Pt substrate Rayleigh wave and the Kr Einstein mode demonstrate strikingly the effect of the hybridization of the two modes: the originally tiny Pt peak increases dramatically, while the Kr peak is pushed slightly toward larger energies. After surpassing the crossover the higher energy loss disappears abruptly. As predicted[52], the two features in the doublet have comparable intensity only quite near the crossover.

Figure 26 shows the dispersion curve of the Kr monolayer obtained from a large number of spectra like those in Fig. 25. The hybridization splitting around the crossing with the substrate Rayleigh wave (solid line) is clearly observed. Also the predicted tiny frequency upshift close to the $\bar{\Gamma}$ point due to the coupling to the substrate vibrations is seen. The observed linewidth broadening is also shown in Fig. 26. As a measure of the broadening, the quantity $\Delta\varepsilon = [(\delta E)^2 - E_I^2]^{1/2}$, with δE the FWHM of the major loss feature and E_I the intrinsic instrumental broadening ($E_I = 0.38$ meV in the present experiment), is plotted as a function of the wavevector. A broadening larger than 0.5 meV is seen, and – as predicted – confined to the region near $\bar{\Gamma}$, where the adlayer mode overlaps the bulk bands of the substrate.

It is noteworthy that the phonon anomaly, due to the dynamical coupling between substrate Rayleigh wave and adlayer mode, is likewise present in the bi- and even the trilayer films[54]. It is only the Q range of the anomaly which

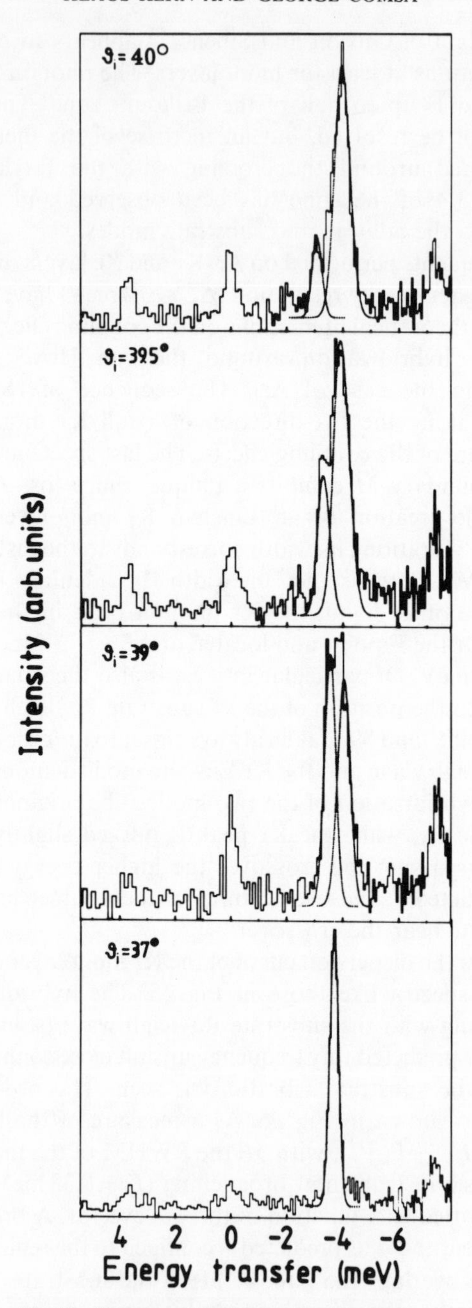

Fig. 25. He energy loss spectra from a Kr monolayer taken along the $\bar{\Gamma}\,\bar{K}_{Kr}$ azimuth. With decreasing incident angle ϑ_i, phonons with larger wavevectors are sampled.

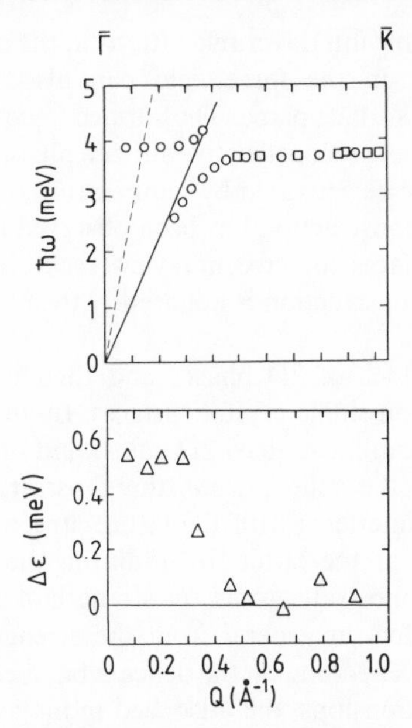

Fig. 26. Experimental dispersion curve of the Kr monolayer and measured line width broadening $\Delta\varepsilon$ of the Kr creation phonon peaks. The solid line in the dispersion plot is the clean Pt(111) Rayleigh phonon dispersion curve and the dashed line the longitudinal phonon bulk band edge of the Pt(111) substrate, both in the $\bar{\Gamma}\ \bar{M}_{Pt}$ azimuth which is coincident with the $\bar{\Gamma}\ \bar{K}_{Kr}$ azimuth.

becomes smaller, and its location shifts towards the zone center together with the location of the intersection between the Kr and the substrate Rayleigh mode. Linewidth broadening, due to radiative damping into the substrate bulk bands, has been found to be still substantial for bilayer films, while the trilayer shows no evidence for additional broadening.

4. SURFACE PHASE TRANSITIONS

4.1. Ordering in two dimensions

Solid surfaces of single crystals provide to some extent the realization of a well-defined two-dimensional (2D) periodic array of atoms. However, the loss of vertical translational invariance at the surface changes the local force field with respect to the bulk forces. As seen in section 3 the charge redistribution is responsible for noticeable interlayer relaxations in the near surface region. In addition, the changes in the force field can also favor lateral atomic rearrangements in the surface plane. The surface 'reconstructs' into a phase with new symmetry. These reconstructive surface phase transitions can either occur spontaneously or be activated by temperature or by small amounts of adsorbates. Surface reconstruction has been observed on a number of metal and semiconductor surfaces (for a recent review, see Refs. 55 and 56). The most thoroughly studied reconstruction is apparently the (7×7) reconstruction of the Si(111) surface[57].

Even closer to the idea of 2D phases and their mutual transitions are adsorbed monolayers on single crystal surfaces. In analogy to bulk matter, these adsorbed layers can form quasi 2D gas, liquid or solid phases[58,59]. Of particular interest are the solid phases; the substrate provides a periodic potential relief which interferes with the lattice structure of the monolayer, inducing modulations in the latter. In addition, the adatoms being of a different kind than the substrate atoms, the strength of the lateral interactions within the adlayer differs in general from the strength of the adsorbate–substrate interaction. Depending on the delicate balance between these forces and on structural relationships, the adsorbed monolayer can form ordered solid structures which are commensurate (in registry) or incommensurate (out of registry) with the substrate.

4.1.1. Critical fluctuations and defects

In contrast to bulk phase transitions, phase transitions at surfaces are quite often high-order transitions. According to Ehrenfest, phase transitions are termed high order if the higher but not the first derivative of the free energy becomes discontinuous; i.e. high-order phase transitions exhibit a continuous change of state from one phase to the other, in contrast to first-order transitions, where a sudden rearrangement occurs.

Continuous transitions are characterized by a critical 'behavior' of a physical observable, which is the order parameter γ of the transition. Above the transition temperature T_c the thermodynamic average of the order parameter $\langle \gamma \rangle$ is zero, indicative of complete loss of long-range order. Below T_c the value of $\langle \gamma \rangle$ is nonzero indicating long-range order, and follows a power law when approaching the critical temperature:

$$\langle \gamma \rangle \sim (T_c - T)^{\beta} \tag{23}$$

The order parameter can be the magnetization in ferromagnetic materials, the electron pair amplitude in superconducting materials, the He^4 amplitudes in superfluid He^4 or the lattice distortion in crystals.

Such continuous phase transitions are conveniently described in a phenomenological Landau free-energy expansion of the order parameter. Since we are dealing with two-dimensional lattices, two-component order parameters have to be choosen, which are parameterized by an amplitude and a phase. While phase fluctuations dominate the low-temperature behavior of two-dimensional phases, amplitude fluctuations dominate at higher temperatures when approaching the critical temperature. Indeed, phase fluctuations, in form of long-wavelength phonons, are responsible for the supression of a genuine long-range order in two-dimensional solid, at all temperatures $T > 0\,\mathrm{K}$. Amplitude fluctuations, which are always present and dominate at the phase transitions at high enough temperatures, appear in two-dimensional systems in the form of defects, in particular as dislocations in two-dimensional solids. Such dislocations (often termed domain walls or solitons) result by adding or removing a half-infinite row of atoms from an otherwise perfect lattice. They play a central role in phase transitions of quasi-two-dimensional systems, in the melting transition as well as in registry–disregistry transitions.

The dominance of fluctuations in lower-dimensional systems can also be understood by simpler arguments. The order of a phase is thermodynamically determined by the free energy, i.e. by the competition between energy and entropy. In three-dimensional systems each atom has a large number of nearest neighbors (e.g. twelve in an f.c.c. crystal); thus the energy term stabilizes an ordered state, local fluctuation being of minor importance. In one dimension, however, each atom has only two nearest neighbors. Here the entropy term dominates the energy term, and even very small local fluctuations destroy the order. In a close-packed two-dimensional system each atom has six nearest neighbors and, depending on temperature, energy and entropy may be in balance. As amplitude fluctuations, i.e. topological defects, can be excited thermally, the two-dimensional systems seem to be ideally suited for studying defect-mediated phase transitions.

4.2. Soliton structure of incommensurate phases

4.2.1. Solitary lattice distortion waves

The essential properties of incommensurate modulated structures can be studied within a simple one-dimensional model, the well-known Frenkel–Kontorova model[60]. The competing interactions between the substrate potential and the lateral adatom interactions are modeled by a chain of adatoms, coupled with harmonic springs of force constant K, placed in a cosine substrate potential of amplitude V and periodicity b (see Fig. 27). The microscopic energy of this model is:

$$H = \sum_n \left[\frac{K}{2}(x_{n+1} - x_n - a_0)^2 \right] + \sum_n V \left[1 - \cos\left(\frac{2\pi}{b} x_n \right) \right] \qquad (24)$$

where x_n is the position of the nth atom. In the absence of the periodic potential

($V = 0$) the distance between nearest-neighbor atoms in the chain is a_0, which, in general, is incommensurate with the substrate lattice constant b. Frank and van der Merwe[61] (FvdM) solved this model analytically within a continuum approximation. They replaced the index n by a continuous variable and x_n by a continuous function $\varphi(n)$. The calculations of FvdM showed that for slightly differing lattice parameters of chain and substrate potential, the lowest energy state is obtained for a system which consists of large commensurate domains, separated by regularly spaced regions of bad fit (Fig. 27). The regularly spaced regions of bad lattice fit are called misfit dislocations, solitons or domain walls. They can be regarded as collective long period lattice distortion waves, which are excitations of the commensurate ground state.

In the continuum approximation of FvdM the microscopic energy can be written:

$$H = \int \left[\frac{Kb^2}{8\pi^2} \left(\frac{\mathrm{d}\varphi}{\mathrm{d}n} - 2\pi\delta \right)^2 + V(1 - \cos(p\varphi)) \right] \mathrm{d}n \tag{25}$$

where $\delta = (a_0 - b)/b$ is the natural misfit of the periodicites of the adatom chain and of the substrate potential, p the commensurability and $\varphi(n) = (2\pi x/b) - 2\pi n$ the phase shift of the chain atom with respect to the potential minimum.

The Hamiltonian, Eq. (25), can be minimized exactly. The ground state satisfies the time-independent sine Gordon equation:

$$\frac{\mathrm{d}^2\varphi}{\mathrm{d}n^2} = pA \sin(p\varphi) \tag{26}$$

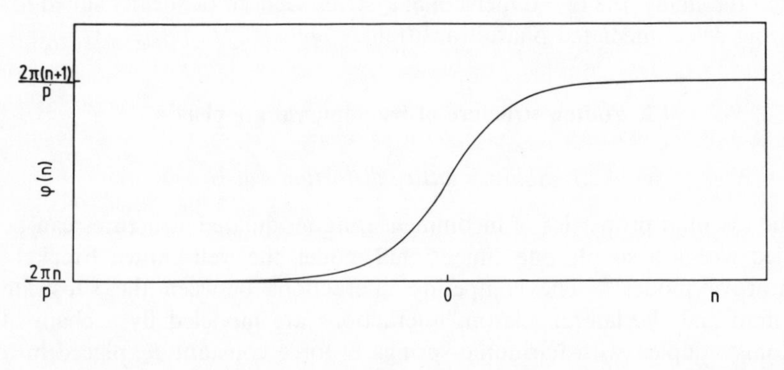

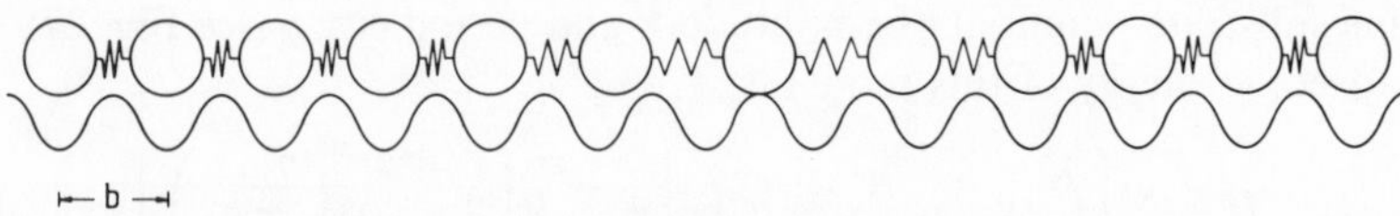

Fig. 27. Single soliton solution of the 1D Frank–van der Merve model of incommensurate monolayers.

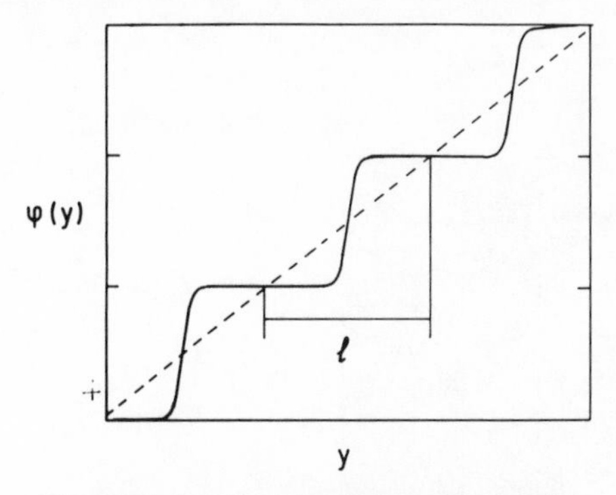

Fig. 28. Soliton lattice solution of the FvdM model with regularly spaced, distance l, domain walls. The dashed line corresponds to the incommensurate phase with negligible potential modulation ($V = 0$). The plateaus represent the commensurate domains.

with

$$\sqrt{A} = 2\pi/b\sqrt{V/K}$$

One solution of this equation is the solitary lattice distortion, the so-called soliton:

$$\varphi(n) = \theta(n) = \frac{4}{p}\arctan\left[\exp\left(pn\sqrt{A}\right)\right] \tag{27}$$

The soliton solution is shown in Fig. 27; it describes a domain wall located at $n = 0$ separating two adjacent commensurate regions. The width of the domain wall in this model is $L_0 = 1/(p\sqrt{A})$. In general, the solutions are regularly spaced solitons (Fig. 28). The distance between solitons, l, is determined by the lateral interaction between neighboring walls which is mediated by the lateral atomic displacements in the chain. At large distances from a wall, $\sin\varphi$ is small and Eq. (26) may be linearized, yielding an exponential decay. Therefore, the repulsive interaction between neighboring walls is also exponential. The phase function of the regularly spaced solitons have the functional form (Fig. 28)

$$\varphi = \varphi_0 + \frac{2\pi f}{p} + \theta(y - fl) \tag{28}$$

with f being the closest integer to y/l and $y = nb$.

4.2.2. The commensurate–incommensurate transition of monolayer Xe on Pt(111)

In two-dimensional systems, walls are lines with finite width. In a triangular lattice there are three equivalent directions. Therefore, domain walls can cross

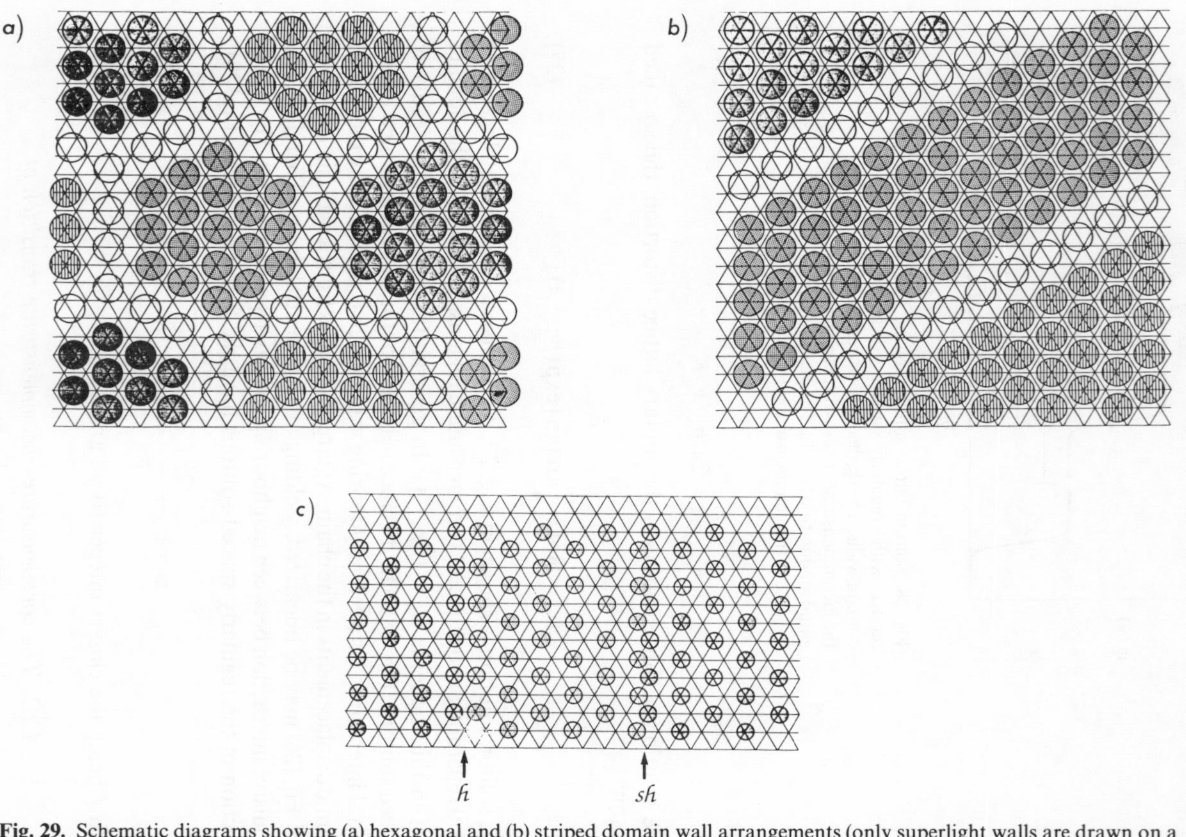

Fig. 29. Schematic diagrams showing (a) hexagonal and (b) striped domain wall arrangements (only superlight walls are drawn on a triangular lattice (e.g. the (111) face of f.c.c. metals)). In incommensurate layers, where the monolayer is compressed with respect to the commensurate lattice, domain walls or either heavy or superheavy (c).

(Fig. 29). Using Landau theory, Bak *et al.*[62] (BMVW) have shown that it is the wall crossing energy Λ which determines the symmetry of the weakly incommensurate phase and the nature of the phase transition:

1. $\Lambda < 0$, i.e. attractive walls. A hexagonal network of domain walls (HI) will be formed at the commensurate (C–I) transition, because the number of wall crossings tends to be as large as possible. This C–HI transition should be first order.
2. $\Lambda > 0$, i.e. repulsive walls. The number of walls tends to be as small as possible, i.e. a striped network of parallel walls (SI) will be formed in the incommensurate region. The C–SI transition should be continuous. The striped phase is expected to be stable only close to the C–I transition. At larger misfits the hexagonal symmetry should be recovered in a first-order SI–HI transition.

The FvdM as well as the BMVW model neglects thermal fluctuation effects; both are $T = 0$ K theories. Pokrovsky and Talapov[63] (PT) have studied the C–SI transition including thermal effects. They found that, for $T \neq 0$ K the domain walls can meander and collide, giving rise to an entropy-mediated repulsive force of the form $F \sim T^2/l^2$, where l is the distance between nearest neighbor walls. Because of this inverse square behavior, the inverse wall separation, i.e. the misfit m, in the weakly incommensurate phase should follow a power law of the form

$$m = \frac{1}{l} \sim \left(1 - \frac{T}{T_c}\right)^{1/2} \tag{29}$$

In a recent He diffraction study[64] we have shown that the adsorption system Xe/Pt(111) is dominated by the existence of a $(\sqrt{3} \times \sqrt{3})$ R30° commensurate phase, shown schematically in Fig. 30. The C-phase has been found to be stable in an extended temperature (62 K–99 K) and coverage range ($\theta_{Xe} \lesssim 1/3$).

The maximum coverage in this $(\sqrt{3} \times \sqrt{3})$ R30° commensurate structure is obviously $\theta_{Xe} = 1/3$ ($\theta_{Xe} = 1$ corresponds to 1.5×10^{15} atoms/cm^2, the density of Pt atoms in the (111) plane). Only one-third of the adsorption sites are occupied, i.e. there exist three energetically degenerate commensurate sublattices. The commensurate Xe lattice being expanded by about 9% with respect to the 'natural' Xe-lattice, the coverge can be increased beyond $\theta_{Xe} = 1/3$. Obviously, above this limit the adatoms cannot all occupy preferred adsorption sites, and the adlayer becomes incommensurate. Due to anharmonic effects[64], the Xe adlayer becomes incommensurate also at $\theta_{Xe} < 1/3$ upon decreasing the temperature below ≈ 62 K. This has been predicted for an analogous case, Kr/graphite, by Gordon and Villain[65], but has not yet observed experimentally in that case.

Before discussing the experimental results of the C–I transition of Xe on Pt(111) in detail, let us have a look at the diffraction patterns expected from a

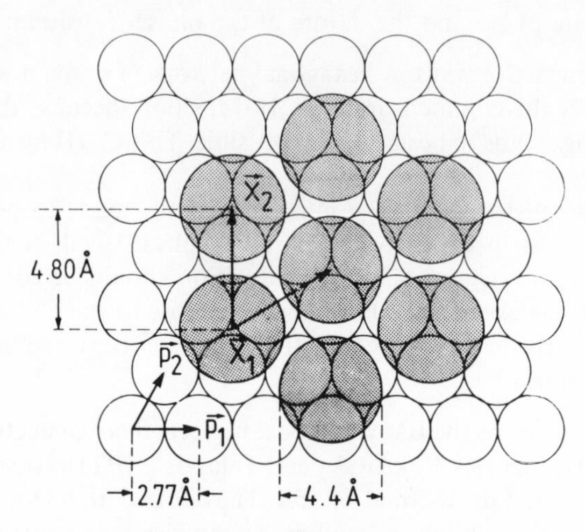

Fig. 30. Commensurate $(\sqrt{3} \times \sqrt{3})R30^{\circ\circ}$ Xe monolayer adsorbed on Pt(111).

striped (SI) and from a hexagonal incommensurate (HI) phase. The structure and the corresponding schematic diffraction patterns are shown in Fig. 31; the diffraction patterns have been calculated for fully relaxed walls, i.e. the SI and the HI phase are in fact uniaxially and uniformly compressed phases, respectively. By inspection of Fig. 31 it is obvious that the various incommensurate structures can easily be identified by their characteristic diffraction patterns.

We discuss first the basic crystallography of the incommensurate Xe phase on Pt(111) as deduced from the measured patterns[66]. Figure 32 shows the $(2,2)_{Xe}$ and $(1,2)_{Xe}$ diffraction features obtained from a Xe layer during the C–I transition induced by cooling below 62 K at constant coverage $\theta_{Xe} \simeq 0.30$. The plots have been obtained by monitoring series of azimuthal scans (i.e. constant Q scans in the reciprocal space). The comparison with Fig. 31 shows that the incommensurate Xe layer on Pt(111) is a striped phase (SI) with a uniaxial compression in the $\bar{\Gamma}\bar{M}$ direction. Indeed, a three-peak structure for the $(2,2)_{Xe}$ diffraction feature, with the doublet located at $Q^{2,2}_{comm} + 0.048\ \text{Å}^{-1}$ and the single peak located at $Q^{2,2}_{comm} + 0.90\ \text{Å}^{-1}$ is observed (with $Q^{2,2}_{comm} = 3.02\ \text{Å}^{-1}$); whereas the $(1,2)_{Xe}$ pattern consists of a single peak at the commensurate position and a shallow doublet with the maximum intensity at about $Q^{1,2}_{comm} + 0.13\ \text{Å}^{-1}$ (with $Q^{1,2}_{comm} = 2.62\ \text{Å}^{-1}$). The observed incommensurability deduced from the well-defined polar location of the peaks in Fig. 32a is $\varepsilon = 0.095\ \text{Å}$ and corresponds to an inter-row distance in the $\bar{\Gamma}\,\bar{M}$ direction of $d_{SI} = 3.91\ \text{Å}$. This results in a misfit $m = 1 - d_{SI}/d_C = 0.059$, where $d_C = 4.80$

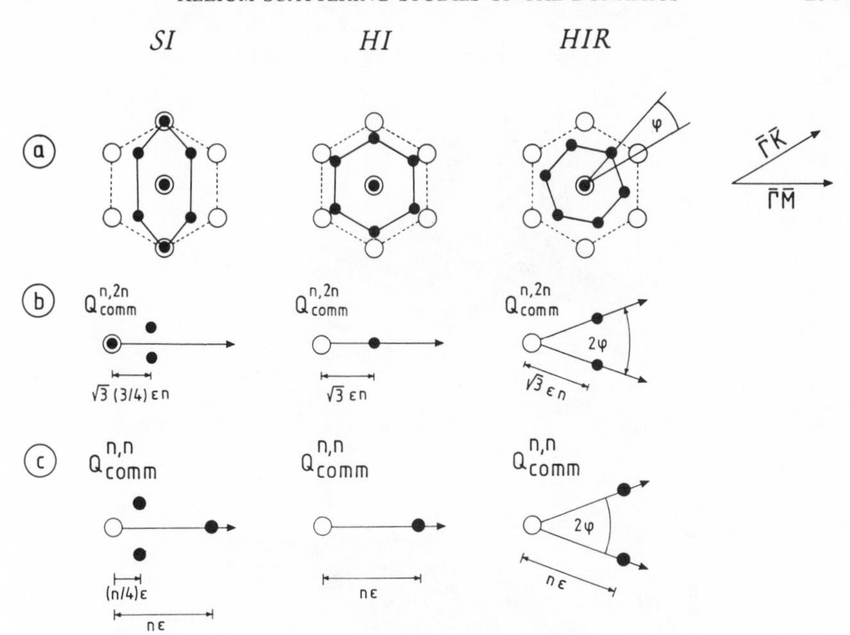

Fig. 31. Schematics of (a) real lattice and (b), (c) the (n, n) and $(n, 2n)$ diffraction features of incommensurate layers. SI – striped incommensurate, HI – hexagonal incommensurate, HIR – hexagonal incommensurate rotated. All phases are assumed to be fully relaxed. ○ denotes the $(\sqrt{3} \times \sqrt{3})$R30° commensurate and ● the incommensurate structures.

$\times \cos 30°$ Å is the inter-row distance of the commensurate Xe structure in the same direction. From the measured polar and azimuthal peak widths in Fig. 32 we can also estimate average domain sizes of the incommensurate layer. For the $\bar{\Gamma}\bar{K}$ direction, i.e. parallel to the walls, we obtain ~ 350 Å and for the perpendicular $\bar{\Gamma}\bar{M}$ direction ~ 50 Å.

The analysis in the last paragraph has shown that the incommensurate Xe layer on Pt(111) at misfits of about 6% is a striped phase with fully relaxed domain walls, i.e. a uniaxially compressed layer. For only partially relaxed domain walls and depending on the extent of the wall relaxation and on the nature of the walls (light, heavy or superheavy) additional statellites in the (n, n) diffraction patterns should appear. Indeed, closer to the beginning of the C–I transition, i.e. in the case of a weakly incommensurate layer (misfits below ~ 4%) we observe an additional on-axis peak at $Q^{2,2}_{comm} + \varepsilon/2$ in the (2, 2) diffraction pattern. In order to determine the nature of the domain walls we have calculated the structure factor for the different domain wall types as a function of the domain wall relaxation[67] following the analysis of Stephens *et al*[68]. The observed additional on-axis satellite is consistent with the occurrence of superheavy striped domain walls; the observed peak intensities indicate a domain wall width of $\lambda \simeq 3$–5 Xe inter-row distances. With

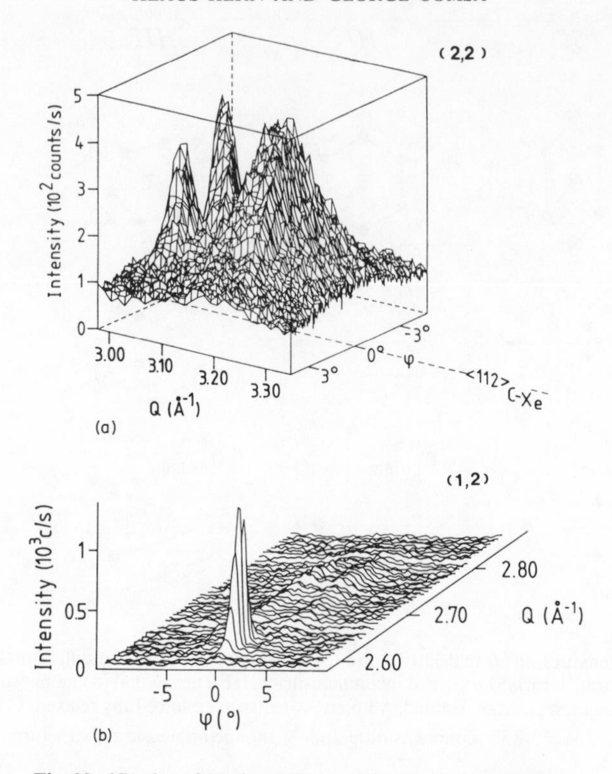

Fig. 32. 3D plot of (a) the $(2,2)_{Xe}$ and (b) the $(1,2)_{Xe}$ diffraction features during the C–I transition at $T = 54$ K ($\theta_{Xe} \approx 0.30$). Q denotes the wavevector in the $\bar{\Gamma}\ \bar{M}$ and $\bar{\Gamma}\ \bar{K}$ directions, respectively, while φ denotes the azimuthal angle.

increasing incommensurability the total length of the domain walls is expected to increase, giving rise to smaller and more numerous commensurate domains[69]. For misfits larger than 4–5%, i.e. where the interwall distance becomes less than three times the wall width, the diffraction pattern of the striped incommensurate layer can no longer be distinguished from an unaxially compressed layer.

In Fig. 33 the misfit during the temperature-induced C–I transition is plotted as a function of the reduced temperature. As long as the phase is weakly incommensurate the data can be fitted by a power law: $m = m_0(1 - T/T_c)^{\beta}$. The best least square fit parameters are $T_c = 61.7$ K, $m_0 = 0.18$ and $\beta = 0.51 \pm 0.04$. The value of β is in agreement with the Pokrovsky–Talapov prediction. Only data points up to misfits of about 4% have been included in the fit. The cutoff $\sim 4\%$ has been choosen in accordance with Erbil *et al.*[70], who have found the $\beta = 1/2$ power law to be valid only in this range for bromine intercalated graphite. For larger values, the misfit variation with reduced temperature is roughly linear; in this region the interwall distance is of

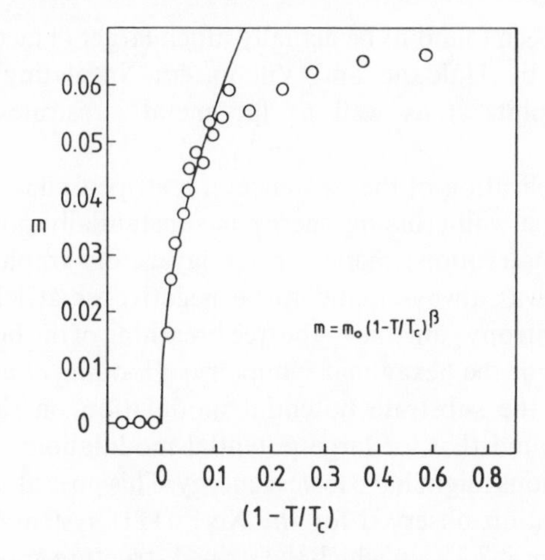

Fig. 33. $\bar{\Gamma}\ \bar{M}$ uniaxial misfit m versus reduced temperature during the C–SI transition. The solid line represents the power law fit (see text).

order of the wall width and thus the Pokrovsky–Talapov theory is not applicable.

Due to the instability of the hexagonal weakly incommensurate phase with respect to the formation of free dislocations (re-entrant melting) the C–HI transition of Kr on graphite[71] (which like Xe/Pt(111) exhibits a $(\sqrt{3} \times \sqrt{3})$ R30° commensurate structure) has been found to be a melting transition. In contrast, the C–SI transition of Xe/Pt(111) is a solid–solid transition with the incommensurability simply related to the domain wall density. According to Coppersmith *et al.*[72] striped structures are stable if the number of sublattices, p (here, $p = 3$), is larger than $\sqrt{8}$, whereas for hexagonal structures the criterion is $p > 7.5 \pm 1.5$. As mentioned, the critical exponent $\beta = 0.51 \pm 0.04$ deduced from the data in Fig. 33 is in agreement with the $\beta = 1/2$ prediction of Pokrovsky and Talapov. The Pokrovsky–Talapov model may essentially be applied to a substrate of uniaxial symmetry; however, the original model calculations are performed for an isotropic substrate and thus applicable to the isotropic Pt(111) substrate. On the other hand, Haldane and Villain[73] pointed out that in the case of rare-gas monolayers on metal surfaces, substrate-induced electric dipole interactions might be responsible for the square root law. Moreover, they inferred that even in the case of an insulating substrate (no induced dipole forces) the square root behavior should be valid, but only for very small misfits ($m < 0.001$!). At present, it is difficult to make a choice between the thermal fluctuation mechanism of Pokrovsky and Talapov and the substrate-induced dipole mechanism of Haldane and Villain. However, it is worth noting that the experimental range of validity of the square root law in 2D striped domain wall phases has been found to be actually much larger (a factor of ≈ 30) than the limit given by Haldane and Villain, for 'insulating' substrates (Br intercalated graphite[46]) as well as for metal substrates (Xe/Cu(110)[74], Xe/Pt(111).[66]

The direct implication of the existence of a striped phase in Xe layers on Pt(111) is that the wall-crossing energy is substantially positive. This is at variance with observations made for Kr layers on graphite[71], where the crossing energy was always found to be negative or at least only slightly positve (so that entropy gain due to the free breathing of the honeycomb lattice is sufficient to favor the hexagonal symmetry). Gooding *et al.*[75] have studied the influence of the substrate potential modulation on the different wall energies. They found that for large potential modulations striped arrays of discommensurations might have lowest energy. This goes along with the large potential modulation observed for the Xe/Pt(111) system[76]. The extended misfit range ($0 < m < 7.2\%$) in which the striped structure appears to be stable, is somewhat puzzling in view of recent theoretical results by Halpin-Healy and Kardar[77]. They have studied the occurrence of striped structures in the 'striped helical Potts lattice gas model'. Their results reveal a strong correlation between the extent of the striped phase regime and the wall thickness. Striped structures in an extended coverage range should appear only for 'sharp' domain walls; with increasing wall thickness this range is expected to shrink substantially. The energy cost due to the wall repulsion seems to be too large for thick walls. They conclude that the wall width of 4–5 inter-rows in Kr monolayers on graphite[78] might be responsible for the absence of a striped phase in this system. The wall width in Xe layers on Pt(111) is of the same size; the coverage range in which the striped phase is found to be stable corresponds in Halpin-Healy and Kardar's calculations to walls widths of 1–2 inter-rows.

When increasing the misfit of Xe monolayers on Pt(111) above 6.5% an additional an-axis peak at $Q^{1,2}_{\text{comm}} + \sqrt{3}\varepsilon$ appears in the (1, 2) diffraction spots (Fig. 34). This marks the transition from the striped to the hexagonal

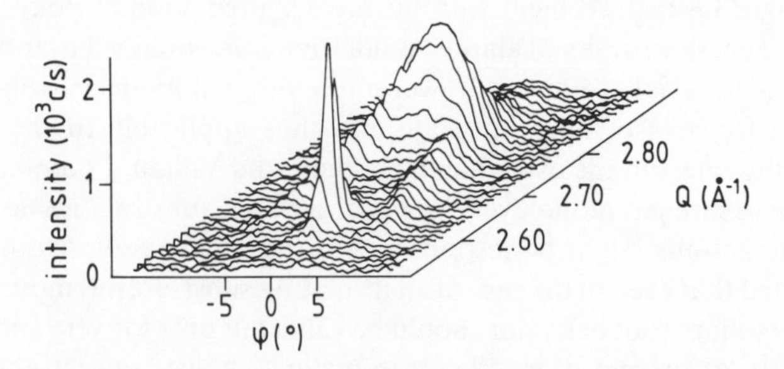

Fig. 34. 3D plot of the $(1,2)_{\text{Xe}}$ diffraction feature of an incommensurate Xe monolayer on Pt(111) at a misfit of 7.0% ($\Theta_{\text{Xe}} \simeq 0.35$, $T = 25\,\text{K}$).

incommensurate phase. Such diffraction patterns composed of a peak at $Q^{1,2}_{\text{comm}}$ and a doublet at $Q^{1,2}_{\text{comm}} + (3/4)\sqrt{3}\varepsilon$ originating from a SI phase, and an on-axis peak at $Q^{1,2}_{\text{comm}} + \sqrt{3}\varepsilon$ originating from a HI phase, are observed in the misfit range $6.5\% \lesssim m \lesssim 7.2\%$, with the HI peak intensity progressively increasing with coverage. We conclude that the SI phase transforms at misfits of $\sim 6.5\%$ to a HI phase in a first-order transition.

Xe on Pt(111) appears to be the first 2D system fully consistent with the BMVW theory, i.e. the first system displaying the full sequence of C → SI → HI transitions with increasing incommensurability. Of course, the BMVW theory is a $T = 0$ K theory neglecting thermal fluctuation effects. However, Halpin-Healy and Kardar[79] have studied recently the various domain wall phases in the framework of a generalized helical Potts model, including finite temperature effects and two species of domain walls. Their results were in general agreement with the BMVW theory; in particular, they pointed out that the C → SI → HI sequence only occurs when assuming repulsive heavy and superheavy wall crossings. This is confirmed by the Xe/Pt(111) system.

4.2.3. The uniaxial soliton reconstruction of Au(111)

It is well known that all low-index clean gold surfaces exhibit reconstructions, in that the outermost layer has always a density larger than the corresponding bulk layers[55,56]. The genuinely rectangular (100) and (110)

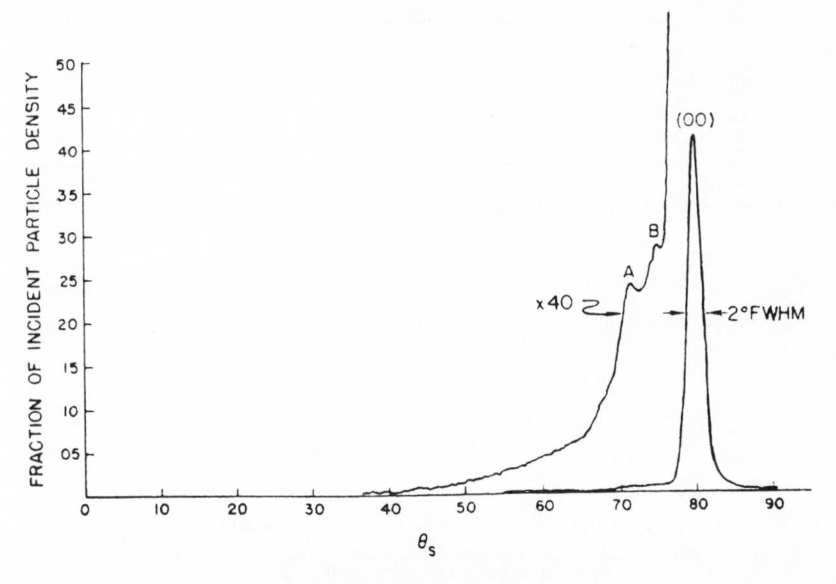

Fig. 35. He diffraction from Au(111) with a primary He beam energy $E_i = 11$ meV ($\vartheta_i = 80°$, $T_S = 145$ K). (*After Ref. 80.*)

outermost Au layers reconstruct to hexagonal close-packed structures, while on the Au(111) surface the density increases by a lateral contraction of the outermost atom layer.

The first experimental evidence for a uniaxial soliton reconstruction of the Au(111) surface was obtained in 1977 by Miller and Horne[80]; unfortunately, they missed the significance of the data. They scattered He beams with energies of 11–63 meV from Au(111) surfaces. Figure 35 shows a polar He diffraction scan with a beam energy of 11 meV, at a fixed incident angle $\vartheta_i = 80°$ obtained

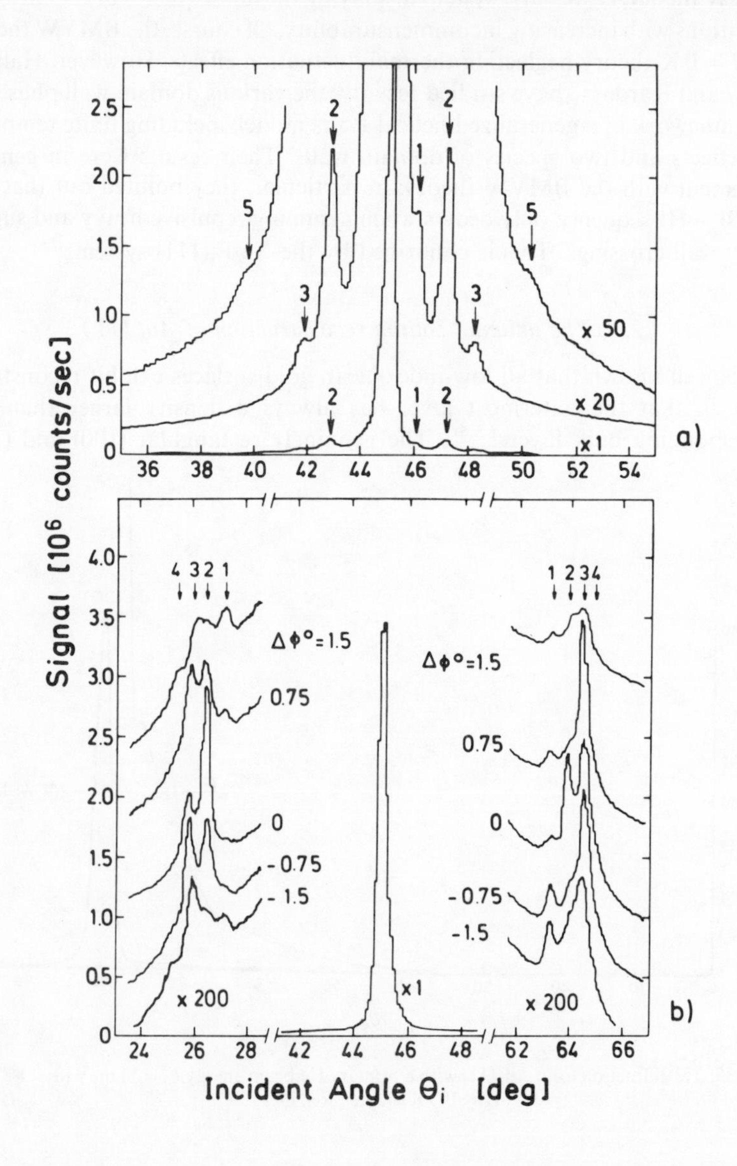

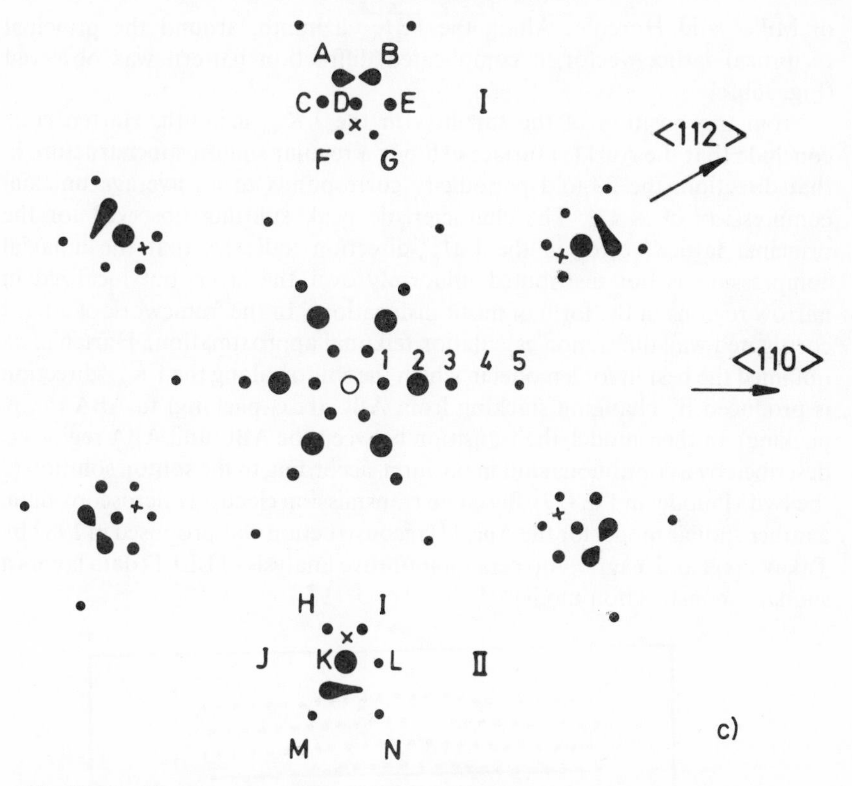

Fig. 36. He diffraction scans taken from Au(111) along (a) the $\bar{\Gamma}\ \bar{K}_{Au}$ azimuth ($E_i = 7.95$ meV) and (b) the $\bar{\Gamma}\ \bar{M}_{Au}$ azimuth ($E_i = 16.6$ meV). (c) Full diffraction pattern shown schematically. (*After Ref. 81.*)

by varying the scattering angle ϑ_f. Despite the relatively poor monochromaticity and angular resolution ($\Delta E/E \approx 0.2$ and angular divergence 0.8°) the polar profile shows two peaks, located at 71.4° (A) and 74.8° (B), corresponding to wavevectors $Q \approx 0.18\,\text{Å}^{-1}$ and $0.09\,\text{Å}^{-1}$. Miller and Horne assigned these peaks to single phonon exchanges between the He atoms and the Au surface.

In 1985 Harten *et al.*[81] repeated these measurements with better monochromaticity and angular resolution ($\Delta E/E \approx 0.02$ and angular resolution $\approx 0.3°$). Their results revealed up to five satellite peaks around the specular beam along the $\bar{\Gamma}\bar{K}_{Au}$ azimuth. Figure 36 shows polar He scans along the $\bar{\Gamma}\bar{K}_{Au}$ and $\bar{\Gamma}\bar{M}_{Au}$ azimuth, taken with beam energies of 7.95 meV and 16.6 meV, respectively. In the $\bar{\Gamma}\bar{K}_{Au}$ scans (a), the satellite peaks are located at wavevectors $Q_n \simeq n0.098\,\text{Å}^{-1}$, with a particularly strong second-order ($n = 2$) peak. This corresponds to a super-periodicity of ≈ 64 Å. This is very close to the value of the reconstruction model given by Takayanagi and Yagi[82]. Note that the peaks 1 and 2 of Harten *et al.*[81] are the peaks A and B in the notation of Miller and Horen[80]. Along the $\bar{\Gamma}\bar{M}_{Au}$ azimuth, around the principal reciprocal lattice vector, a complicated diffraction pattern was observed (Fig. 36b, c).

From the position of the satellites in the $\bar{\Gamma}\bar{K}_{Au}$ azimuth, Harten *et al.* conclude that the Au(111) surface exhibits a regular soliton superstructure in that direction; the 23-fold periodicity corresponds to an average uniaxial compression of $\sim 4\%$. The characteristic peak splitting observed for the principal lattice vector in the $\bar{\Gamma}\bar{M}_{Au}$-direction indicates that the uniaxial compression is not distributed uniformly over the layer, but localized in narrow regions in the form of misfit dislocations. In the framework of a hard corrugated wall diffraction calculation (eikonal approximation), Harten *et al.* obtained the best fit for a model in which the soliton along the $\bar{\Gamma}\bar{K}_{Au}$-direction is produced by changing stacking from ABC (f.c.c. packing) to ABA (h.c.p. packing). In their model, the transition between the ABC and ABA regions is described by a continuous shift in position, according to the soliton solution of the FvdM model in Eq. (27). Based on transmission electron microscopy data, a rather similar model for the Au(111) reconstruction was proposed in 1982 by Takayanagi and Yagi[82]. A recent quantitative analysis of LEED data favors a similar reconstruction model[83].

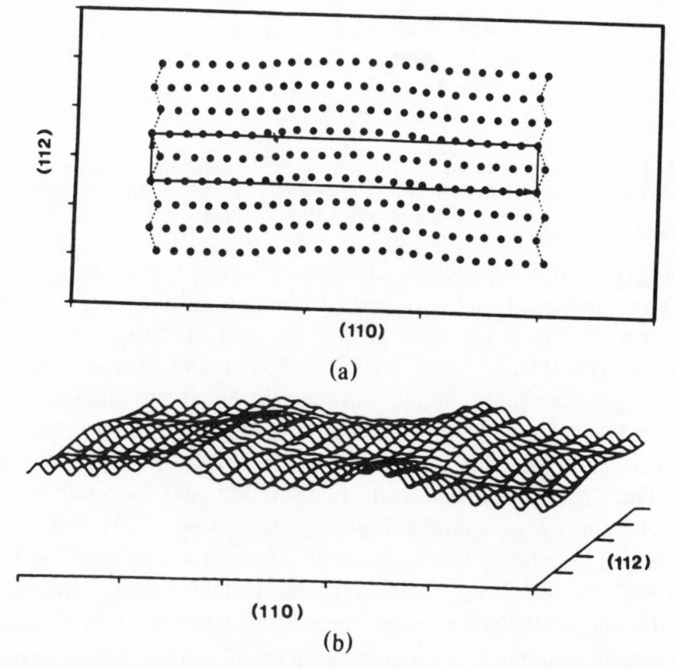

Fig. 37. (a) Atomic arrangement (top view) of the Au(111) surface atoms in the double sine Gordon model, (b) corrugation of the corresponding unit cell. (*After Ref. 84.*)

TABLE 3
Surface energy of reconstructed Au(111) surfaces according to the 'glue model' of Ercolessi *et al.*[85].

Surface arrangement	Surface energy (meV $Å^{-2}$)
Au(111) bulk, unrelaxed	105.1
Au(111) relaxed, non-reconstructed	96.1
Au(111) $(7 \times \sqrt{3})$	90.4
Au(111) $(11 \times \sqrt{3})$	88.1
Au(111) $(23 \times \sqrt{3})$	90.2

El-Batanouny *et al.*[84] have recently analyzed the He diffraction data of Harten *et al.* in a refined soliton model. For the Au(111) surface, the A and C sites are not expected to be degenerated; a continuation of the Au-bulk structure would favor C-sites. To account for this nondegeneracy, El-Batanouny *et al.* have introduced the double-sine Gordon (DSG) equation to model the Au(111) reconstruction. Using molecular dynamics techniques they have solved this model and obtained an atomic arrangement of the surface atoms shown in Fig. 37. Hard corrugated wall diffraction calculations based on this atomic arrangement agree well with the experimental data.

The reconstruction of the Au(111) surface has also been subject to theoretical studies. Ercolesi *et al.*[85] have explored the reconstruction in a phenomenological many-body Hamiltonian, an approach which is known as the 'glue model'. In this model, the many-body forces minimize classically the non-directional cohesive forces of the filled d-bands, appearing as the main term in enforcing optimal coordination of all atoms. The procedure is empirical in nature and adjusted to reproduce the macroscopic properties of solid and liquid gold, as well as surface formation energies. The optimum structure of the surface is obtained by an 'annealing' procedure based on molecular dynamics techniques. Within this model, the (111) surface of gold does exhibit reconstruction. In qualitative agreement with the experiment, the surface energy is minimized by a soliton-like reconstruction, with a smooth transition from f.c.c. to h.c.p. stacking, with a unit cell length of $(L \times \sqrt{3})$. However, in contrast to the experimental findings ($L = 23$) the theoretical optimal value is $L = 11$ (see Table 3).

4.3. Surface reconstruction and soft phonons, W(100)

A basic concept in the reconstruction theory of solid surfaces is the soft phonon approach of displacive structural transitions. An essential property of these structural phase transitions is the existence of an order parameter which

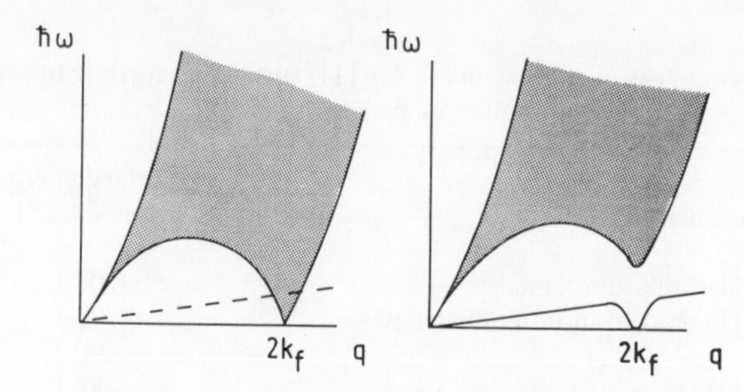

Fig. 38. 1D model of metal surface reconstruction. The coupling between discrete phonon modes (dashed line) and the electron–hole pair continuum (hatched) produces lattice instability.

represents a collective displacement of the atoms. At the reconstructive transition the surface lattice distorts via the softening of the corresponding phonon (the order parameter diverges), i.e. the configuration of atomic displacements changes continuously during the transition.

Several authors have suggested that these lattice instabilities may arise as a result of 'giant Kohn anomalies'[86,87]. The electronic energy of a metal surface may be lowered by periodic lattice distortion waves of wavevector $Q_{\mathrm{LDW}} = 2k_{\mathrm{f}}$, where k_{f} is the Fermi wavevector of the surface electrons. This is exemplified in Fig. 38 where the continuum of electron–hole pair excitations $\omega(q)$ with its characteristic low-energy regime around the Fermi vector is shown. The dashed line represents a longitudinal surface phonon. The phonon vibrations couple to the electrons for coincident phonon and electron–hole pair creation, giving rise to a mode repulsion which depresses the phonon energy near $2k_{\mathrm{f}}$ and creates a gap in the electron–hole pair band. In the case of sufficient coupling strengths, the phonon frequency becomes zero and the surface can reconstruct by freezing this particular lattice distortion.

The strength of the lattice instability near the Fermi vector depends on the magnitude of the electron–phonon coupling and on the phase space available for electron–hole pair excitation around $2k_{\mathrm{f}}$. Thus, a reconstructive surface phase transition has to fulfill the following requirements in order to be ascribed to an electronically driven lattice instability:

1. Surface states or resonances have to be present near the Fermi energy in the electronic surace band structure.
2. The Fermi surface must have large regions of flat parallel areas, i.e. it must exhibit strong nesting properties.
3. The symmetry of the electronic surface bands has to allow for strong electron-phonon coupling to distortions of appropriate symmetry.

The W(100) surface is suspected to be a representative for this kind of

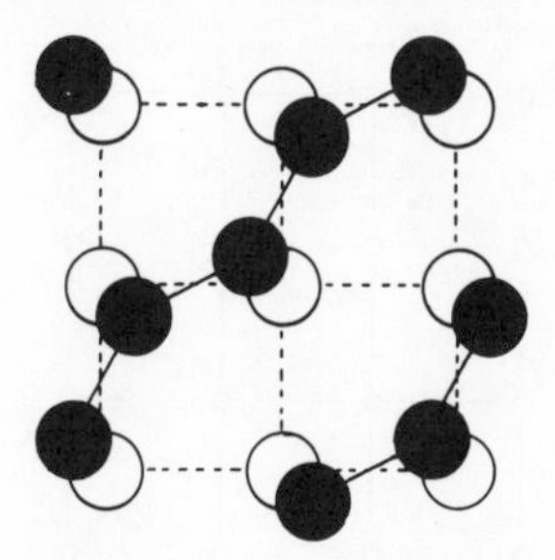

Fig. 39. Schematics of the W(001) p2mg surface reconstruction; the Debe–King model.

electronically driven reconstructions. Already in 1971 Yonehama and Schmidt[88] observed with LEED that the clean W(100) reconstructs upon cooling below room temperature. At low temperature, the reconstructed surface exhibits a sharp $c(2 \times 2)$ LEED pattern, i.e. extra Bragg spots in the (1/2, 1/2) position, which become weak and diffuse upon approaching the transition temperature.

A qualitative structural model of the reconstructed $c(2 \times 2)$ W(100) surface was first proposed by Debe and King[89] on the basis of symmetry arguments. Figure 39 shows this reconstruction model. The surface atoms exhibit only in-plane displacements along diagonal directions. A subsequent LEED structurel analysis of Barker *et al.*[90] supported this picture. In a more recent quantitative LEED analysis, Walker *et al*[91] deduced a lateral displacement of 0.16 Å at 200 K.

More recently, the reconstruction of the clean W(100) surface has also been studied by He diffraction[92,93]. These studies reveal a complex behavior during the transition. Only at temperatures below ~ 240 K sharp diffraction spots *centered* at the (1/2, 1/2) positions are observed. In the temperature range between ~ 400 K and ~ 240 K broad superstructure spots are observed which progressively shift to the (1/2, 1/2) position upon cooling. Lapujoulade and Salanon[92] explain this behavior in the framework of a domain wall model: reconstructed domains of various sizes are separated by dense domain walls, which disappear continuously upon cooling.

Detailed electronic energy-band calculations[94] have revealed the existence of appropriate surface states near the Fermi energy, indicative of an electronically driven surface instability. Angle-resolved photoemission studies, however, showed that the Fermi surface is very curved and the nesting is far from perfect[95]. Recently Wang and Weber[96] have calculated the surface phonon dispersion curve of the unreconstructed clean W(100) surface based on the first principles energy-band calculations of Mattheis and Hamann[94].

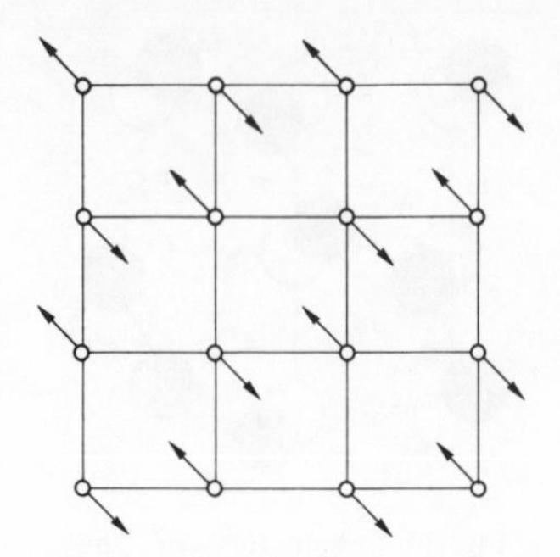

Fig. 40. Displacement pattern of the W(100) $\bar{M}_5$ surface phonon.

They found a whole bunch of soft phonons, which are primarily horizontally polarized, near the zone boundaries between $\bar{M}$ and $\bar{X}$. The most unstable mode they observed is the $\bar{M}_5$ phonon, the displacement pattern of which is shown in Fig. 40; note the similarity between this pattern and the reconstruction model in Fig. 39. According to Wang and Weber, these soft phonons are caused by electron–phonon coupling between the surface phonon modes and the electronic $\bar{\Sigma}_2$ surface states at the Fermi surface. They attributed the predominant M_5 phonon instability to an additional coupling between $d(x^2 - y^2)$ and $d(xy)$ orbitals of the Σ_2 states.

The $\bar{M}_5$ phonon instability as the cause of the $c(2 \times 2)$ reconstructionwas also predicted in the pioneering work of Tosatti and coworkers[97] and in a more recent 'frozen phonon' total energy calculation of Fu *et al.*[98]. In the approach of Tosatti *et al.* the reconstruction mechanism is treated by incorporating the electronic driving forces in extra forces between surface atoms, i.e. studying the reconstruction essentially as a lattice dynamical problem by introducing empirical force constants between nearest neighbors. For values of the surface force constants, corresponding to increased lateral attraction in the surface plane, the $\bar{M}_5$ phonon frequency becomes imaginary, i.e. the W(100) surface can reconstruct via the observed lattice distortion. In contrast to the Wang–Weber model, which includes long-range interatomic interactions, the model of Tosatti *et al.* is based on the short-range nature of the interaction.

More recently, Ernst *et al.*[93] have studied the surface phonon dispersion of W(100) along the $\bar{\Gamma}\bar{M}$-azimuth by means of high-resolution inelastic He scattering. Besides the Rayleigh phonon, they observed an additional acoustical low-frequency mode. This mode resulted in rather broad energy losses in the He time-of-flight spectra and has an unusual dispersion. Its energy increases initially with increasing wave-vector, reaches a maximum at about 2/5 of the Brillouin zone and softens upon further increase of the wavevector; it

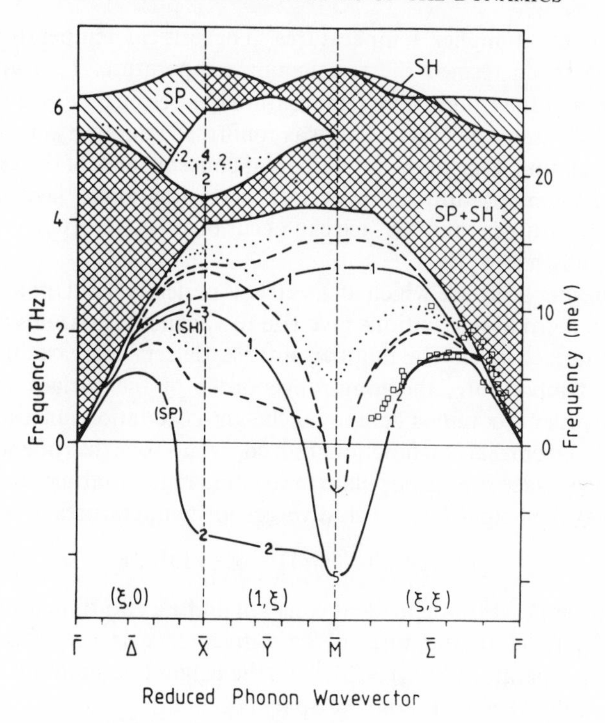

Fig. 41. Surface phonon dispersion of the unreconstructed W(100) surface. The theoretical dispersion curves are from Ref. 96; the data (□) are due to Ernst *et al.*[93].

eventually reaches zero energy near the zone boundary. Upon lowering the temperature, i.e. initiating the reconstruction, the Rayleigh mode does not change, while the soft phonon mode lowers slightly its energy. This behavior is consistent with the assertion that the soft mode is an in-plane mode, which might be connected with the reconstruction mechanism. Indeed, the Rayleigh mode being a vertical mode, we expect neither an anomalous dispersion nor a softening, because these forces are not important in driving the reconstruction. In Fig. 41 we have included the experimental data of Ernst *et al.*[92] in the theoretical dispersion curve of Wang and Weber[96]; the agreement between experiment and theory is reasonable.

4.4. Surface roughening

In their pioneering work on crystal growth Burton, Cabrera and Frank[99] predicted that on an atomic length scale the equilibrium structure of a crystal surface should exhibit a transition from a smooth state at low temperatures to a rough surface at higher temperatures. The critical temperature of this transition has been termed the roughening temperature, T_R. Burton *et al.* suggested that at the roughening temperature the free energy associated with the creation of a step vanishes[99]. This was confirmed later by Swendsen[100] in a detailed calculation. One of the fundamental consequences of the existence of a roughening temperature for a certain crystallographic face below the melting temperature is that this face can occur on an equilibrium crystal only at temperatures below T_R.

Let us consider a surface which at $T = 0$ K is perfectly flat. Upon raising the temperature, thermal fluctuations give rise to vacancies, adatoms atoms and steps in the surface layer. The number of these 'defects' increases until, at the roughening temperature, the long-range order of the surface disappears. Long-range order is confined here to the 'height-correlation function' and not to the positional correlation function. Indeed, even above the roughening temperature, the surface atoms populate in average regular lattice sites. It is the fluctuation of the height $h(r)$ which diverges for temperatures $T > T_R$[101,102]:

$$\langle [h(\vec{r}') + \vec{r}) - h(\vec{r})]^2 \rangle \propto C(T) \ln(\vec{r}) \tag{30}$$

where C is a temperature-dependent constant and $\vec{r}$ a two-dimensional vector in a plane parallel to the surface. This divergence is very weak. At the roughening temperature $C(T_R) = 2/\pi^2$ [102]; the height fluctuation is one lattice spacing for a distance of 139 lattice spacings.

The roughening transition has also been studied by computer simulation methods[103]. Figure 42 shows characteristic configurations of a f.c.c. (100) surface in the simple solid-on-solid (SOS) model, calculated by Gilmer[103]. The roughening temperature in this model corresponds to a parameter $kT/\Phi = 0.6$.

The first indication for the existence of a roughening transition was obtained by Jackson and Miller[104] who studied the crystal shape of chloroethane and ammonium chloride. Above 370 K and 430 K, respectively, they observed a drastic change in crystal morphology, which might be interpreted as roughening. Similar observations for adamantane have been reported by Pavlovska[105].

The first first direct experimental evidence for a roughening transition was reported in 1979. Several groups[106,107] have studied the thermal behavior of the basal plane of a hexagonal close-packed ^{4}He crystal. In a beautiful experiment Balibar and Casting[108] obtained for this surface a roughening temperature of $T_R \approx 1.2$ K.

More recently, the question of thermal roughening has also been addressed in the study of metal surfaces. Detailed He diffraction studies from the high Miller index (115) surface of Cu and Ni proved the existence of a roughening transition on these surface. These studies were performed by means of He scattering. Let us make first two short comments.

The microscopic mechanism which leads to the roughening of a low and a

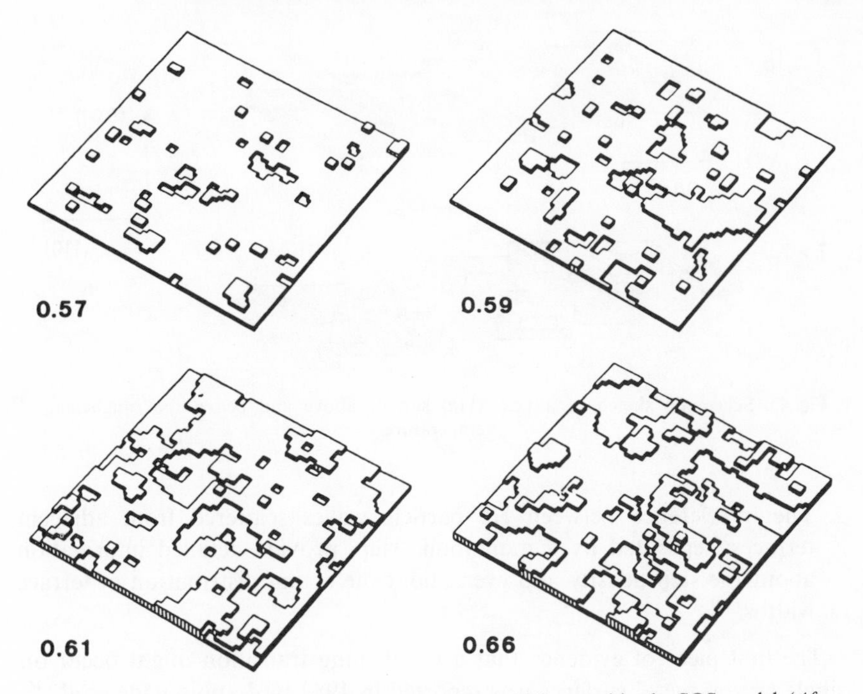

Fig. 42. Typical equilibrium configurations of a f.c.c. (100) crystal in the SOS model. (*After Ref. 103.*)

high Miller index surface is expected to be essentially different. Indeed, as already mentioned, a low indexed surface – which at $T = 0$ K is perfectly flat – fulfills the roughening condition, Eq. (30), when the free energy for the creation of a step becomes zero. In contrast, on a high indexed surface – which at $T = 0$ K is already stepped – Eq. (30) can be fulfilled also without the creation of new steps. It appears that the proliferation of kinks is sufficient to roughen the surface (Fig. 43). Indeed, the ensuing meandering of the step rows, in conjunction with the mutual repulsion between these rows, leads also to the divergence of the 'height-correlation function'. Thus, the roughening temperature of high indexed surfaces might be substantially lower than that of the low indexed ones.

Most studies of metal surface roughening have been performed with He scattering because two specific features make this probe particularly suited for the study of the roughening of single crystal surfaces:

1. Due to the large total cross-section of defects (about 150 Å^2 for a monovacancy on Pt(111)) He scattering is extremely sensitive to the presence of defects. In addition, this large cross-section enables one to study the lateral distribution of defects[6].

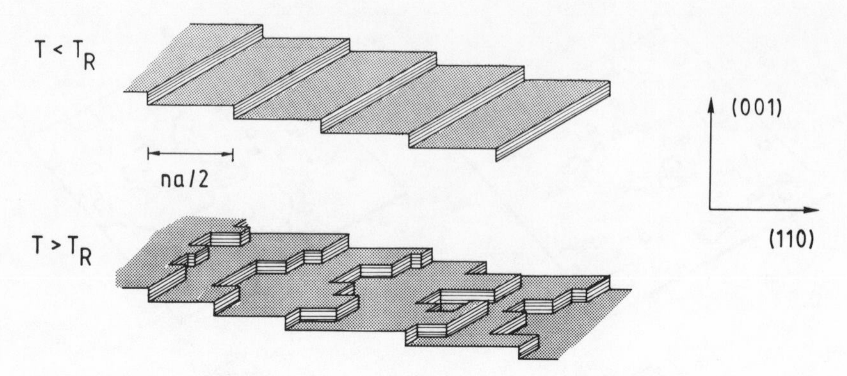

Fig. 43. Schematic sketch of a f.c.c. (11*n*) surface above and below the roughening temperature.

2. The interference between He particle waves scattered from adjacent terraces, separated by a monatomic step, provide detailed information about the step density and even about the actual distribution of terrace widths[109].

The first piece of evidence that a roughening transition might occur on single crystal metal surfaces was reported in 1982 by Lapujoulade *et al.*[110]. They observed a dramatic drop in the He intensities coherently scattered from the Cu(115) surface upon increasing the temperature above ~400 K (Fig. 44). This behavior contrasted with the generally observed, much weaker decrease of the coherent intensity, accounted for by the Debye–Waller factor[111]. Lapujoulade and coworkers claimed that this anomalous behavior was due to thermal roughening and assigned the temperature at which the data deviate from the Debye behavior to the roughening temperature. Such anomalous behavior, however, may have other causes (see below), but this assignment happened to be correct.

More sound experimental arguments for the existence of a roughening transition of high Miller index surfaces were reported three years later both by Engel's and Lapujoulade's groups[112–114]. In these studies, the thermal behavior of the He diffraction peak line shapes have been analyzed in the framework of simple roughening models, developed by den Nijs *et al.*[115] and by Villain *et al.*[116] Since the height correlation disappears when approaching T_R and since the diffraction intensity is simply related to the pair correlation function, the diffraction peak shape is expected to undergo characteristic changes at the roughening transition. Both theoretical models predict power-law lineshapes for the He diffraction peak. In particular, for the tail of the specular (0, 0) peak they infer that

$$I(Q_{\parallel}) \propto Q_{\parallel}^{\eta} \tag{31}$$

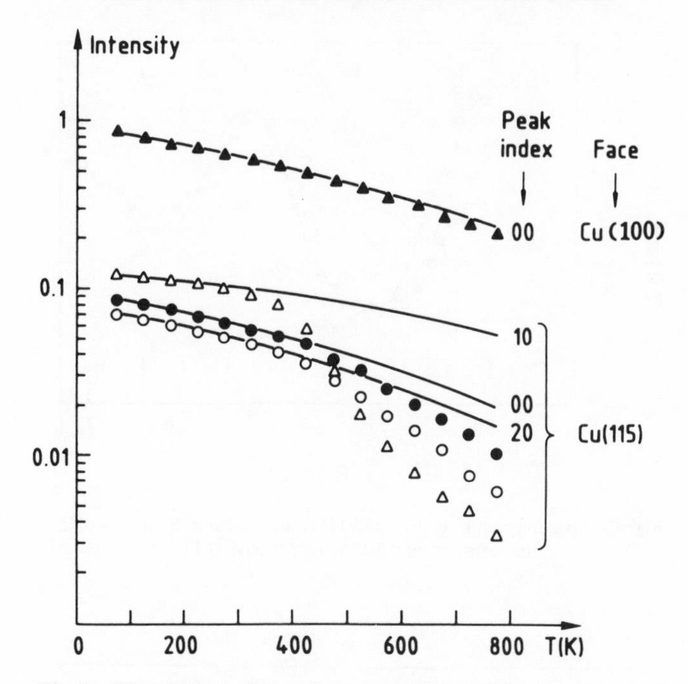

Fig. 44. Thermal dependence of coherently diffracted He intensities from Cu(100) and Cu(115). (*After Ref. 112.*)

with $\eta = (2x_R - 2)$, $Q_\parallel$ being the parallel momentum transfer and x_R the so-called roughening exponent. In the non-roughened state of the surface, the roughening exponent is expected to be $x_R = 0$ ($\eta = -2$), while it takes a value $x_R = 1/2$ ($\eta = -1$) at T_R and increases further with the temperature above T_R. We should note that this dependence is expected to be valid for elastic scattering only; the influence of inelastic scattering on the diffraction lineshape has not been considered so far in theory. Since the roughening temperature usually occurs well above the surface Debye temperature, inelastic phonon scattering strongly influences the lineshape, and has to be accounted for. In the study of Ni(115) and Ni(113), Conrad *et al.*[113,117] have corrected for this inelastic influence.

In Figs. 45 and 46 we summarize the results obtained by these authors for the Ni(115) and Ni(113) surface. Figure 45 shows the peak width (FWHM) of the specular He beam scattered from the Ni(115) surface along the $(\bar{5}, \bar{5}, 2)$ azimuth as a function of the incident angle θ_i. The oscillation characterizes the step density. The lineshapes of the specular He peak are shown in Fig. 46 in a log-log plot for Ni(113) along the $(\bar{3}, \bar{3}, 2)$ direction with the temperature as parameter. The profiles suggest the expected power-law behavior and the variation of the power exponent η (i.e. the slope) with temperature is obvious.

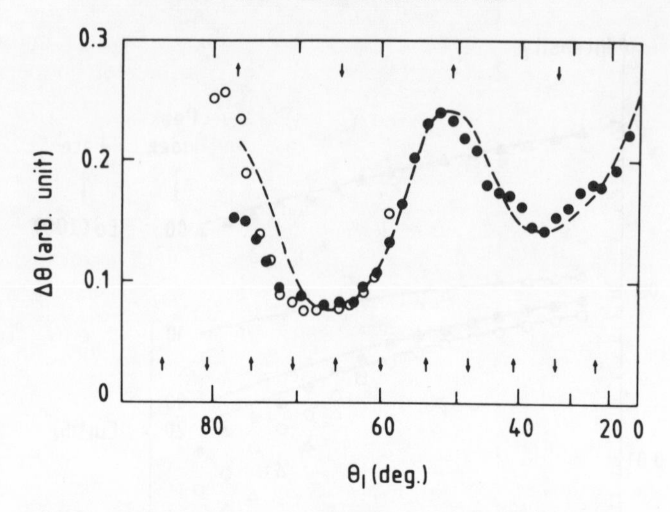

Fig. 45. Specular peak width (FWHM) vs incident angle for He scattering from Ni(115). (*After Ref. 113.*)

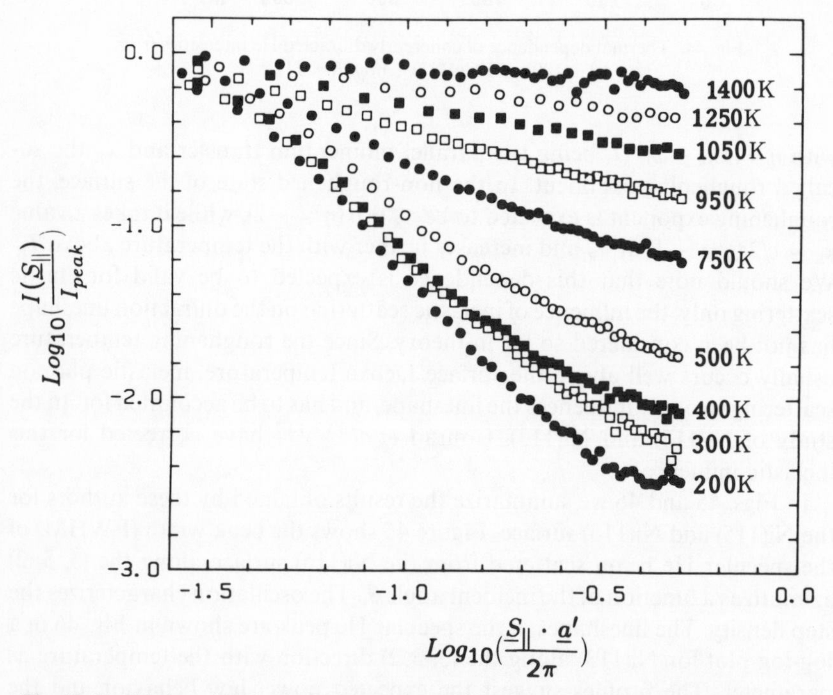

Fig. 46. Double-logarithmic plot of the lineshapes of the specular He intensity from a Ni(113) surface along the $(\bar{3}, \bar{3}, 2)$ direction with the temperature as parameter. (*After Ref. 117.*)

The data being obtained with an energy integrating detector, Conrad *et al.* tried to account for inelastic effects, by assuming a wave vector-independent multiphonon scattering and a one-phonon scattering probability proportional to $Q_{\|}^{-1}$. Then the roughening temperature was obtained from the corrected data via the roughening exponent $x_R = 1/2 \rightarrow \eta = -1$ at T_R. For the (113) surface of Ni they deduced a value of $T_R \approx 750 \pm 50$ K and for the (115) surface of the same metal they obtained $T_R \approx 450 \pm 50$ K. By using a somewhat similar procedure Lapujoulade and coworkers analyzed the (113) and (115) surface of Cu, obtaining roughening temperatures of 720 ± 50 K and 380 ± 50 K, respectively. The observed decrease of T_R with increasing terrace width is in agreement with the microscopic roughening mechanism of high indexed surfaces mentioned above. Indeed, in view of their mutual repulsion, it is easier for the step rows to meander when they are further apart.

The data analysis in the work of Conrad *et al.* and Lapujoulade *et al.* relies on statistical mechanical models assuming power-low lineshapes for the He diffraction peaks. In particular they assume a power exponent $x_R = 0 \rightarrow \eta = -2$ for a perfect flat surface. We have recently studied the lineshape of the specularly scattered He peak from an almost ideal (less than 0.1% defects) Pt(111) surface at low temperatures. By using energy-resolved He scattering, we were able to discriminate directly between purely elastic scattered atoms and those which have undergone an inelastic event, and thus to avoid more or less justified corrections. A log-log plot of the lineshape of the purely elastic component of the specular beam obtained with an 18 meV incident beam and a Pt surface temperature of 110 K is shown in Fig. 47. The data show the characteristic power-law behavior, but with the unexpected exponent $\eta \simeq -2.76$. Practially the same value for η ($\simeq -2.86$) has been observed also in the case of Cu(110). We have yet no explanation for this unusual exponent, but it raises at least some questions concerning the analysis of He scattering data measured without energy discrimination.

The discussion on the experimental evidence for the existence of roughening on low indexed surfaces is not yet settled. Let us consider for example the Cu(110) surface. More than ten years ago it had been noticed that the intensities in the photoemission spectra taken from Cu(110) decrease rapidly with temperature above ~ 500 K[118]. Similar effects have been seen recently in low-energy ion scattering[119], in X-ray diffraction[120] and in thermal He scattering[110,121]. The dramatic intensity decrease observed in all cases above 450–500 K could not be accounted for by simple Debye–Waller effects. While Lapujoulade *et al.*[110] and Fauster *et al.*[119] proposed as explanation either anharmonic effects or some kind of disorder, Mochrie[120] concluded categorically – without qualitative additional evidence – that he was observing the roughening transition. He even tentatively identified the temperature at which 'the intensity has fallen essentially to zero' (870 K) with T_R. A He specular intensity measurement on Cu(110) versus temperature performed in

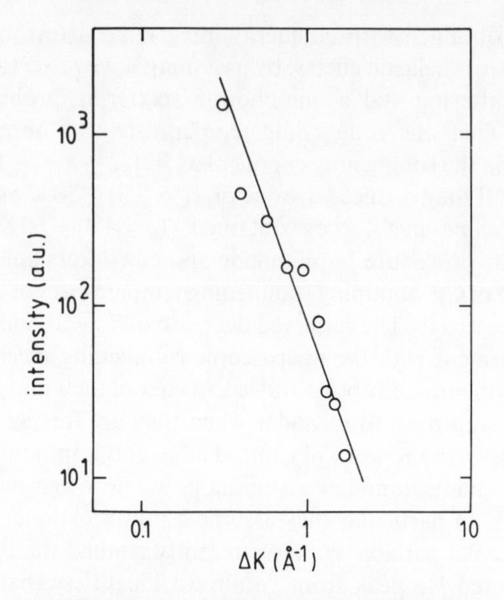

Fig. 47. Double-logarithmic plot of the lineshape of the purely elastic specular He intensity from a Pt(111) surface along the $\bar{\Gamma}\,\bar{M}$ azimuth. The primary beam energy was 18.3 meV.

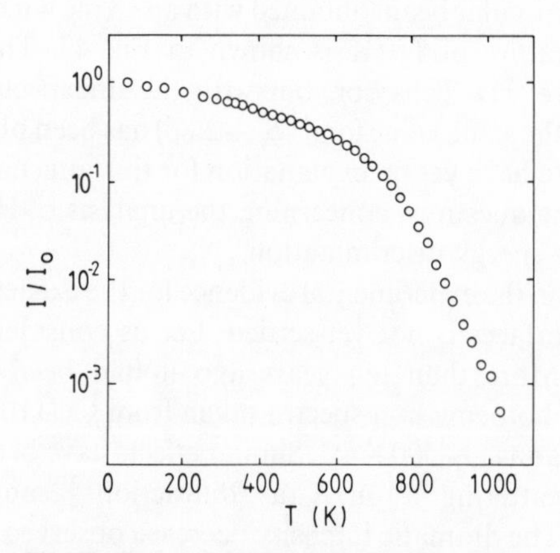

Fig. 48. Thermal dependence of the He specular peak height from Cu(110). The primary beam energy was 18.3 meV, and $\theta_i = \theta_f = 45°$.

our laboratory shows (Fig. 48) that also above 870 K the intensity continues to drop (at 1000 K it is already one order of magnitude lower) and that there is no sign of saturation even above 1000 K. Whether the intensity becomes 'essentially zero' appears to depend on the dynamical range of the instrument, and is not a criterion for the choice of value of T_R. Zeppenfeld *et al.*[121] have analyzed very recently in detail the energy distribution of the scattered He atoms in the whole temperature range up to 1000 K. The various scattering components (elastic, one phonon and multiphonon) were clearly discriminated. The analysis shows that the anomalous intensity drop above ~ 500 K can be ascribed to multiphonon interactions, i.e. the strong increase of the mean square wave displacement of the surface atoms. Thus, the intensity drop is due to the enhancement of the anharmonicity and not necessarily to topological effects (no roughening up to 1000 K).

Acknowledgements

We are indebted to many colleagues all around the world, who by their contributions, lectures and discussions helped us to put together this review; especially to those who have accepted the reproduction of their results. We are particularly grateful to Rudolf David, Bene Poelsema and Peter Zeppenfeld for their decisive contribution to the developments originating in this laboratory. For illuminating discussions we wish to thank in particular to Harald Ibach and Jaques Villain.

For the typing and retyping to always new changes of the manuscript we are very much indebted to Maria Kober.

References

1. Johnson, T. H., *Phys. Rev.*, **37**, 847 (1931).
2. Engel, T., and Rieder, K. H., *Springer Tracts in Modern Physics*, Vol. 91, Springer, Berlin, 1982.
3. Kern, K., David, R., Palmer, R. L., and Comsa, G., *Phys. Rev. Lett.*, **56**, 2823 (1986).
4. Lahee, A. M., Manson, J. R., Toennies, J. P., and Wöll, Ch., *Phys. Rev. Lett.*, **57**, 471 (1986).
5. Frenken, J. W., Toennies, J. P., and Wöll, Ch., *Phys. Rev. Lett.*, **60**, 1727 (1988).
6. Comsa, G., and Poelsema, B., *Appl. Phys.*, **A38**, 153 (1985); Poelsema, B., and Comsa, G., *Springer Tracts in Modern Physics*, to be published.
7. Kern, K., David, R., and Comsa, G., *Rev. Sci. Instr.*, **56**, 369 (1985).
8. David, R., Kern, K., Zeppenfeld, P., and Comsa, G., *Rev. Sci. Instr.*, **57**, 2771 (1986).
9. Sköld, K., *Nucl. Instr. Meth.*, **63**, 114 (1968).
10. Comsa, G., David, R., and Schumacher, B. J., *Rev. Sci. Instr.*, **52**, 789 (1981).
11. Verheij, L. K., and Zeppenfeld, P., *Rev. Sci. Inst.*, **58**, 2138 (1987), and references therein.
12. Allen, R. E., Aldredge, G. P., and de Wette, F. W., *Phys. Rev.*, **B4**, 1661 (1971).
13. Lord Rayleigh, *Proc. London Math. Soc.*, **17**, 4 (1887).
14. Matthews, H., *Surface Wave Filters*, Wiley, New York, 1977.
15. Ho, K. M., and Bohnen, K. P., *Phys. Rev. Lett.*, **56**, 934 (1986).
16. Finnis, M. W., and Heine, V., *J. Phys.*, **F4**, L37 (1974).

17. Smoluchowski, R., *Phys. Rev.*, **60**, 661 (1941).
18. Jiang, P., Marcus, P. M., and Jona, F., *Solid State Communications*, **59**, 275 (1986), and references therein.
19. Badger, R. M., *J. Chem. Phys.*, **2**, 128 (1934); **3**, 710 (1935).
20. Baddorf, A. P., Lyo, I. W., Plummer, E. W., and Davis, H. L., *J. Vac. Sci. Technol.*, **A5**, 782 (1987).
21. Copel, M., Gustafsson, T., Graham, W. R., and Yalisove, S. M., *Phys. Rev.*, **B33**, 8110 (1986).
22. Baddorf, A. P., and Plummer, E. W., to be published.
23. Manson, J. R., and Celli, V., *Surf. Sci.*, **24**, 495 (1971).
24. Weare, J. H., *J. Chem. Phys.*, **61**, 2900 (1974).
25. Brusdeylins, G., Doak, R. B., and Toennies, J. P., *Phys. Rev. Lett.*, **44**, 1417 (1980); **46**, 437 (1981).
26. Bledsoe, J. R., and Fisher, S. S., *Surf. Sci.*, **46**, 129 (1974).
27. Horne, J. M., and Miller, D. R., *Phys. Rev. Lett.*, **41** 511 (1978).
28. Armand, G., and Manson, J. R., *Surf. Sci.*, **80**, 532 (1979).
29. Bortolani, V., Franchini, A., Garcia, N., Nizzoli, F., and Santoro, G., *Phys. Rev. B.*, **28**, 7358 (1983).
30. Levi, A. C., and Suhl, H., *Surf. Sci.*, **88**, 221 (1979).
31. Feuerbacher, B., and Willis, R. F., *Phys. Rev. Lett.*, **47**, 526 (1981).
32. Ibach, H., *Chemistry and Physics of Solid Surfaces V*, Springer, Berlin, 1985, p. 455.
33. Harris, J., and Liebsch, A., *J. Phys. C*, **15**, 2275 (1982).
34. Brusdeylins, G., Rechtsheimer, R., Skofronick, J. G., Toennies, J. P., Benedek, G., and Miglio, L., *Phys. Rev. Lett.*, **54**, 466 (1985).
35. Zeppenfeld, P., Kern, K., David, R., and Comsa, G., *Phys. Rev.* **B38**, 12 329 (1988).
36. Stroscio, J. A., Persson, M., Bare, S. R., and Ho, W., *Phys. Rev. Lett.*, **54**, 1428 (1985).
37. Persson, M., Stroscio, J. A., and Ho, W., *Physica Scripta*, **36**, 548 (1987).
38. Lehwald, S., Wolf, F., Ibach, H., Hall, B. M., and Mills, D. L., *Surf. Sci.*, **192**, 131 (1987).
39. Svensson, E. C., Brockhouse, B. N., and Rowe, J. M., *Phys. Rev.*, **155**, 619 (1967).
40. Doak, R. B., Harten, U., and Toennies, J. P., *Phys. Rev. Lett.*, **51**, 578 (1983).
41. Harten, U., Toennies, J. P., and Wöll, Ch., *Faraday Discuss. Chem. Soc.*, **80**, 137 (1985).
42. Harten, U., Toennies, J. P., Wöll, Ch., and Zhang, G., *Phys. Rev. Lett.*, **55**, 2308 (1985).
43. Kern, K., David, R., Palmer, R. L., Comsa, G., and Rahman, T. S., *Phys. Rev.*, **B33**, 4334 (1986).
44. Neuhaus, D., Joo, F., and Feuerbacher, B., *Surf. Sci.*, **165**, L90 (1986).
45. Jayanthi, C. S., Bilz, H., Kress, W., and Bededek, G., *Phys. Rev. Lett.*, **59**, 795 (1987).
46. Bortolani, V., Franchini, A., Nizzoli, F., and Santoro, G., *Phys. Rev. Lett.* **52**, 429 (1984).
47. Kern, K., David, R., Palmer, R. L., Comsa, G., He, J., and Rahman, T. S., *Phys. Rev. Lett.* **56**, 2064 (1986).
48. Kern, K., David, R., Palmer, R. L., Comsa, G., and Rahman, T. S., *Surf. Sci.*, **178**, 537 (1986).
49. Neuhaus, D., Joo, F., and Feuerbacher, B., *Phys. Rev. Lett.*, **58**, 694 (1987).
50. Gibson, K. D., and Sibener, S. J., *Phys. Rev. Lett.*, **55**, 1514 (1985).
51. Kern, K., David, R., Palmer, R. L., and Comsa, G., *Phys. Rev. Lett.*, **56**, 2823 (1986).

52. Hall, B. M., Mills, D. L., and Black, J., *Phys. Rev.*, **B32**, 4932 (1985).
53. Gibson, K. D., and Sibener, S. J., *Faraday Discuss. Chem. Soc.*, **80**, 203 (1985).
54. Kern, K., Zeppenfeld, P., David, R., and Comsa, G., *Phys. Rev.*, **B35**, 886 (1987), to be published.
55. Englesfield, J. E., *Prog. Surf. Sci.*, **20**, 105 (1985).
56. Heinz, K., in *Kinetics of Interface Reactions* (Ed. M. Grunze and H. J. Kreuzer), Springer, Berlin, 1987, p. 202.
57. Binnig, G., and Rohrer, H., *Phys. Blätter*, **43**, 282 (1987).
58. Sinha, S. K. (Ed.), *Ordering in Two Dimensions*, North Holland, Amsterdam, 1980.
59. Dash, J. G., and Ruvalds, J. (Eds.), *Phase Transitions in Surface Films*, Phenum, New York, 1980.
60. Kontorova, T., and Frenkel, Ya. I., *Zh. Eksp. Teor. Fiz.*, **89**, 1340 (1938).
61. Frank, F. C., and van der Merwe, J., *Proc. Roy. Soc.* A, **198**, 205 (1949).
62. Bak, P., Mukamel, D., Villain, J., and Wentowska, K., *Phys. Rev.*, **B19**, 1610 (1979).
63. Pokrovsky, V. L., and Talapov, A. L., *Sov. Phys. JETP*, **51**, 134 (1980).
64. Kern, K., David, R., Palmer, R. L., and Comsa, G., *Phys. Rev. Lett.*, **56**, 620 (1986).
65. Gordon, M. B., and Villain, J., *J. Phys.*, **C18**, 3919 (1985).
66. Kern, K., David, R., Zeppenfeld, P., Palmer, R. L., and Comsa, G., *Solid State Comm.*, **62**, 361 (1987).
67. Zeppenfeld, P., Kern, K., and Comsa, G., *Phys. Rev.* **B38**, 3918 (1988).
68. Stephens, P. W., Heiney, P. A., Birgeneau, R. J., Horn, P. M., Moncton, D. E., and Brown, G. S., *Phys. Rev.*, **B29**, 3512 (1984).
69. Abraham, F. F., Rudge, W. E., Auerbach, D. J., and Koch, S. W., *Phys. Rev. Lett.*, **52**, 445 (1984).
70. Erbil, A., Kortan, A. R., Birgenau, R. J., and Dresselhaus, M. S., *Phys. Rev.* **B28**, 6329 (1983).
71. Fain, S. C., Chinn, M. D., and Diehl, R. D., *Phys. Rev.*, **B21**, 4170 (1980); Moncton, D. E., Stephens, P. W., Birgeneau, R. J., Horn, P. M., and Brown, G. S., *Phys. Rev. Lett.*, **46**, 1533 (1981).
72. Coppersmith, S. N., Fisher, D. S., Halperin, B. I., Lee, P. A., and Brinkman, W. F., *Phys. Rev.*, **B25**, 349 (1982).
73. Haldane, F. D. M., and Villain, J., *J. Physique*, **42**, 1673 (1981).
74. Jaubert, M., Glachant, M., Bienfait, M., and Boato, G., *Phys, Rev. Lett.*, **46**, 1679 (1981).
75. Gooding, R. J., Joos, B., and Bergersen, B., *Phys. Rev.*, **B27**, 7669 (1983).
76. Kern, K., David, R., Zeppenfeld, P., and Comsa, G., *Surf. Sci.*, **195**, 353 (1988).
77. Halpin-Healy, T., and Kardar, M., *Phys. Rev.*, **B34**, 318 (1986).
78. D'Amico, K. L., Moncton, D. E., Specht, E. D., Birgeneau, R. J., Nagler, S. E., and Horn, P. M., *Phys. Rev. Lett.*, **53**, 2250 (1984).
79. Halpin-Healy, T., and Kardar, M., *Phys. Rev.*, **B31**, 1664 (1985).
80. Miller, D. R., and Horne, J. M., *Proc. III Int. Conf. Solid, Surf.* (Eds. R. Dobrozemsky *et al.*), Vienna, 1977, p. 1385.
81. Harten, U., Lahee, A. M., Toennies, J. P., and Wöll Ch., *Phys. Rev. Lett.*, **54**, 2619 (1985).
82. Takayanagi, K., and Yagi, K., *Jpn. Inst. Met.* **24**, 337 (1983).
83. Moritz, W., private communication.
84. El-Batanouny, M., Burdick, S., Martini, K. M., and Stancioff, P., *Phys. Rev. Lett.*, **58**, 2762 (1987).
85. Ercolessi, F., Bartolini, A., Garofalo, M., Parrinello, M., and Tosatti, E., *Surf. Sci.*, **189/190**, 636 (1987).

86. Peirls, R. E., *Quantum Theory of Solids*, Clarendon, Oxford, 1955.
87. Kohn, W., in *Many Body Physics* (Eds. C. Dewitt, and L. Balian), Gordon and Breach, New York, 1968, p. 353.
88. Yonehama,K., and Schmidt, L. D., *Surf. Sci.*, **25**, 238 (1971).
89. Debe, M. K., and King, D. A., *Phys. Rev. Lett.*, **39**, 708 (1977).
90. Barker, R. A., Estrup, P. J., Jona, F., and Marcus, P. M., *Solid State Commun.*, **25**, 375 (1978).
91. Walker, J. A., Debe, M. K., and King, D. A., *Surf. Sci.*, **104**, 405 (1981).
92. Lapujoulade, J., and Salanon, B., *Surf. Sci.*, **173**, L613 (1986).
93. Ernst, H. J., Hulpke, E., and Toennies, J. P., *Phys. Rev. Lett.*, **58**, 1941 (1987).
94. Mattheiss, L. F., and Hamann, D. R., *Phys. Rev.*, **B29**, 5372 (1984).
95. Holmes, M. I., and Gustafsson, T., *Phys. Rev. Lett.*, **47**, 443 (1981).
96. Wang, X. W., and Weber, W., *Phys. Rev. Lett.*, **58**, 1452 (1987).
97. Fasolino, A., Santoro, G., and Tosatti, E., *Phys. Rev. Lett.*, **44**, 1684 (1980).
98. Fu, C. L., Freeman, A. J., Wimmer, E., and Weinert, M., *Phys. Rev. Lett.*, **54**, 2261 (1985).
99. Burton, W. K., Cabrera, N., and Frank, F. C., *Phil. Trans. Roy. Soc.*, **243A**, 299 (1951).
100. Swendsen, R. W., *Phys. Rev.*, **B17**, 3710 (1978).
101. Chin, S. T., and Weeks, J. D., *Phys. Rev.*, **B14**, 4978 (1976).
102. van Beijeren, H., and Nolden, I., in *Structure and Dynamics of Surfaces* II (Eds. W. Schommers, and P. van Blanckenhagen), Springer, Berlin, 1986, p. 259.
103. Gilmer, G. H., *Science*, **208**, 4442 (1980).
104. Jackson, K. A., and Miller, C. E., *J. Crys. Growth*, **40**, 169 (1977).
105. Pavlovska, A., *J. Cryst. Growth*, **46**, 551 (1979).
106. Balibar, S., Edwards, D. O., and Laroche, C., *Phys. Rev. Lett.*, **42**, 782 (1979).
107. Avron, J. E., Balfour, L. S., Kuper, C. G., Landau, J., Lipson, S. G., and Schulman, L. S., *Phys. Rev. Lett.*, **45**, 814 (1980).
108. Balibar, S., and Castaing, B., *Surf. Sci. Rep.*, **5**, 87 (1985).
109. Verheij, L. K., Lux, J., and Poelsema, B., *Surf. Sci.*, **144**, 385 (1984), and references therein.
110. Lapujoulade, J., Perreau, J., and Kara, A., *Surf. Sci.*, **129**, 59 (1983).
111. Idiodi, J., Bartolani, V., Franchini, A., and Santaro, G., *Phys. Rev.*, **B35**, 6029 (1987).
112. Lapujoulade, J., *Surf. Sci.*, **178**, 406 (1986).
113. Conrad, E. H., Aten, R. M., Kaufman, D. S., Allen, L. R., Engel, T., den Nijs, M., and Riedel, E. K., *J. Chem. Phys.*, **84**, 1015 (1986); **E85**, 4657 (1986).
114. Fabre, F., Gorse, D., Lapujoulade, J., and Salanon, B., *Europhys. Lett.*, **3**, 737 (1987).
115. den Nijs, M., Riedel, E. K., Conrad, E. H., and Engel, T., *Phys. Rev. Lett.*, **55**, 1689 (1985); **E57**, 1279 (1986).
116. Villain, J., Grempel, D. R., and Lapujoulade, J., *J. Phys.*, **F15**, 809 (1985).
117. Conrad, E. H., Allen, L. R., Blanchard, D. L., and Engel, T., *Surf. Sci.*, **187**, 265 (1987).
118. Williams, R. S., Wehner, P. S., Stöhr, J., and Shirley, D. A., *Phys. Rev. Lett.*, **39**, 302 (1977).
119. Fauster, Th., Schneider, R., Dürr, H., Engelmann, G., and Taglauer, E., *Surf. Sci.*, **189/190**, 610 (1987).
120. Mochrie, S. G. J., *Phys. Rev. Lett.*, **59**, 304 (1987).
121. Zeppenfeld, P., Kern, K., David, R., and Comsa, G., *Phys. Rev. Lett.* **62**, 63 (1989).

Advances in Chemical Physics
Edited by K. P. Lawley

GAS–SURFACE REACTIONS: MOLECULAR DYNAMICS SIMULATIONS OF REAL SYSTEMS*

DONALD W. BRENNER

Naval Research Laboratory, Washington, D.C. 20375, USA

and

BARBARA J. GARRISON

Department of Chemistry, The Pennsylvania State University, University Park, PA 16802, USA

CONTENTS

*We would like to dedicate this chapter to Don E. Harrison, Jr., a dear friend and colleague, who died during the final stages of the preparation of this work. Don's first paper on computer simulations of sputtering was published in 1964, long before many of us even dreamed of performing molecular dynamics on real systems. Eleven years ago Don graciously agreed to collaborate and even gave us his computer code – a beginning of an interaction that led to what we believe has been interesting science.

1. INTRODUCTION

In the not too distance past, surface science could have been described as an emerging field of research that borrows science and engineering techniques to study surface-related phenomena. This field has sufficiently matured, however, to the point that it has spurred new techniques which are unique to the study of surfaces, and which have in turn been adapted for studying technologically important processes. Examples of fields that have been impacted by surface science include heterogeneous catalysis, bio-engineering, tribology, and semiconductor device fabrication. In fact, any technological process which involves an interface has probably benefited in some way from techniques which were developed specifically to study surfaces.

Central to the understanding of surface-related phenomena has been the study of gas–surface reactions. A comprehensive understanding of these reactions has proven challenging because of the intrinsic many-body nature of surface dynamics. In terms of theoretical methods, this complexity often forces us either to treat complex realistic systems using approximate approaches, or to treat simple systems with realistic approaches. When one is interested in studying processes of technological importance, the latter route is often the most fruitful. One theoretical technique which embodies the many-body aspect of the dynamics of surface chemistry (albeit in a very approximate manner) is molecular dynamics computer simulation.

To cover all of the contributions which molecular dynamics has made to surface science would be an almost impossible task. Instead, this review is intended as a brief survey of several areas of surface science in which we have been associated. In a sense, this chapter can be considered as a guided tour through a small (but we hope interesting) area of theoretical surface science.

The phrase in the title 'real systems' is somewhat ambiguous and should be better defined. The theory of chemical processes is an extremely complex topic which is intertwined with other subjects, especially applied mathematics. It can, however, be conveniently divided into two steps – the development of theoretical techniques followed by their application. The former category is especially tied to applied mathematics, and will not be discussed in detail in this chapter. The second category, while relying on the first, is as equally important and often provides the ultimate test of a theoretical method. Such a test includes both the severity of any approximations used in the theory, and its ability to understand and predict phenomena which are of practical importance. It is the latter part of the test which will be emphasized in this chapter, and so 'real systems' are defined here as those which lead to an enhanced understanding of technologically important processes. This is contrasted with systems, for example, which may be sufficiently simple to be used to test the assumptions used in a theoretical method or to test a particular experimental technique, but are of limited practical importance.

1.1. Why molecular dynamics?

Molecular dynamics simulations yield an essentially exact (within the confines of classical mechanics) method for observing the dynamics of atoms and molecules during complex chemical reactions. Because the assumption of equilibrium is not necessary, this technique can be used to study a wide range of dynamical events which are associated with surfaces. For example, the atomic motions which lead to the ejection of surface species during keV particle bombardment (sputtering) have been identified using molecular dynamics, and these results have been directly correlated with various experimental observations[1]. Such simulations often provide the only direct link between macroscopic experimental observations and microscopic chemical dynamics.

In its pure classical form, molecular dynamics is straightforward to carry out. One starts with given initial conditions for the system of interest. These conditions include atomic positions and velocities as well as a given interaction potential. For example, if one were simulating a sputtering event, the initial conditions might correspond to a collection of atoms which comprise the solid surface, another collection of atoms adsorbed on top of the surface, and an incoming energetic particle (Fig. 1). The atomic dynamics are then determined by numerically solving a set of classical equations of motion. Various aspects of the dynamics, such as reaction mechanisms and product distributions, can then be determined by examining the motion of the atoms during the simulation (Fig. 2). The numerical details of this type of simulation can be found in a large number of excellent reviews[1–7].

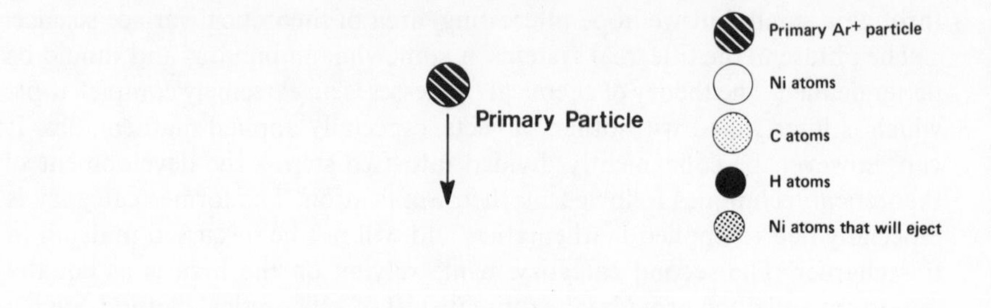

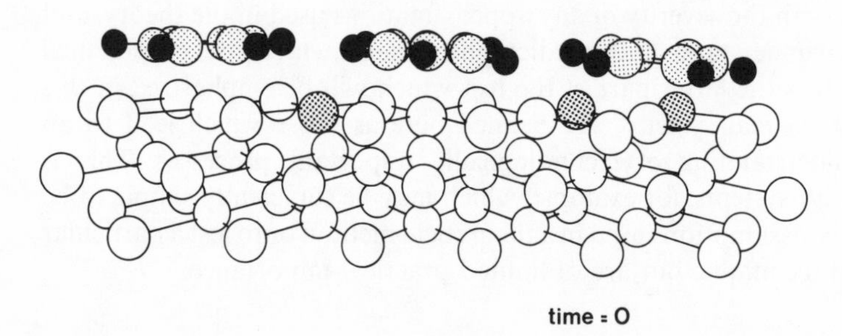

Fig. 1. Initial position of atoms before Ar^+ ion bombardment of a layer of C_6H_6/Ni(001).

Molecular dynamics is an extremely powerful technique for understanding the atomic-scale dynamics of chemical processes. This is because it is sufficiently simple that large numbers of atoms can be treated, and because a minimal number of approximations are required. Specifically, all that is assumed is the validity of classical mechanics for the given problem and a potential-energy surface. The former approximation, although never totally true, is reasonably well understood. For example, classical mechanics describes the dynamics of heavy particles better than light particles. Hence, if one were interested in understanding the motion of hydrogen atoms, it would be understood that the classical mechanical method involves severe approximations. If one were interested in modeling the translational motion of silicon atoms, however, classical mechanics would be adequate. Potential surfaces, on the other hand, are only known well for a very few systems, and the influence which the interaction potential has on the results of a simulation is often unknown. One must therefore weigh results against the potential used. This concern is compounded for computer simulations of more than a few atoms, because the additional factor of available computer resources often demands that simple potentials be used. Because of the important role which interaction potentials play in computer simulations of real systems, they will be emphasized throughout this article.

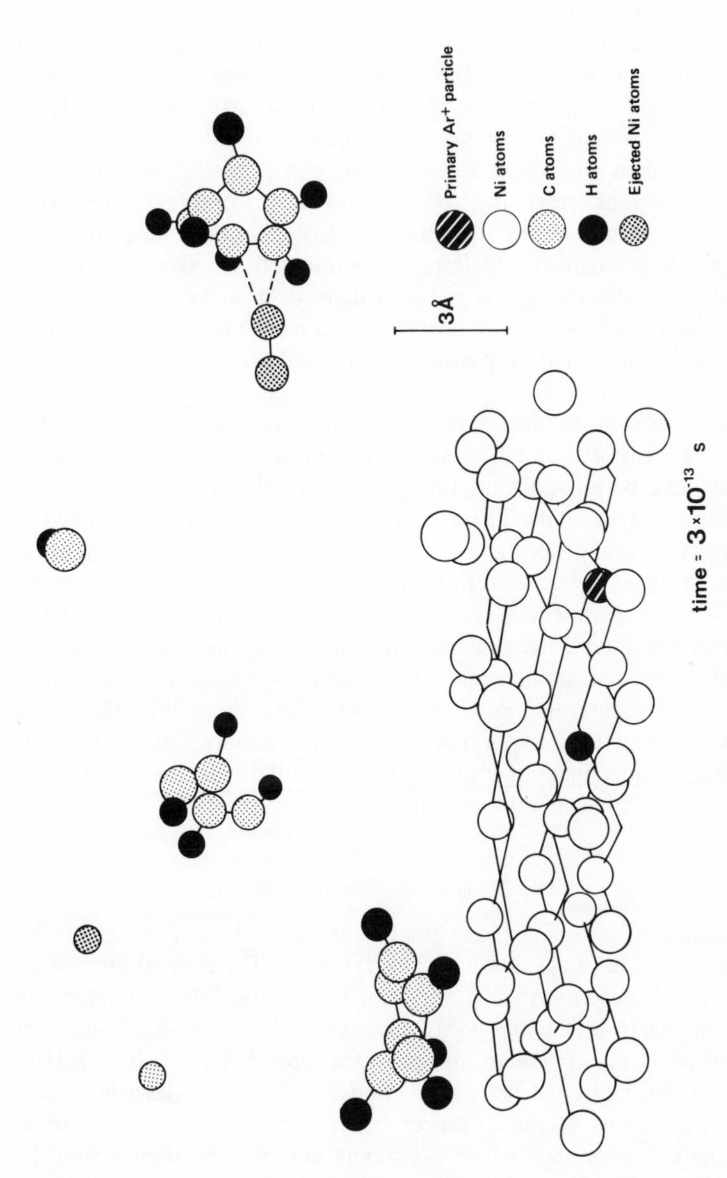

Fig. 2. Final positions of the atoms 0.3 ps after the Ar^+ ion impact.

1.2. Interaction potentials

When the results of a molecular dynamics study are being judged, the question, 'Is the potential realistic?' is often asked. This can be the incorrect approach to evaluating the validity of a computer simulation. The appropriate question to be addressed should be, 'Is the potential used appropriate for the phenomena being modeled?'. In the same vein, it is common for a potential developed to model one property of a system to be arbitrarily extended to phenomena for which it may be inappropriate. In this way, interaction potentials are often misjudged. For example, pair potentials with tails that mimic the oscillations present in an electron gas due to ion cores can be used to understand the properties of bulk metals[8]. It is obvious that these potentials, however, would not realistically describe the interaction between three metal atoms, where an electron gas is not well defined. It is the rare interaction potential which works well for all properties of a particular system and so one needs to understand why a particular potential works well for a given property.

Computer simulation demands that interatomic potentials and their derivatives be easily evaluated. Hence, it is common practice to fit available theoretical and experimental data, such as the energetics of static structures, to simple analytic expressions. These expressions are often further refined by comparing dynamical properties calculated from a simulation with quantities measured during an experiment. For example, Garrison *et al.* have compared simulated and experimental sputtering results to refine an interaction potential for rhodium[9]. Because different energy regimes sample different sections of a potential, simulations can be effectively categorized in terms of the range of atomic interactions which the dynamics explore. With these ideas in mind, we develop the discussion in this section by dividing interactions into three regimes – short, medium and long range – and discuss how each can be effectively modeled in a computer simulation.

1.2.1. Short-range interactions

For atomic collisions of energies greater than typical bond energies (a few eV), strong repulsive forces dominate the dynamics. The general form of the potentials which describe these forces can be qualitatively understood by recalling that electron densities decay approximately exponentially outward from the nuclei. Hence, any relationships which depend on the radial structure of the electronic charge density can often be effectively described using exponential functions of the distances between atoms. Using this simple concept, analytic potentials which accurately describe the strong repulsive forces can be developed.

The repulsion between atomic cores arises primarily from two contributions. First, the 'bare' nuclei interact via simple pair-additive electrostatic repulsions, where the potential is given by

$$V_{nn} = Z_1 Z_2 e^2/R. \tag{1}$$

The quantities Z_1 and Z_2 are the charges of the two nuclei, R is their scalar separation and e is the electronic charge. To account for the shielding of the nuclear charge by the negative charge distribution of the surrounding electrons, this interaction is modified so that the effective nuclear charges are reduced as the interatomic distance increases. The form for this modified nuclear repulsion can be written as

$$V_{rep} = V_{nn} F(R) \tag{2}$$

where $F(R)$ is a 'screening' function[10]. If the core electronic shells do not appreciatively distort during bonding, a radially symmetric electronic screening function can be used. Bohr first proposed a simple screening function of the form

$$F(R) = \exp(-R/a) \tag{3}$$

where the quantity a is either given by

$$a = a_0(Z_1^{2/3} + Z_2^{2/3})^{-1/2} \tag{4}$$

where a_0 is the Bohr radius, or a may be used as an adjustable parameter[11].

While the physical significance of Eq. (3) is apparent based on the orbital decay mentioned above, the single exponential decays too rapidly and so it is valid only for atomic separations of up to a few tenths of an angstrom. A more generally useful form of the screening function is the Molière potential, given by the screening function[10]

$$F(R) = 0.35\exp(-0.3R/a) + 0.55\exp(-1.2R/a) + 0.1\exp(-6.0R/a). \tag{5}$$

The quantity a can either be given by the equation

$$a = 0.885a_0(Z_1^{1/2} + Z_2^{1/2})^{-2/3} \tag{6}$$

(called the Firsov screening) length, or a may also be taken as an adjustable parameter. A typical modification is to reduce the Firsov value by a factor of 0.8. This form was originally obtained as an analytic fit to the Thomas–Fermi calculated screening function, although the parameter a is often empirically adjusted to fit experimental results. A more detailed discussion of these screening functions, as well as a variety of others, may be found in the book by Torrens[10]. The Molière potential with an adjustable screening length displays realistic decay behavior over a range of energies, and so it is generally deemed to be useful.

The second source of repulsion comes from the interaction when filled electronic shells overlap and is due to the exclusion principle. Based again on the exponential decay of electron densities, it would be appropriate to assume that an exponential function of atomic distances could realistically describe this interaction. Furthermore, based on the assumption that the filled electronic shells maintain their spherical symmetry during an atomic collision, radial pair potentials provide a simple analytic description of this interaction. A popular interaction of this type is the Born–Mayer potential of the form

$$V_{bm} = A \exp(-BR) \tag{7}$$

where A and B are adjustable parameters[12]. This potential is realistic when the inner electronic shells do not overlap significantly, and so it is only useful for low-energy collisions. It is often used, however, as an adjustable repulsion which balances a many-body attraction (see below).

In conclusion, the repulsive interactions arise from both a screened coulomb repulsion between nuclei, and from the overlap of closed inner shells. The former interaction can be effectively described by a bare coulomb repulsion multiplied by a screening function. The Molière function, Eq. (5), with an adjustable screening length provides an adequate representation for most situations. The latter interaction is well described by an exponential decay of the form of a Born–Mayer function. Furthermore, due to the spherical nature of the closed atomic orbitals and the coulomb interaction, the repulsive forces can often be well described by pair-additive potentials. Both interactions may be combined either by using functions which reduce to each interaction in the correct limits, or by splining the two forms at an appropriate interatomic distance[10].

1.2.2. Medium-range interactions

Medium-range interactions can be defined as those which dominate the dynamics when atoms interact with energies within a few eV of their molecular binding energies. These forces determine a majority of the physical and chemical properties of surface reactions which are of interest, and so their incorporation in computer simulations can be very important. Unfortunately, they are usually many-body in nature, and can require complicated functional forms to be adequately represented. This means that severe approximations are often required when one is interested in performing molecular dynamics simulations. Recently, several potentials have been semi-empirically developed which have proven to be sufficiently simple to be useful in computer simulations while still capturing the essentials of chemical bonding.

Solid surfaces lie at the interface of two historically distinct regimes. On the one hand, a surface can be thought of as a perturbation on a crystalline solid. Hence ideas based on the properties of condensed matter can be used to develop interaction potentials. For example, in a bulk metal the concept of a free electron gas is well developed, and simple potentials based on these ideas have been extended to include surfaces[13–20]. Unfortunately, these ideas are not well developed in the context of angle-dependent interactions, and so they are not exceptionally useful for describing covalent bonding.

The alternate approach to developing interaction potentials is to consider the solid surface as a very large molecule. One can then apply theoretical techniques based on gas-phase reaction ideas. The simulation of real systems, however, often requires that both reactive adsorbed atoms as well as a large number of substrate atoms be explicitly treated, and so these techniques rapidly become computationally infeasible. It is apparent that to simulate the general situation, bonding ideas from both regimes should be used. This breakdown does, however, provide a useful format within which to discuss intermediate-range interaction potentials, and so it will be used to illustrate potentials which are in current use in simulations of gas–surface interactions.

(a) Gas-phase approaches

Surfaces provide a unique environment for promoting chemistry which would not otherwise occur in the gas phase. For example, surfaces can provide a heat bath, source (or sink) of electrons, or simply a physical surrounding which brings together species which otherwise would have a low probability of meeting. If the substrate atoms do not move appreciably during the physical event of interest, and the number of atoms of interest is not too large, then few-body gas-phase potentials can be modified to model gas–surface interactions. A large number of few-body formalisms have been developed over the years[21,22] and so rather than present an overview of all of these ideas, this discussion will emphasize general modifications used to model surface chemistry.

The spirit of this type of approach is to consider the entire surface as one body of a few-body reaction. An effective first approximation is to treat the surface as rigid, and to write all contributions of the surface to the potential energy as a periodic function of the surface lattice vectors. For example, Morse parameters which control the binding of an atom to a surface can be made a function of the position of the adsorbate within the surface unit cell[23]. The advantage of this approximation is that the sum of the interactions between an adsorbed atom and each substrate atom is reduced to a single effect. A disadvantage is that because the substrate atoms are considered rigid, coupling of the reaction to thermal motion of the lattice is not included. This means, for example, that energy cannot be removed from a reaction without adding additional forces, and long-lived exothermic chemical bonding is not possible.

A correction to the rigid lattice approximation is to write the potential as a contribution commensurate with the lattice vectors as above, and to add an additional term which depends on the displacement of substrate atoms from their lattice sites. For example, restoring forces between atoms and their atomic sites combined with correction terms to the adsorbate–surface interactions which account for the displacement of the substrate atoms have been used to modify rigid-lattice potentials[24,25]. The advantage of this approach is that the rigid-lattice potentials, which are often simpler to fit, are easily modified to allow coupling to a heat bath.

Finally, the symmetry constraint can be removed by considering a pair sum over substrate atoms as a single contribution to the many-body energy. For example, the periodic contribution of the substrate can be replaced by a sum of contributions from each individual substrate atom[26,27]. This allows the study of the effect of features such as amorphous surfaces, steps and defects on surface reactivity, while still retaining a potential derived from a rigid lattice. These types of potentials, however, can become time consuming in their evaluation, and can therefore be inconvenient for use in large-scale computer simulations.

(b) Condensed-phase approaches

For condensed phases of bulk metals, the binding energy can be divided into repulsions between nuclei (see above) and the interaction of the positively charged nuclei with an electron gas. Within this breakdown, the motion of the nuclei can be determined by pair-additive forces with the addition of volume-dependent terms arising from the pressure of the electron gas[8]. While computer simulations based on these types of interactions have been carried out[28], volume-dependent interactions are difficult to define unambiguously for surfaces.

An alternate approach, which has proven to be extremely useful for metals, has been developed by Daw, Baskes and Foiles[13,29] (and to a lesser extent, by Ercolessi, Tosatti and Parrinello[15–17]). Called the embedded atom method (EAM) (or the glue model by the second group), the interactions in this approach are developed by considering the contribution of each individual atom to the local electron density, and then empirically determining an energy functional for each atom which depends on the electron density. This circumvents the problem of defining a global volume-dependent electron density.

While the embedded atom method has been formally derived by Daw and Baskes[13], the functions used in computer simulations are typically empirically determined. The description presented here will therefore treat this approach as an empirical method. The first step in determining the potential is to define a local electron density at each atomic site in the solid. A simple sum of atomic electron densities has proven to be adequate, and so in most cases a sum of free atom densities is used[13,29]. The second step is to determine an embedding function which defines the energy of an atom for a given electron density. Finally, the attractive contribution to the binding energy produced by the function is balanced by pairwise additive repulsion interactions (see above). The expression for the total binding energy is given by

$$E_{\text{tot}} = \sum_i F_i(\rho_i) + \tfrac{1}{2}\sum_i \sum_j \Phi(R_{ij}) \tag{8}$$

where ρ_i is the electron density at each atomic site, $F_i(\rho_i)$ is the embedding function, and $\Phi(R_{ij})$ is the pair term arising mainly from the core–core repulsions. The function ρ_i is given by the expression

$$\rho_i = \sum_j \rho(R_{ij}) \tag{9}$$

where the quantity $\rho(R_{ij})$ is the contribution of electron density to site i from atom j, and is a function of the distance R_{ij}.

If free atom electron densities are used in the sum (9) above, the embedding function is left to determine the properties of the condensed phase. An accurate determination of this function is therefore important for modeling realistic systems. The approach commonly used is to fit this function to a large number of properties. For example, experimentally determined values of the lattice constant, sublimation energy, elastic constants and vacancy formation energies are often combined with theoretically determined relations such as the universal equation of state to provide an extensive database[29]. This formalism has proven to provide both a realistic and easily evaluated potential which is suitable for describing a large range of properties of various pure metals and alloys[9,13–20,30–32]. Examples of the application of the EAM to surfaces will be given below. The approach, however, is not sufficiently developed at the present time to model covalent bonding, although some progress has made by the introduction of angle-dependent electron densities[33]. For covalent interactions, which determine the properties of semiconductors, other approaches based on condensed-matter ideas have been developed which have been used to model surface chemistry.

A covenient starting point for developing covalent interactions is to write the energy as a many-body expansion of the form

$$E_{\text{tot}} = V_2 + V_3 + V_4 + \ldots \tag{10}$$

where the first term represents a sum over pairs of atoms, the second term represents a sum over triples of atoms, etc. A well-known example of this type of expansion, if restricted to atomic displacements near equilibrium, is the valence-force field. While this expression is exact if all terms are included, computational restrictions demand that it be truncated. In most applications, it is truncated at three-body interactions. This is partly for computational convenience and partly because the three-body term can be written in the form of a bond bend – a concept which is physically appealing. While this approach is well developed for few-body, gas-phase reactions[21,22], it has only recently been extended to condensed phases.

Silicon has been the test case for potentials of this type, and so this discussion will be restricted to this element, although examples of other systems have very recently become common. The most widely used silicon potential was developed by Stillinger and Weber[34]. The interactions used are composed of a sum of two-body and three-body terms, with the three-body interactions serving to destabilize the sum of the pair terms when bond angles are not tetrahedral. The parameters for this potential were determined by reproducing the binding energy, lattice stability and density of solid silicon, and also by reproducing the melting point and the structure of liquid silicon. Although this potential was originally developed to model liquid–solid properties, subsequent studies have demonstrated that it also provides a good description of the Si(001) surface[35,36]. The wide applicability of this potential can be considered a testament to the care (and computer time) invested by Stillinger and Weber in its development.

A simpler potential of the form of Eq. (10) has been used by Pearson *et al.* to model Si and SiC surfaces[37]. The two-body term is of the familiar Lennard–Jones form while the three-body interaction is modeled by an Axilrod–Teller potential[38]. The physical significance of this potential form is restricted to weakly bound systems, although it apparently can be extended to model covalent interactions.

Brenner and Garrison introduced a potential which was derived by rewriting a valence force expression so that proper dissociation behavior is attained[39]. Because the equations were extended from a set of terms which provided an excellent fit to the vibrational properties of silicon, this potential is well suited for studying processes which depend on dynamic properties of crystalline silicon. For example, Agrawal *et al.* have studied energy transfer from adsorbed hydrogen atoms into the surface using this potential[40].

While these potentials have been successful in modeling dynamic processes on silicon surfaces, the many-body expansion as applied in this case suffers from several drawbacks. Because all of the three potentials above have been fit to properties of the crystalline silicon solid, they implicitly assume tetrahedral bonding. Atoms on the surface of silicon are known to exhibit nontetrahedral hybridizations, and so the results for surfaces are at best uncertain. Also, none of these potentials reproduce accurately the properties of the Si_2 diatomic molecule. This again inhibits a complete description of surface reactions.

A related potential form, which was primarily developed to reproduce structural energetics of silicon, was introduced by Tersoff[41,42] and was based on ideas discussed by Abell[43]. The binding energy in the Abell–Tersoff expression is written as a sum of repulsive and attractive two-body interactions, with the attractive contribution being modified by a many-body term.

The stability of the diamond lattice is achieved by modifying the attractive pair terms according to local coordination, so that the atomic binding energy is at a minimum when each atom has four nearest neighbors. The parameters in this potential were determined by fitting to the properties of the Si_2 molecule, and by reproducing the binding energies and lattice constants of several crystal structures of silicon.

While the original many-body expression introduced by Tersoff (given in Ref. 41) reproduced the energy of a number of structures, it was seriously flawed because it does not give the diamond structure as the global potential minimum. Furthermore, dynamic properties of crystalline silicon are not well described, with the optic mode frequencies being too high and the nearest-neighbor radial distribution being too narrow[42,44]. A modification of Tersoff's original potential expression which was suggested by Dodson eliminated the former flaw in Tersoff's expression[45], although it is unclear as to whether the dynamic properties are improved[42]. Finally, Tersoff has introduced a different expression which appears to have remedied both problems[42]. The Tersoff potential is expected to yield a good overall expression for silicon because it correctly describes the isolated dimer, and because it is fit to nontetrahedral structures. This means that it should provide an adequate description of silicon surfaces, although thorough testing is still being carried out[42].

Additional silicon potentials have been introduced, but they appear to be cumbersome for studying large numbers of atoms[46]. Also, several of the silicon potentials mentioned above have been modified to represent germanium[47–49]. These potentials are discussed elsewhere and will not be presented here.

1.2.3. Long-range interactions

The field of long-range, or intermolecular interactions is an extremely interesting and extensive topic in itself. In the context of surfaces, these forces are responsible for physisorption and can play a major role in governing processes such as rotational excitation of NO due to scattering from Ag(111)[50–53]. For the simulations discussed here, however, these interactions do not play a major role, and so discussions of this topic are deferred to a number of review articles[54,55].

The remainder of this chapter is devoted to describing the results of computer simulations which have used the ideas discussed above. The overall goal of these studies is to describe and understand phenomena which depend for the most part on bonding ('medium-range') interactions. For example, simulations of the reaction of small molecules on metal surfaces are discussed in section 3.1, where bond formation occurs at thermal energies. The major drawback for using simulations to study these types of processes is that the interactions are not well known, and where a reasonable estimate can be made, analytic functional forms are complicated. Two solutions to this problem can be imagined. First, one could develop simple schemes for handling 'medium-range' interactions which still capture the essence of chemical bonding. Simulations using this type of approach will be described in section 3. The second solution would be to perform simulations where information which depended on bonding interactions was obtained, but where the dynamics probed regions of the potential which are well described using simple potentials. This is the essence of the simulations described in the next section.

2. SIMULATIONS RELYING ON SHORT-RANGE INTERACTIONS: ION-INDUCED SPUTTERING

In secondary-ion mass spectrometery (SIMS) and its sister technique fast atom bombardment mass spectrometry (FABMS), a surface is bombarded with energetic particles, and the kinetic energy of the particles converts substrate and chemisorbed atoms and molecules to gas-phase species. The ejected (or sputtered) material is subsequently interrogated using various analytical tools, such as lasers and mass spectrometers, to indirectly deduce information about the initial surface. The relationships between sputtered material and the surface, however, are not always clear, and erroneous conclusions are easily made. Computer simulations have demonstrated that a fundamental understanding of the sputtering process is required to interpret experimental data fully[1].

SIMS has traditionally been used in several regards. First, it has proven to be an important analytical tool for studying surface structures. For example, the identity as well as the velocity and spatial distributions of the sputtered material can reflect the bonding and local structure of the initial surface[1]. Often this information is not easily obtained with other analytical techniques. It has also been discovered that if biomolecules are adsorbed onto a surface, FABMS can be used to convert these molecules into gas-phase species with limited fragmentation[56]. This allows large molecules with atomic masses in the range of 1000–20,000 AMU to be studied on a molecule-by-molecule basis. Finally, SIMS can be used to study energy transfer between the bombarding ions and the solid. For example, recent measurements of energy- and angle-resolved neutral (EARN) data[57–59] have provided experimental information with which analytic[60,61] and simulated[9,62] sputtering calculations can be compared.

The material sputtered from surfaces on which molecules have been chemisorbed generally consists of a collection of single atoms, strongly bound molecules, and weakly bound molecular clusters, any of which may be neutral or charged. It is often assumed, perhaps naively, that the identity and the local environment of molecules on the surface can be directly deduced from the sputtered species. This implies that a molecule, with bond energies on the order of a few eV, can remain intact during the sputtering process. This is despite the fact that the kinetic energy of the bombarding particles is orders of magnitude larger than bond energies. This curious result, along with the total picture of the sputtering process, can be understood with the use of computer simulations.

2.1. The ejection process

As discussed in section 1, a unique feature of computer simulations is that they provide a microscopic picture of atomic motion. In the case of ion-induced sputtering, they can reveal general ejection mechanisms as well as enhance the interpretation of experimental results for specific systems. In the sputtering simulations, a set of atomic positions and velocities are chosen which mimic experimental conditions[1]. In particular, the substrate is modeled by a microcrystallite which is typically five to six atomic layers deep, with each layer containing 100 to 150 atoms. When desired, chemisorbed molecules and atoms are placed at predetermined binding sites on the substrate. The dynamics are initiated by impacting the surface with an energetic particle (see Fig. 1). The kinetic energy of the impacting particle is chosen so that a relatively small number (500–1000) of substrate atoms is sufficient to effectively model the ejection process. Typical kinetic energies used are on the order of a keV. One unique feature of the sputtering simulations in contrast to other simulations of solids and liquids is the boundary conditions on the sides and bottom of the crystallite. There can easily be one or two very energetic (20–500 eV) particles that reach the edge of the crystallite. These particles in a real solid penetrate further into the bulk, damaging the sample along their path. By enlarging the crystallite in the simulations it can be verified that these energetic particles do not substantially contribute to the ejection of atoms and molecules into the gas phase. Thus these atoms are truncated from the simulation once they leave the side or bottom edge. One should *not* use periodic boundary conditions as it is nonphysical to have the energy enter the other side of the crystal. Likewise the generalized Langevin[63–65] prescription or a rigid layer would cause reflection of the energy back into the crystal, again a nonphysical phenomenon.

Once the initial and boundary conditions are specified, the classical equations of motion are integrated as in any other simulation. From the start of the trajectory, the atoms are free to move under the influence of the potential. One simply identifies reaction mechanisms and products during the dynamics. For the case of sputtering, the atomic motion is integrated until it is no longer possible for atoms and molecules to eject. The final state of ejected material above the surface is then evaluated. Properties of interest include the total yield per ion, energy and angular distributions, and the structure and stability of sputtered molecules. In order to further mimic experimental conditions, many trajectories are evaluated by choosing an ensemble of impact points for the energetic particle within the surface unit cell. The experiments with which the simulations are compared are performed so that the majority of the bombarded surface is undamaged[1]. This makes direct comparisons between the simulated and experimental results possible.

Simulations of this type have demonstrated that the ejected species arise from energetic collisions which sample primarily the short-range repulsive part of the potential. This means that simple pairwise additive interactions can be used to describe much of the dynamics of interest without introducing overly severe approximations. This is opposed to the situation, for example, where surface damage due to the incoming ion is being examined. In this case more complicated potentials should be warranted. The sputtering simulations have used pairwise additive Molière and Born–Mayer interactions combined with attractive terms such as Morse potentials. The attractive interactions are incorporated to model the overall cohesive energy, with their exact form again not exerting a large influence on much of the dynamics of interest. Incorporation of more realistic potentials and the influence which these potentials have on the sputtering process will be discussed in section 3.2.

Molecular dynamics simulations have yielded a great deal of information about the sputtering process. First, they have demonstrated that for primary ion energies of a few keV or less, the dynamics which lead to ejection occur on a very short timescale on the order of a few hundred femtoseconds. This timescale means that the ejection process is best described as a small number of direct collisions, and rules out models which rely on many collisions, atomic vibrations and other processes to reach any type of 'steady state'. Within this same short-timescale picture, simulations have shown that ejected substrate atoms come from very near the surface, and not from subsurface regions.

These simulations have also shed light on whether the gas-phase molecules observed arise from molecules which were originally bound on the surface. In other words, are the detected molecules a direct picture of the surface? Computer simulations have clearly demonstrated that molecules such as CO and C_6H_6 can survive the sputtering process without necessarily fragmenting (Fig. 2). This is due in part to two reasons. First, the roughly equivalent effective sizes of metal substrate atoms and adsorbed hydrocarbon molecules facilitates the ejection of molecules without significant fragmentation. For example, the hard-sphere diameter for a single Rh atom in a crystal is 2.7 Å and the van der Waals diameter for a four-atom CH_3 group is 4 Å[66]. This means that a moving Rh atom can 'see' this group as a single species, and can eject it without dissociating the CH bonds. The second reason for the intact ejection of large molecules is that they contain many vibrational degrees of freedom which can absorb kinetic energy that would otherwise lead to fragmentation. This contribution is especially important for the intact ejection of large biomolecules mentioned above[56].

The fact that molecules can survive the ion impact is encouraging for correlating gas-phase species with surface structure. Unfortunately, sputtered atoms and molecules have also been observed in the simulations to recombine into bound molecules in the near-surface region. This mechamism can lead to the detection of stable molecules which were not present on the surface, and can complicate the interpretation of experimental results. For example, the observation of NiCO and Ni_2CO clusters from the sputtering of a CO covered Ni surface could be interpreted as indicating that chemisorbed CO molecules are bound both to single Ni atoms (atop binding sites) as well as pairs of Ni atoms (bridge binding sites)[67]. This could be an erroneous interpretation because recombination processes can also lead to the formation of multiply bonded CO molecules[68]. The recombination mechanism, however, does not completely inhibit the ability to interpret gas-phase structures because it only occurs between species that are originally in close proximity on the surface. The general picture which simulations yield is that the identity of sputtered material can be used to identify surface species, but that one must be cautious in how far the interpretations can be taken.

One important technique which has been confirmed by simulation is the ability to deduce local bonding arrangements on the surface from the angular distribution of ejected species. The local geometry surrounding an adsorbed molecule is anisotropic. That is, along some directions a moving atom would quickly encounter a substrate atom while in other directions it would encounter an open space, perhaps between two substrate atoms along $\phi = 0°$

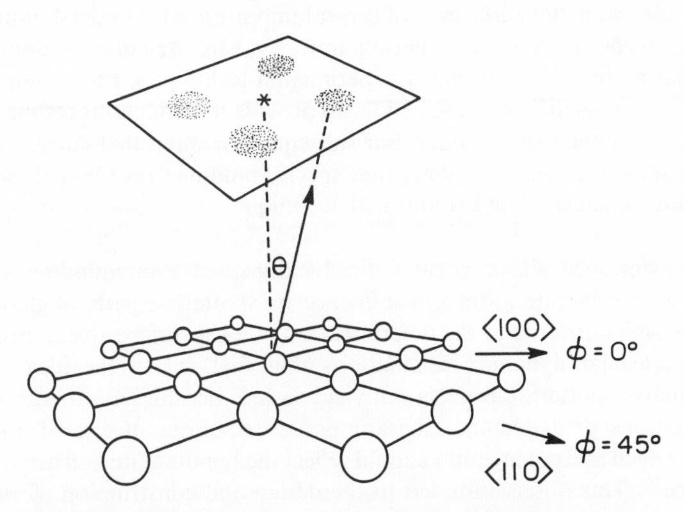

Fig. 3. Schematic representation of the angular distribution of ejected atoms from the (001) crystal face of a face-centered cubic metal. The polar angle is θ and the azimuthal angle $\phi = 45°$ corresponds to the close-packed row of surface atoms. (*From Ref. 1.*)

of Fig. 3. This symmetric arrangement serves to direct the motion of ejected species into regions where open surface channels exist. The collective effect over the sputtering of many surface atoms is that the magnitude of the detected gas-phase species is enhanced for particular orientations. Computer simulations have been shown to agree with experimental results in regards to the angular distributions, and have confirmed this channeling mechanism[69]. Furthermore, experiments have demonstrated that in most cases the angular enhancement is more dramatic for particles which eject with high kinetic energies. Again, computer simulations agree with this general conclusion, and suggest that the reason for this effect is that the higher-energy particles eject early in the collision cascade when most of the local structure is still well defined. Specific studies of this kind are discussed below.

The use of molecular dynamics simulations for understanding and interpreting SIMS experiments has been quite successful. This is despite the use of rather crude potentials, and is possible because the dynamics of interest depend primarily on the crystal structure, which is reflected in the short-range repulsion between atoms. But just as one must be cautious about carrying the interpretation of SIMS results too far, care should be taken in implying too much from this type of simulation. With the general picture of sputtering outlined above, this discussion is continued below by considering some specific details of various SIMS simulations.

2.2. Angular distributions and surface structure

The determination of the equilibrium binding sites of atoms and small molecules on metal surfaces is of central importance for understanding gas–surface reactions. Such information, however, can often only be obtained by correlating the results of many experimental techniques. For example, low-energy electron diffraction (LEED) can provide unambiguous results regarding the symmetry of a surface, but subsequent complicated current–voltage (I–V) analysis is needed to determine specific binding sites. Often these results are not conclusive, and additional techniques are required to confirm a result.

As mentioned above, because the local geometry surrounding a chemisorbed or substrate atom can influence its sputtering path, angle-resolved SIMS can contribute to the determination of local surface structures. It is in fact molecular dynamics calculations that first led to the idea that the destructive sputtering process can yield surface bonding information[70]. The calculations predicted that the angular distributions of ejected adsorbate species such as oxygen atoms should reflect the bonding site and height on the surface[70]. This observation led to the design and construction of an angle-resolved secondary ion mass spectrometry (SIMS) apparatus[71].

Continued interplay between experiment and theory has prompted the development of a new method to measure the energy- and angle-resolved neutral (EARN) atom distributions[57–59,72]. In this technique, ejected neutral atoms are ionized by a laser above the surface. Because the ionization of the sputtered neutral atoms takes place at a different point in the trajectory than the atoms which are ionized during the SIMS processes, an applied potential bias serves to separate the two types of ejected species. This latest development now allows direct comparisons between predicted and measured particle trajectories. Previous to the EARN experiments, the SIMS experiments measured *ion* distributions while the calculations described the *neutral* distributions. The ability to make detailed and direct comparisons between the molecular dynamics model and the measured energy and angle distributions of neutral particles that eject during keV ion bombardment has provided the impetus for significantly expanding the scope and applications of the molecular dynamics technique. Below are given four examples of how combined experiment and molecular dynamics calculations have provided insight into bonding geometries on surfaces. The first two are EARN experiments on clean and oxygen-covered Rh(111) in which neutral Rh atoms are detected, and the latter two are SIMS experiments in which ions are detected.

2.2.1. Clean Rh(111)

The first detailed comparison between EARN data and molecular dynamics simulations was on Rh(111). The initial simulations were performed using a pairwise additive potential[73]. The polar angle distributions from both the calculations and experiment are shown in Fig. 4. The azimuthal angles are defined at the top of the figure. Examination of the pair-potential results (right-hand side of the figure) reveals overall semiquantitative agreement between the calculated and measured distributions. The peak intensity along the azimuthal direction $\phi = -30°$ is greater than along $\phi = +30°$ or $0°$ for all secondary-particle energy ranges. As the particle energy increases, the $\phi = +30°$ intensity increases relative to the $\phi = 0°$ intensity. Both theory and experiment find that the intensity at $\theta = 0°$ relative to the peak intensity ($\theta = 25$–$40°$) increases as the Rh atom energy increases. The agreement between the measured and calculated distributions is quite remarkable as pair potentials were used, and no adjustment of the parameters in conjunction with this experimental data was performed. Obviously, the pair potential description is a reasonable first approximation to describe the ejection events. The reason for this agreement is that much of the ejection process is dominated by the surface structure and not the details of the interaction potential. Of note is that the pair potential does very well at reproducing the azimuthal anisotropy. There are discrepancies between the experimental and calculated results which can be removed by using a many-body EAM interaction potential in the simulations. These results are discussed in section 3.2.

It is very difficult experimentally to determine the positions of the second-

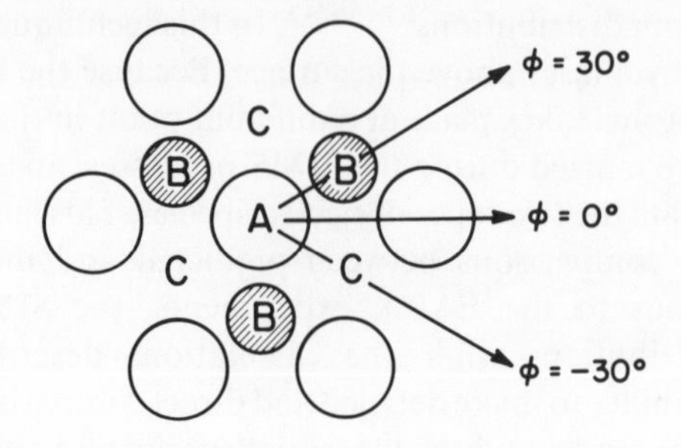

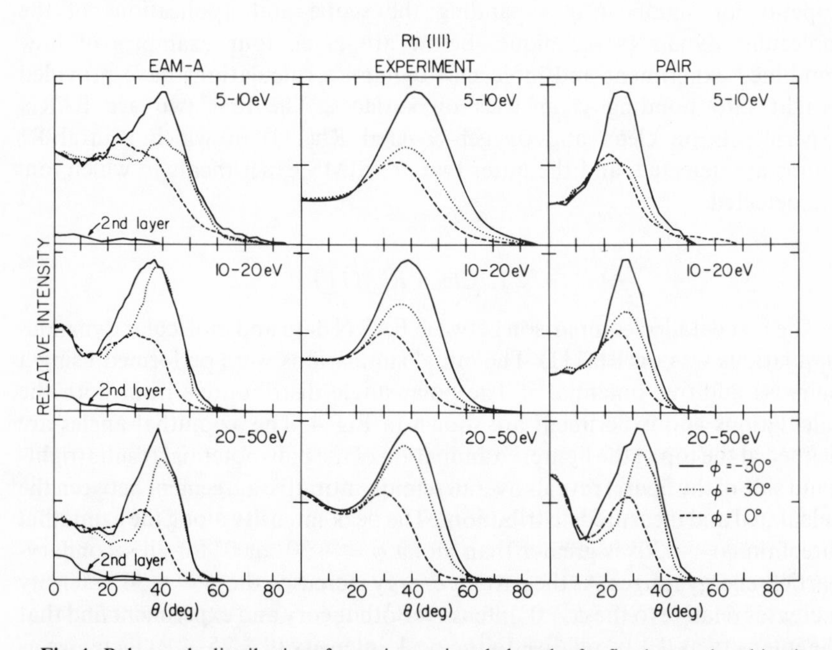

Fig. 4. Polar angle distributions for various azimuthal angles for fixed secondary kinetic energy of the Rh atoms. In each frame the data are normalized to the $\phi = -30^\circ$ peak intensity. For the calculated data the full width at half maximum (FWHM) of the resolution is 15° in the polar angle. A constant solid angle is used in the histogramming procedure. The experimental resolution is approximately the same. The surface normal corresponds to $\theta = 0^\circ$ and $\theta = 90^\circ$ is parallel to the surface. The curve marked '2nd layer' is the polar distribution along $\phi = -30^\circ$ for the ejected second-layer atoms. The azimuthal directions are defined above. Open circles designate first layer atoms and shaded circles second-layer atoms. The letters A, B, and C designate possible adsorption sides for oxygen atoms. (*From Ref.* 9.)

layer atoms in the Rh(111) (or any face-centered cubic(111)) face. Note that there is a six-fold rotation axis (Fig. 4) if only the surface atoms are considered but that if the second-layer atoms are considered there is a three-fold rotation axis. With LEED it is difficult to determine the placement of the second-layer atoms without a detailed I–V analysis. With the EARN experiment in conjunction with the simulations the determination of the absolute crystal orientation is straightforward. The sputtered intensity along $\phi = -30°$ is greater than along $\phi = +30°$. This ambiguity in placement of the second-layer atoms carries over to the placement of adsorbates on the surface. For example, oxygen atoms tend to bond in high-coordination sites, for example, the three-fold hollow sites on Rh(111). However, there are two such sites, one where the oxygen atom would be directly above a second-layer atom (B-site) and one where the oxygen atom would not be above a second-layer atom (C-site). In the following section we show how the EARN distributions in conjunction with the calculations can determine both the site and coverage of oxygen on Rh(111).

2.2.2. Oxygen-covered Rh(111)

Oxygen adsorbs atomically on Rh(111) at room temperature in an ordered overlayer structure that yields a (2×2) LEED pattern[74]. There are at least two structural unknowns for this system – one that involves coverage and one that involves adsorption site. A simple (2×2) structure with one adsorbate atom per unit cell results in a 0.25 monolayer coverage, that is, one oxgen atom per four Rh surface atoms. However, if there are three domains of a 0.5 monolayer coverage in an arrangement where each domain would yield a (1×2) pattern, the resulting overlapping LEED pattern looks like a (2×2). Thus the adsorbate coverage is not apparent from the LEED pattern. Second, oxygen is generally believed to adsorb in three-fold hollow sites on the Rh(111) surface. As shown in Fig. 4 there are two three-fold hollow sites on the (111) surface, one above a second-layer atom (B-site) and one not above a second-layer atom (actually above a third-layer atom (C-site)). To distinguish between these two three-fold sites experimentally is extremely difficult. In the EARN distributions intuition indicates that if an oxygen atom is adsorbed in a C-site then it should preferentially attenuate ejection in the $\phi = -30°$. If it is adsorbed in the B-site, however, it would preferentially attenuate ejection in the $\phi = +30°$ direction.

The EARN distributions of Rh atoms ejected from O/Rh(111)-(2×2) have recently been measured[75]. In addition simulations were performed for the oxygen atoms in both of the three-fold sites and for both the 0.25 and 0.5 monolayer coverages. Experimentally the Rh atom yield decreases by about a factor of two with the oxygen adsorption. To reproduce this effect in the simulations the 0.5 monolayer coverage was required. In the experimental

EARN distributions the yield along the azimuth $\phi = -30°$ was preferentially reduced with respect to $\phi = +30°$. In agreement with intuition, the calculations confirm that the oxygen atom resides in the C-site. Thus from the cooperation of EARN experiments and computer simulations the coverage and nature of the adsorption site of O/Rh(111) has been determined. It will be of interest to see if other surface structure techniques can be used to confirm these specific surface structures.

2.2.3. *Pyridine versus benzene*

In addition to the enhancement of sputtering yields for specific azimuthal angles, structure in the polar (defined from the surface normal) angular distributions can also be used to determine bonding geometries. When benzene and pyridine molecules are adsorbed at low coverages on a Ag(111) surface, molecule–surface bonding can occur through the molecular π-cloud. This results in the molecules lying flat on the surface. As the molecular coverage is increased, however, the molecules become 'crowded' and repel one another. For the case of benzene, the π-bonding is stronger than the repulsive intermolecular forces, and the molecules remain flat on the surface. Pyridine, however, can also σ-bond to the surface through the N atom, and so the rings can lift from the surface while leaving the N atom σ-bonded[76]. The net effect of increasing pyridine coverage is therefore thought to lead to a 'standing' orientation of the molecules. It was believed that this orientation would effect the angular distributions of sputtered molecules, and so a combined computer simulation–SIMS study was undertaken[77–79].

Computer simulations of the sputtering of benzene and pyridine molecules suggest that the molecular orientation influences the sputtering dynamics in two respects[79]. First, it was observed that the π-bonded rings often eject intact, while the upright σ-bonded rings have a very small cross-section for molecular desorption. The second effect observed is that polar-angle distributions are sharper for the σ-bonded molecules than for the π-bonded molecules. This is a result of the channeling of ejected molecules upward by their upright neighbors, a process which does not occur for the π-bonded molecules. Furthermore, it appeared that particles ejected with low kinetic energy showed sharper distributions than those ejected with higher kinetic energies. This is different from the case of azimuthal-angle enhancements, where the higher-energy particles reflect the symmetry of the initial environment of the sputtered species better than low-energy particles. The reason for this difference is that the upright molecules are easily moved, and so molecules with high kinetic energy distort the local environment more than low-energy particles, and thus have broader polar distributions.

The SIMS experiments were performed by sputtering both adsorbed benzene and pyridine at various coverages from Ag(111) surfaces[77,78]. For increasing coverages of pyridine, and presumably increasing 'upright angle', a decrease in the intensity of sputtered intact molecules was observed. This is consistent with the dynamics predictions, where again decreased molecular ejection was observed for standing configurations of pyridine over the π-bonded benzene molecules. In addition, experimental polar-angle distributions showed agreement with the dynamics results. For the sputtering of benzene and low coverages of pyridine, the angular distributions were relatively broad. As the pyridine coverage was increased, however, the polar distributions for the ejected pyridine molecules became more sharply peaked, with low-energy molecules showing slightly sharper peaks than those ejected with high energies. Both of these effects were predicted by the dynamics calculations, and are easily understood based on the molecular orientation of the molecules on the surface.

2.2.4. CO on Ni(7 9 11)

Steps and defects are thought to exert a large influence on the chemical reactivity of surfaces, and so an understanding of the role which they play in the binding of adsorbates is important. When surfaces containing atomic steps are sputtered, it would be expected that the azimuthal angle of the incident ion beam would affect the sputtering results. For example, an ion beam that was angled so that the bottom of a step was bombarded would be expected to 'peal off' more surface atoms than one that was angled to bombard the top of a step. This would lead to an enhancement of the ion yield for certain azimuthal angles of the ion beam. Similarly, if adsorbates showed preferential bonding to the top or bottom of a step, the sputtering yield or molecular clusters would also show an azimuthal-angle dependence of the ion beam. To test this hypothesis, a combined SIMS and computer simulation study of the adsorption of CO on a stepped Ni(7 9 11) surface was undertaken[80].

The Ni(7 9 11) surface explored in this study was composed of terraces with (111) orientation five atomic rows wide separated by steps of monatomic height with (110) orientation. An ensemble of surface science techniques have suggested that adsorbed CO molecules on this surface preferentially bond to sites at the bottom of the steps at low coverages and temperatures[81,82]. At higher temperatures and coverages, however, the CO molecules are thought to bind both at the steps and on the terraces. At the bottom of the step two bridge sites are accessible, with one being two-fold coordinate and the other three-fold coordinate. The occupation of both sites had been suggested based on previous studies.

In the SIMS experiment the CO covered Ni surface was sputtered using an ion beam whose azimuthal angle could be varied[80]. The ion yields for various ejected species were found to vary with sputtering angle, with the greatest sensitivity being shown by sputtered $NiCO^+$ clusters. For low coverages, the

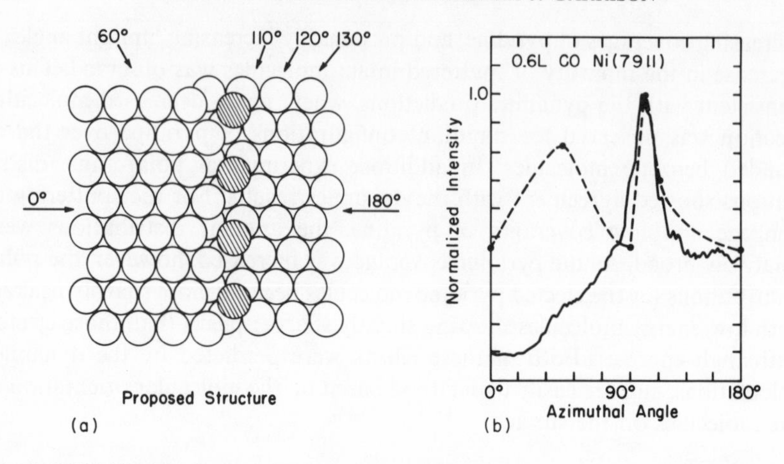

Fig. 5. Ni(7 9 11) with adsorbed CO. (a) Proposed structure. The open circles represent Ni atoms and the shaded circles are CO molecules. The numbers refer to the incident azimuthal angle of the primary ion. (b) $NiCO^+$ intensity versus azimuthal angle of the bombarding Ar^+ ion. The nickel surface was exposed to 0.6 L of CO. The solid line represents the experimental data and the dashed line results from the classical dynamical calculation. The additional peak at $\phi = 60°$ is not yet understood. (*From Ref. 1.*)

yield of the $NiCO^+$ ions with the ion beam aligned towards the bottom of the step (180° in Fig. 5) was found to be twice that with the beam aligned down the step (0° in Fig. 5). This result confirmed the concept of using aligned ion beams to study the structure of adsorbates at steps. An additional feature of the results was a sharp peak in the $NiCO^+$ distribution for the Ar^+ azimuthal angle of incidence of $\sim 120°$ (Fig. 5). This feature disappeared for higher coverages, and so appeared to be a signature of the binding site of the CO molecules.

Computer simulations were performed for CO molecules adsorbed at the two proposed sites at the bottom of the step and for CO molecules adsorbed in bridge sites on the terrace[80]. The experimental results for low CO coverages could only be reproduced for the CO molecules adsorbed at the two-fold bridge site at the bottom of the step. However, no distinct formation mechanism for the sputtered NiCO species which contributed to the signature peak could be discerned. The only general trend seemed to be that these species were not composed of CO and Ni atoms which were originally bound on the surface, but formed via recombination in the near-surface region.

As demonstrated in this section, the agreement between simulation and experimental results for keV particle bombardment of solids is remarkable. This is especially true when one considers the rather crude potentials used in the calculations. To understand the reason for this agreement, the underlying features of the dynamics should be reviewed. The surface structures which are deduced from experiment are clearly caused by complicated many-body interactions. 'Guesses' for these structures enter into the simulations as initial conditions, and if left to evolve under the influence of the pair-additive potentials, the atoms would undoubtedly collapse into nonphysical configurations. The sputtering process, however, occurs much faster than these relaxations (10^{-13} s), and so the repulsive collisions dominate the dynamics. Again, for this regime pair-additive potentials are often adequate. Since the final atomic positions and velocities of the sputtered material reflect the initial configurations, structures which depend on complicated many-body interactions are deduced using simple short-range forces. This is the key to understanding the SIMS and EARN simulation results.

Several topics which are important to the keV particle bombardment processes have not been discussed here. Foremost is the ionization probability for sputtered species. In the traditional SIMS mode, only those species which eject charged are detected whereas in the simulations the motions of neutral species are determined. This complication is currently being overcome in two ways. First, models which include ionization effects have been incorporated into dynamics simulations[83,84]. While this is the obvious approach, the appropriate manner in which to treat the ionization problem is not clear. The second approach is to ionize sputtered neutral atoms after ejection in the experiment, and to study the properties of these ions. The significant progress recently made using lasers to post-ionize sputtered neutral atoms was discussed above and results deduced in this way have led to an enhanced understanding of ion-surface energy transfer[57–59]. A second topic which has not been addressed here is the influence which the surface image charge has on the trajectories of sputtered ions. This is a relatively simple effect to study, and simulations have been performed which mimic these forces[85]. Finally, the enhanced agreement with experiment through the use of more realistic potentials has not been explored. Studies of this type, along with comparisons with data for post-ionized sputtered neutral atoms, are discussed in the next section.

3. SIMULATIONS RELYING ON REALISTIC MEDIUM-RANGE INTERACTIONS

In this section examples of classical dynamics simulations where the kinetic energies of the atoms involved are on the order of typical bond energies are discussed. This energy regime is more difficult to model than the higher-energy dynamics discussed in the previous section because the many-body nature of the forces strongly governs the dynamics. Most chemical reactions of interest, however, take place in this energy regime, and so considerable effort has gone into studies of this type.

In section 3.1, reactions of diatomic molecules with metal surfaces are discussed. These studies, although perhaps not sufficiently complicated to directly address processes of technological interest, have produced considerable insight into the dynamics of gas–surface reactions. Simulations of metal surfaces where more realistic interactions are required than are used in the gas–surface studies are presented in section 3.2. This is followed in section 3.3 by a discussion of simulations of reactions on the surfaces of covalently bonded solids. These final studies are particularly suited for addressing technologically relevant processes due to the importance of semiconductor technology.

3.1. Reaction of gas-phase molecules with metal surfaces: modified LEPS potentials

The exchange between the gas-phase and chemisorbed states of small molecules plays a vital role in such technologically important fields as heterogeneous catalysis and corrosion. The dynamics involved in these processes, however, are not currently well understood. Molecular-beam studies combined with classical trajectory calculations have proven to be a successful tool for understanding the underlying features of atomic-scale motion in the gas phase. The extension of these techniques to surfaces has also helped in elucidating the details of gas–surface reactions.

One formalism which has been extensively used with classical trajectory methods to study gas-phase reactions has been the London–Eyring–Polanyi–Sato (LEPS) method[86,87]. This is a semiempirical technique for generating potential energy surfaces which incorporates two-body interactions into a valence bond scheme. The combination of interactions for diatomic molecules in this formalism results in a many-body potential which displays correct asymptotic behavior, and which contains barriers for reaction. For the case of a diatomic molecule reacting with a surface, the surface is treated as one body of a three-body reaction, and so the two-body terms are composed of two atom–surface interactions and a gas-phase atom–atom potential. The LEPS formalism then introduces adjustable potential energy barriers into molecule–surface reactions.

Although the theoretical roots of this technique are very well established, it is more often used as a flexible surface which can be adjusted to fit either exprimental data or data established by better electronic-structure methods. The LEPS formalism has also been extensively used to explore the relationships between the potential energy surface and the details of chemical dynamics[87]. Because of the widespread use of this potential for studying gas-phase reactions, the specific form of the equations will not be discussed here. The interested reader is instead referred to references which discuss this approach in more detail[23,86,88].

3.1.1. *Hydrogen–metal reactions*

Exchange reactions of $H + H_2$(or H_3) have provided the testing ground for theoretical methods which are used to understand gas-phase chemical dynamics[89]. Interest in modeling the reaction of hydrogen with metal surfaces is therefore not unexpected. In addition, hydrogen often plays an important role in reactions associated with catalysis, so studies of this type also have practical application.

Several researchers have used modified forms of the LEPS potential (see section 1.2 for a discussion of the modifications) to study the dynamics of H_2 on the surfaces of various metals. Initial studies of this type were restricted to rigid surfaces, and the parameters in the LEPS surfaces were either determined by fitting to available experimental or theoretical data, or systematically varied to produce potential energy surfaces with specific properties.

In a series of studies, McCreery, Wolken and coworkers have used a LEPS potential to model reaction of H_2 and HD with the W(001) surface[23,90,91]. The substrate in these studies was restricted to be rigid, and Morse functions were used for the hydrogen–surface and H_2 two-body interactions. The parameters in the Morse functions were determined for single hydrogen atoms adsorbed on the tungsten surface by fitting to extended Hückel molecular orbital (EHMO) results, and the H_2 Morse parameters were fit to gas-phase data. The Sato parameter, which enters the many-body LEPS prescription, was varied to produce a potential barrier for the desorption of H_2 from the surface which matched experimental results.

Using their H_2–W potential, McCreery and Wolken have modeled the recombination and desorption of two H atoms that are initially adsorbed on the surface[90]. These simulations were initiated by dividing 3.5 eV of translational energy between the two adsorbed atoms. Classical equations of motion were then integrated until desorption was observed, or it was determined that both atoms would remain on the surface. They reported that 3.4% of the atomic collisions resulted in the desorption of an H_2 molecule. An analysis of the initial translational energies of two hydrogen atoms showed that this mode was enhanced for cases where the two atoms had similar energies. They also reported that 4.6% of the collisions resulted in the desorption of only one of the hydrogen atoms. This mode of reaction was enhanced when one of the atoms possessed almost all of the initial energy. Desorption of both hydrogen atoms into isolated gas-phase species was not possible, since the total energy was insufficient to reach this state.

Energy and angular distributions for the desorbed species were also analyzed. The polar angular distributions for all species were reported to be peaked toward the surface normal, and did not have the cosine form which is indicative of thermal processes. This was shown to result because atoms which approach at large impact parameters along the surface do not react. This orientation would lead to desorption with large polar angles, and so this region of the distribution was depleted. They also showed that the distribution of energy among the degrees of freedom of the desorbed H_2 molecules matched those predicted by statistical theories[92,93].

Elkowitz, McCreery, and Wolken have also examined the reaction of a gas-phase hydrogen atom with a hydrogen atom initially adsorbed on the tungsten surface using the same LEPS function[91]. They report that for an initial energy of the incoming hydrogen atom of 0.44 eV, 1.3% of the collisions resulted in the desorption of a hydrogen molecule, while 12.8% led to the desorption of one of the atoms. Furthermore, they report that in a vast majority of cases the atom initially on the surface remains adsorbed unless molecular desorption occurs, and the reflection of the incoming atom tends to be unaffected by the presence of the surface species. Unlike the combination and desorption of the two atoms initially on the surface, they report that the energy distribution of the desorbed H_2 molecules is not well described by statistical theories.

McCreery and Wolken have also extended their model by exploring the effect of various potential energy surfaces on the dynamics of hydrogen atoms on a metal surface, and by extending the LEPS prescription so that three atoms adsorbed on the surface could be modeled[88]. Varying heights of the potential barrier were used to explore the relationships between the chemical dynamics of reaction and the potential surface for the chemisorption of H_2 molecules[94,95]. For large barrier heights, they found that translational energy was more effective in promoting the chemisorption of the H_2 molecule than was internal energy. The topology of the potential surface was further changed in a study of the combination and desorption of H_2 from this surface by varying both the barrier heights and relative exothermicity of the reaction[23]. In both of the dynamics studies, the populations of various energy modes and their relationships to the chemical dynamics agreed with trends predicted by gas-phase studies, and so the application of gas-phase models to gas–surface reactions was verified.

Other researchers have used LEPS plus rigid-surface potentials to study the reaction of H_2 with metal surfaces. Gelb and Cardillo have used a LEPS potential to model the reaction of gas-phase H_2 molecules with Cu(001) and Cu(011) surfaces[96–98]. Their studies have suggested that a 'rough' surface (that is, one with a high barrier for surface diffusion) is needed to match experimental molecular-beam results[98]. Furthermore, they reported that increased translational and vibrational energy of the incoming H_2 molecule enhances the probability of dissociative adsorption, while rotational energy was ineffective in promoting reaction.

In efforts to improve upon the LEPS scheme outline above, other prescriptions for the single atom–surface interaction have been formulated. The initial studies using the LEPS approach modeled the atom-surface interaction as a two-body term where the parameters used in the function are dependent on the position of the atom within the surface until cell. These interactions are not transferable between different crystal faces, and they do not allow the study of amorphous surfaces or surfaces which contain defects. Purvis and Wolken have addressed this problem by proposing that the atom–surface two-body term that enters the LEPS formalism be replaced with a sum of two-body terms between the adsorbed atom and the surface atoms[27]. This modification allows the study of surfaces which do not possess periodicity. Lee and DePristo have further extended the LEPS formalism by replacing the two-body atom–surface interaction with other more realistic many-body interactions[99,100]. These relatively complicated functions combined with the LEPS formalism have produced potential energy surfaces which are transferable between crystal faces, and which can be adjusted to reproduce dynamical results.

Another advance in the LEPS description of H_2–metal surface dynamics has been the introduction of moving substrate atoms[101–103]. The use of a static surface can be justified on the basis of the difference in mass between the hydrogen atoms and the metal atoms. The substrate atoms are sufficiently heavy compared to the hydrogen atoms that they do not have sufficient time to move during much of the dynamics of interest. For multiple collisions, however, the timescale of reaction becomes longer, and including substrate atoms which can adjust to the influence of the motion of the molecule can be important. This motion can change the features of the potential energy surface, and conclusions drawn from rigid surface studies may be affected. Furthermore, energy transfer between the H_2 molecule and the substrate is not incorporated with rigid surfaces, and so moving substrate atoms are necessary to produce long-lived trapping of an incoming adsorbate on the surface.

The initial studies which incorporated moving substrate atoms used correction terms to the rigid-surface LEPS potential functions, and did not specifically include temperature effects in the dynamics of the substrate[101,103]. Further studies, however, incorporated the generalized Langevin equations[63–65] into the equations of motion governing the dynamics of the substrate. This advance introduced a realistic temperature-regulated response of the substrate to the surface dynamics without significantly increasing the number of atoms explicitly entering the simulation. It also reduces the number of approximations which may affect dynamical results, and allows the study of systems where the mass of the reacting surface atoms is comparable to the atoms in the substrate.

Studies of H_2 have proven the feasibility of using the LEPS formalism to study gas–surface reactions, and have indicated that relationships between the potential surface and chemical dynamics derived from gas-phase studies can be generalized to reactions with surfaces. Reactions of H_2, however, represent simple systems compared even to other diatomic molecules, and extensions to other more complicated reactions are rare. A few studies of other diatomic molecules interacting with metal surfaces have been undertaken, and the results of these studies have helped in elucidating the dynamics of complicated surface reactions.

3.1.2. Nitrogen, oxygen and carbon monoxide–metal reactions

The chemical reactions of N_2, O_2 and CO on metal surfaces are the ones that have received the most attention both from experimentalists and theorists. Kara and DePristo have examined the dissociative chemisorption of N_2 on the W(011) surface using the combination many-body atom–surface interactions plus LEPS potential formalism[104] which was developed by Lee and DePristo to model H_2 adsorption[100]. For this system, molecular beam studies have demonstrated that the probability of dissociative chemisorption scales with the normal component of the initial translational energy for large incoming polar angles, but it scales with total energy for angles between 0° and 45°[105]. This is different from most other cases, where the probability for reaction depends on the normal component at all angles. Kara and DePristo demonstrated that the energy scaling for dissociative chemisorption could be understood based on the width of the transition state at the potential energy barrier. For a narrow transition state, their results show that the probability for dissociative chemisorption scales with total kinetic energy. This is because a restricted atomic configuration is required to overcome the barrier, and so the energy must redistribute (with a sufficient amount of energy entering the bond stretching mode) for reaction to occur. For a wider barrier, however, the molecule is less restricted, and so the normal contributions to the translational energy is directly converted to the stretching of the N—N bond. This leads to dissociation of the bond and chemisorption. This study demonstrates the ability of the LEPS formalism to be adjusted so that the characteristics of reaction can be correlated with features of the potential energy surface.

Another system which displays complex behavior is the reaction of oxygen on silver. For the Ag(011) surface, O_2 displays molecular adsorption at substrate temperatures below 185 K[106–109], and dissociative chemisorption at substrate temperatures between 185 K and 600 K[106–113]. Furthermore, overlayers of chemisorbed oxygen atoms produce $(n \times 1)$ structures, where n varies from 7 at low coverage to 2 at the highest coverage[111–113]. The low coverage (7×1) surface structure is composed of rows of oxygen atoms that are separated by 20 Å. This implies a long-range repulsive force between the atoms along the direction between rows, and a shorter-range interaction between atoms within the rows. Both experimental and theoretical studies have suggested that the physical origin of this interaction is a transfer of electron density from the solid to the adsorbates which leaves the oxygen atoms with a net negative charge[106,108,109,114,115]. This charge apparently results in an anisotropic long-range repulsion between adsorbed atoms.

In an effort to model this system, Lin and Garrison modified the LEPS prescription to include a long-range anisotropic force between adsorbed oxygen atoms[116]. The oxygen–surface bonding interactions were modeled by a Morse function which has parameters that depend on the position of the oxygen atom in the surface unit cell. The oxygen–oxygen two-body interaction was also modeled using a Morse function. The anisotropic contribution to the potential was incorporated into the two-body oxygen–oxygen antibonding interaction that enters into the LEPS formalism. A long-range $1/R$ (where R is the scalar distance between atoms) dependency was used for the component of this interaction in the direction between rows, while a shorter-range exponential function was used in the direction within the rows. This resulted in an oxygen–oxygen interaction which produces the $(n \times 1)$ structures for varying coverages. Also, because it enters the antibonding two-body interaction (instead of the bonding interaction) in the LEPS prescription the long-range interaction does not change the gas-phase characteristics of the O_2 molecule.

The parameters in the O_2-Ag LEPS potential were determined by fitting to a variety of experimental data, and the surface was restricted to be rigid. Two different potential energy surfaces were tested, both of which were consistent with the available data, but which differed in bonding characteristics for the O_2 molecule. For the first potential function, classical dynamics trajectories showed that both molecular and atomic adsorption were possible, while for the other the reaction of O_2 resulted only in dissociative chemisorption. Both potentials, however, displayed sticking probabilities which are strongly dependent on the orientation of the molecule with respect to the surface. Although the results are somewhat inconclusive due to the lack of sufficient data to determine a unique potential surface, this study did demonstrate that the LEPS formalism can be modified to include long-range electrostatic interactions. This suggests that the LEPS approach can be sufficiently flexible that the dynamics of complicated systems can be modeled.

The reaction of CO with platinum is an example of another complicated and technologically important system which has been studied using the LEPS formalism. Tully has modeled the abstraction of a chemisorbed C atom from the Pt(111) surface by a gas-phase O atom using a LEPS potential and a dynamic surface[117]. This abstraction reaction is exothermic, and so Tully was interested in the partitioning of the energy released by the reaction. The parameters entering the potential surface were determined from a variety of experimental measurements, and realistic temperature effects were included by using a generalized Langevin equation to govern the motion of the lattice atoms.

The classical trajectories were initiated with the C atom on the lattice near the lowest-energy binding site, and the O atom heading toward the surface aimed in the vicinity of the C atom. The exothermicity of the reaction combined with the small barriers in the potential surface resulted in a large reaction probability. Furthermore, Tully reported that most of the energy of reaction is carried away by the desorbing molecule rather than being deposited into the surface. He reports an approximate ratio of energy deposited into translational, vibrational and rotational modes of the CO molecule of 2:2:1. Although he cautions that uncertainties in the potential energy surface makes this ratio inconclusive, variations of the potential within reasonable limits are reported to still result in the majority of the released energy being carried away by the CO molecule.

The modification of theoretical gas-phase reaction techniques to study gas–surface reactions continues to hold promise. In particular, the LEPS formalism appears to capture a sufficient amount of realistic bonding characteristics that it will continue to be used to model gas–surface reactions. One computational drawback of the LEPS-style potentials is the need to diagonalize a matrix at each timestep in the numerical integration of the classical equations of motion. The size of the matrix increases dramatically as the number of atoms increases. Many reactions of more direct practical interest, such as the decomposition of hydrocarbons on metal surfaces, are still too complicated to be realistically modeled at the present time. This situation will certainly change in the near future as advances in both dynamics techniques and potential energy surfaces continue.

3.2. Metals: the embedded-atom method

The structure and dynamics of clean metal surfaces are also of importance for understanding surface reactivity. For example, it is widely held that reactions at steps and defects play major roles in catalytic activity. Unfortunately a lack of periodicity in these configurations makes calculations of energetics and structure difficult. When there are many possible structures, or if one is interested in dynamics, first-principle electronic structure calculations are often too time consuming to be practical. The embedded-atom method (EAM) discussed above has made realistic empirical calculations possible, and so estimates of surface structures can now be routinely made.

Prior to the development of the EAM[13–17] a majority of simulations of metals and metal surfaces used pair-additive potentials. These potentials, however, rarely yield a good description of most metals. For example, if the parameters entering a typical pair potential are determined from an isolated diatomic molecule, the binding energy for the metal is generally overestimated, and the lattice constant is usually too small. To circumvent this problem, pair terms can be fit to properties of the bulk metal without regard to the isolated dimer molecules[118]. Although lattice constants and binding energies can be reproduced this way, other properties (particularly those of small clusters) are often meaningless. For example, pair potentials generally predict an outward expansion of surface layers, while surface relaxations for most metals involve a contraction.

The EAM has been widely used to predict and understand the structure of the surfaces of metals. One system that has been thoroughly studied is gold. Ercolessi, Parrinello and Tosatti (and also Dodson[19]) have used the EAM to determine a likely structure for the reconstruction of the Au(100) surface[15,16]. Using molecular dynamics to relax the surface, Ercolessi *et al.* determined that a (35 × 5) reconstruction of this surface is the lowest energy structure. This observation is very similar to a (5 × 1) structure reported by Dodson[19]. This reconstruction is a result of a rearrangement of surface atoms into a more densely packed configuration, and also involves a relaxation of the first few atomic planes below the surface. Both Ercolessi and Dodson report quantitative agreement with experimental observations, including scanning-tunneling microscopy results[119].

Garofalo, Tosatti and Ercolessi have also studied the structure of the Au(110) surface using the same method[17]. For this surface, their molecular dynamics studies gave a (1 × 2) missing row geometry as the lowest-energy structure. This result is in agreement with both experimental evidence[120–125] and other theoretical calculations[126,127]. They also found that other similar missing row structures are very close in energy, and that the appearance of these other structures at finite temperatures may account for additional experimental observations.

The EAM has been used to study the surface structure of other metals and metal alloys. For example, Daw has suggested that a missing row configuration is also the likely structure for the (2 × 1) reconstruction of the Pt(110) surface[14]. Studies have also been made of the surface structures of various alloys, where for example surface segregation of one constituent over the other has been observed[20,128–130]. In addition to studies of specific systems, the EAM formalism is also sufficiently general that it has been used to understand trends in surface reconstructions among various metals[131,132].

The EAM method has proven to provide structures and energetics at thermal energies which agree with experiment for a large number of metals. Simulations of atomic dynamics at higher energies, however, have been fewer in number. In section 2 the case was made that pair potentials are generally adequate for describing the short-timescale dynamics involved in sputtering. Some of the results of the simulations, however, are very dependent on the potential. For example, the yield of Rh_2 ejected from Rh(111) is impossible to determine using pair potentials that describe only the bulk energetics. This has created interest in using the EAM formalism to determine potential functions which can be used at a variety of energies, and which describe both few-atom clusters as well as the bulk metals[32,133,134].

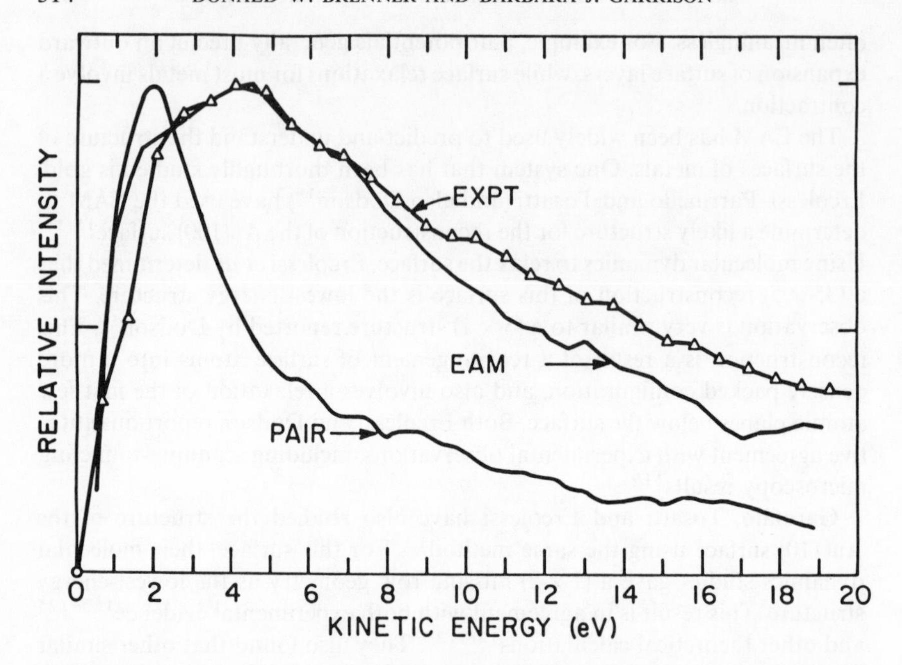

Fig. 6. Experimental and calculated angle-integrated kinetic energy distributions. In all cases the curves are peak normalized. (*Modified from Ref. 9.*)

As shown in Fig. 4, molecular dynamics calculations using pairwise-additive potentials do quite well at reproducing and explaining the angular distributions of Rh atoms ejected from Rh(111). The question is then, are many-body potentials necessary or is the ejection process dominated by crystal structure and thus the repulsive wall of the potential? Shown in Fig. 6 are measured and calculated (with pairwise-additive potentials) angle-integrated energy distributions[9]. The curves do not agree nor could they be made to agree with any reasonable variation of the parameters. In addition, the calculated peaks in the polar distributions (Fig. 4) were 5–10° closer to the surface normal than the experimental ones.

A preliminary fit of the embedding function and the core repulsive term was made to the properties of Rh metal in order to determine if the EAM description of the interaction better predicts the EARN data of Rh atoms ejected from Rh(111) than the pair potentials[9]. The most dramatic change in the predicted distributions arises in the angle-integrated energy distributions. As shown in Fig. 6 the experimental and calculated distributions using the EAM interaction are in excellent agreement while the calculated distribution using pair potentials is quite different from the experimental curve. The peak in the polar angle distributions as calculated from the EAM are also found to increase by about 10° from those predicted by the pair potentials (Fig. 4). The agreement between the EAM and the experimental energy distributions is better than one could have hoped, and the polar distribution correction is in the right direction.

Is the better agreement fortuitous or is there a sound basis for it? The pair-potential description in the surface region has been thought to be inadequate but the detailed data that exposed the nature of the deficiencies was not available. There are several differences between the EAM and pair potentials. First the surface binding energy of the EAM potential is larger (5.1 eV) than that of the pair potential (4.1 eV). Of note is that both potentials were fit to the bulk heat of atomization of Rh (5.76 eV). The peak position in the energy distribution is proportional to the binding energy[135], thus it is logical that the peak in the EAM energy distribution occurs at a higher value than for the pair potential. In addition to the larger binding energy at the equilibrium site, the EAM potential is relatively flat in the attractive portion of the entire surface region. There is more than a 4 eV attraction for the ejecting atom even above a neighboring atom, while the pair potential has only ~ 1 eV overall attraction. Thus particles that eject at more grazing angles will experience a larger attraction to the surface in the EAM potential than in the pair potential. This will tend to make the peak in the energy distribution shift to larger energies, and will also pull the particles away from the surface normal and move the peak in the polar distribution (Fig. 4).

Recently the previously developed Rh(111) EAM potential has been employed to model the ejection process from Rh(331), a stepped surface that consists of (111) terraces three atoms wide with a one-atom step height. In this surface there are atoms that are both more and less coordinated than on the (111) surface. The agreement between the experimental and calculated angular distributions is excellent[136]. This same EAM potential was used for the Rh interactions in the O/Rh(111) study discussed in section 2.

In a similar study, Lo *et al.* have compared the characteristics of atoms sputtered from copper surfaces in simulations which used both pair-additive potentials and EAM potentials[137]. Significant differences were found for many properties of interest, including the peak in the energy distributions. Although adjustment of the potentials to fit experimental data was not attempted, this study concluded that many-body potentials are required to realistically model much of the sputtering process.

The EAM approach appears to provide a formalism within which realistic potentials which describe atomic dynamics can be developed. It should also provide a method for realistically incorporating adsorbates into dynamics simulations. Both of these applications can be considered significant advances, and will help molecular dynamics simulations to continue to contribute to the understanding of technologically important processes.

3.3. Silicon: covalent many-body potentials

There has been recent widespread interest in simulating semiconductors. This has been especially true for silicon, and to a lesser extent for germanium. Prior to 1984, no general potential energy expressions were available which could be used to model the chemical dynamics of semiconductors. Between 1984 and 1986, at least five different expressions were introduced which can successfully model condensed phases of silicon[34,37,39,41,46]. These potential energy schemes, which were discussed in section 1.2, have made possible the use of molecular dynamics to study atomic-scale motion on semiconductor surfaces.

Reactions that occur on the surface of covalent solids have a complexity that is not as prevalent in metals. Many metal surfaces, especially the close-packed faces, retain the same geometry and bonding arrangement as would be present in the bulk phase. Most semiconductor surfaces, however, undergo reconstructions in which the surface atoms move significant distances from the bulk terminated positions. For example, the Si(001) surface, if bulk terminated, would have each atom bonded to two other silicon atoms in the second layer (Fig. 7). There would be two dangling bonds each with one electron, and the

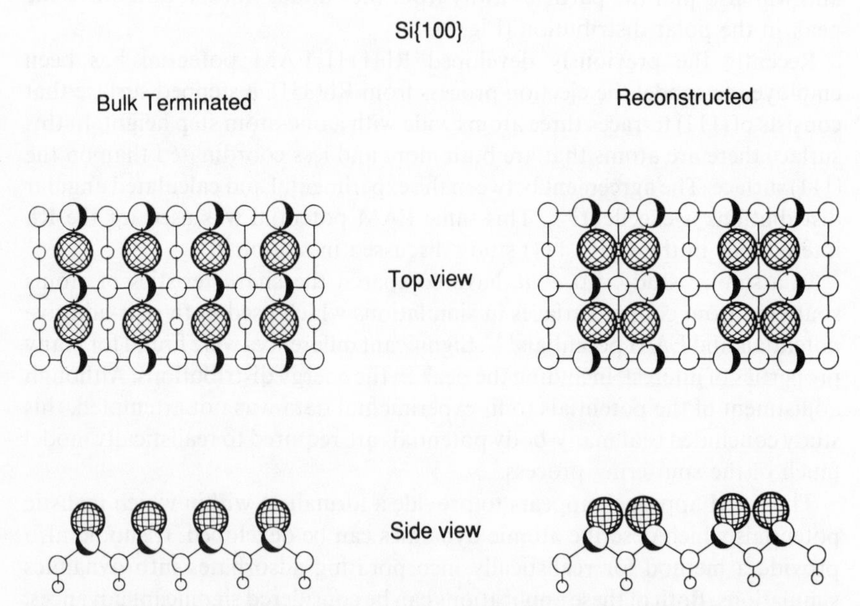

Fig. 7. Si(001). Bulk terminated atom positions, top and side views. (2 × 1) dimer reconstruction, top and side views. In all cases, the first-layer atoms are hatched, the second-layer atoms are shaded and deeper-layer atoms are smaller. (*Drawn by Tracy Schoolcraft.*)

nearest-neighbor distance on the surface would be ~3.84 Å. Since Si prefers higher coordination environments two surface atoms react and form a bond (i.e. 'dimerize') – with a distance of ~2.4 Å (Fig. 7)[138–146]. The Si(111) surface reconstructs with a unit cell 7 times that of the bulk terminated surface with a reconstruction thought to involve vacancies, adatoms and five and seven-membered rings[146–152]. If one is going to realistically model reactions on these surfaces, then the reconstructions must be incorporated into the simulation.

Two popular means of growing silicon single crystals are molecular beam epitaxy (MBE) and chemical vapor deposition (CVD)[153–156]. In MBE a beam of Si atoms from an oven source impinges on a Si crystal surface with the goal being to grow additional crystalline layers. Since the timescale of this process is approximately a layer per minute, there is ample time for diffusion of adsorbed atoms, reactions to remove surface reconstructions, and growth of subsequent layers. In CVD a gas of silanes (SiH_4, Si_2H_6) is present above the Si crystal surface and reactions occur between the gas and surface which result in the growth of a single crystal. Again numerous processes including diffusion, removal of the reconstruction, and crystal growth can and do occur.

3.3.1. Silicon on silicon

As described above, silicon crystals can be grown from a variety of gas sources. Because the rate of growth can be modulated using these techniques, dopants can be efficiently incorporated into a growing crystal. This results in control of the atomic structure of the crystal, and allows the production of samples which have specific electronic properties. The mechanisms by which gas-phase silicon species are incorporated into the crystal, however, are still unclear, and so molecular dynamics simulations have been used to help understand these microscopic reaction events.

(a) Surface diffusion

The growth rate of silicon crystals by either MBE or CVD is relatively slow, and so there is ample time for adsorbed atoms and molecules to diffuse to energetically more favorable sites. Experimental rates of diffusion of silicon on silicon and the activation barriers, however, are not known well. Experimental estimates of the activation barrier for silicon atoms diffusing on the Si(111) surface have ranged from 58 kcal/mole when the pyrolysis of silane is used to produce surface silicon atoms[157], to 4.6 kcal/mole for the direct deposition of silicon atoms under ultrahigh vacuum[158]. Furthermore, a comparison of these values with other silicon crystal faces has not been available. Because data for surface diffusion is necessary to model various aspects of semiconductor production, there has been interest in using molecular dynamics as a method of confirming and characterizing experimental observations.

In an effort to understand silicon surface diffusion, NoorBatcha, Raff and Thompson have employed molecular dynamics to model the motion of single silicon atoms on the Si(001) and Si(111)surfaces[159]. Morse functions are used for the pair forces, with the parameters being determined by the heat of sublimation. Because different forces were used for the diffusing and substrate atoms, the incorporation of gas-phase species into the crystal could not be directly modeled. Nonetheless, they were able to explore the characteristics of adsorption and diffusion for single atoms.

Using classical trajectories, NoorBatcha *et al.* determined a sticking coefficient of 0.96 for adsorption of Si atoms on the (001) surface[159], and effective activation energies of 3.63 kcal/mole and 2.43 kcal/mole for diffusion on the Si(001) and Si(111) surfaces, respectively[159,160]. The calculated sticking coefficient is in agreement with the experimentally determined value of near unity[162], and the activation energies suggest that the experimental number of 4.6 kcal/mole for diffusion on the (111) surface is the more accurate value. They also found diffusion on the Si(001) surface to be highly anisotropic, with atoms diffusing along channels in the surface. Similar modes of anisotropic diffusion have been proposed as being responsible for the occurrence of steps with specific orientations during the deposition of silicon atoms on the Si(001) surface[163].

In a subsequent study, NoorBatcha *et al.* varied the valence-force parameters used for the lattice interactions to evaluate the effect of the vibrational properties of the crystal on diffusion characteristics[161]. Using three sets of lattice potential parameters, they determined a range of effective activation barriers for diffusion of 3.63 kcal/mole to 7.47 kcal/mole on the Si(001) surface. This range encompasses the experimental estimate of 4.6 kcal/mole for the Si(111) surface, and further suggests this value as the more accurate experimental estimate.

In a similar study, Khor and Das Sarma studied the diffusion of Si, Si_2 and Si_3 on the (001), (011), and (111) surfaces of silicon[164]. In their study, the forces on all of the atoms were determined by the Stillinger–Weber potential[34]. For single atoms on all surfaces, they report an upper bound for diffusion of 4.8×10^{-5} cm^2/sec at 1600 K. This value is significantly less than both the range determined by NoorBatcha of 2.031×10^{-4} to 15.8×10^{-4} cm^2/sec[159–161], and the experimental estimate of $\sim 10^{-3}$ cm^2/sec[158]. They also report that diatomic Si_2 molecules diffuse more readily than single atoms on the Si(111) surface.

Despite the insights which the dynamics have provided into surface diffusion, additional studies will be required to fully characterize surface dynamics. In particular, the role of surface reconstructions on diffusion has not been fully explored, and additional studies of the relationship between anisotropic diffusion and step stability are currently needed.

(b) Epitaxial growth

Another area in which molecular dynamics has been used is in the study of the dynamics and structure of vapor-deposited crystals[165–167]. The main drawback to using molecular dynamics to study vapor-phase deposition is that epitaxial growth is an intrinsically slow process. Typical growth rates for techniques which employ molecular beams are on the order of monolayers per minute, while present timescales accessible to molecular dynamics stretch from picoseconds to nanoseconds. In a recent article by Abraham, the amount of CPU time on a Cray supercomputer required to simulate modest realistic growth conditions using a Lennard–Jones potential is estimated to be seven months[4]. If one of the silicon potentials currently available were substituted for the pair-additive interactions, this estimate would increase dramatically. Nonetheless, it is expected that dynamics simulations can provide important microscopic information about the growth process, and so studies of silicon growth at high deposition rates have been undertaken.

Gossmann and Feldman have employed a combination of low-energy electron diffraction (LEED) and high-energy ion scattering to experimentally probe the structure of the (001) and (111) surface of silicon during the deposition of silicon atoms[168]. This combination of experimental techniques provides a very good characterization of surface structures. This is because LEED probes the long-range symmetry of the first few atomic layers, while ion scattering provides information on the short-range structure of all atomic layers which are displaced from bulk positions[169]. In their studies, molecular beams were used to deposit controlled amounts of silicon atoms on substrates which were maintained at various constant temperatures. This allowed intermittent determinations of surface structures during the growth process.

As discussed above, the Si(001) surface is reconstructed into dimers, a side view of which is shown in Fig. 7. In addition to the reconstruction of the surface atoms there is significant distortion of the subsurface region. This distortion blocks channels which enter into the bulk, and causes an excess of backscattered ions over what would be expected from atomic rows in the ideal crystal. This local distortion is the property monitored by backscattered ion intensities. The dimer pairs also tend to form in rows, which is the source of the (2×1) LEED pattern observed for this surface.

For the deposition of silicon on Si(001) and Si(111) surfaces, Gossmann and Feldman determined that epitaxial growth occurs for substrate temperatures maintained over 570 K and 640 K, respectively[168]. Above the epitaxial temperature the growth is single crystal, while below this temperature the growth is amorphous. This difference in epitaxial growth temperature between the two faces was ascribed by Gossmann and Feldman to be due to the different surface reconstructions, where the (111) surface presumably requires a higher temperature to 'unreconstruct' during deposition than the (001) surface. They also observed that for the deposition of up to ~ 3 monolayers on the (001) surface at low temperatures, the (2×1) symmetry of the surface was lost, yet the number of backscattered ions remained constant. This is in contrast to epitaxial growth at high substrate temperatures, where both sharp diffraction patterns and constant ion-scattering signals were reported for all coverages. Furthermore, as additional monolayers were deposited, the number of backscattered ions increased and the surface symmetry changed to (1×1), i.e. the same as the bulk terminated surface.

Gossmann and Feldman proposed that the initially constant backscattered ion intensities were a result of two effects. First, as the silicon was being deposited, channels into the lattice were being filled. This would result in an increased number of backscattered ions. At the same time, however, they proposed that the reconstruction was being disordered, and so the channels which were originally blocked by the subsurface distortion were opening. This would result in a decreasing number of backscattered ions. The net result of these two effects would be an almost constant backscattered-ion intensity. Furthermore, because approximately three surface layers were distorted due to the reconstruction, this cancellation would occur for the deposition of approximately three monolayers, as observed experimentally. The conclusions of the study were that for high substrate temperatures (above 560 K), epitaxial growth occurred. For low substrate temperatures, however, deposited atoms form amorphous or polycrystalline layers, with the ion-scattering results suggesting that these layers reorder the initial reconstruction.

Two molecular dynamics studies of the gas-phase deposition of silicon atoms on the silicon (001) reconstructed surface have been reported. In a pair of simulations which used the Stillinger–Weber potential[34], Gawlinski and Gunton modeled the gas-phase growth of silicon by depositing eight monolayers of silicon atoms on reconstructed (001) substrates[165]. In one simulation, the substrate was maintained at a low temperature of 250 K, while in the other the substrate was maintained at a high temperature of 1500 K. These two temperatures were chosen because they are below and above the epitaxial growth temperature reported by Gossmann and Feldman. After the low-temperature deposition, they observed that the surface of the initial substrate (which was buried under the deposited atoms) remained relatively unchanged, while for deposition on the high-temperature substrate the initial surface reconstruction was disordered. Furthermore, they observed that for the low-temperature deposition, amorphous overlayers resulted, while for deposition on the higher-temperature substrate enhanced ordering of the surface layers was apparent. Despite the very high deposition rate used in this study, the difference between the structures produced at the two growth temperatures reflected a difference in growth modes as suggested by Gossmann and Feldman's study. This result also confirmed the idea that molecular dynamics can be successfully used to study crystal growth.

In a different approach to this problem, Brenner and Garrison used molecular dynamics to examine the chemical mechanisms which lead to reordering of the atom-pairing reconstruction during atom deposition[166]. This simulation incorporated a dissociative valence-force field potential[39] and consisted essentially of a high-temperature anneal of $1\frac{1}{2}$ monolayers of silicon atoms which had been deposited on a silicon (001) reconstructed surface.

During the dynamics, two modes of surface dimer opening ('unreconstruction') were observed. In the first mode the distance between the two atoms in a dimer pair would intermittently change between that corresponding to the reconstruction, and the distance corresponding to the bulk terminated surface. This mechanism, termed an unstable opening, resulted when a surface dimer was surrounded by several randomly positioned atoms. Because the atoms surrounding the open dimer were not in lattice sites, this mode of dimer opening was proposed as being the initiation of an amorphous overlayer which reorders the reconstruction. Also, since diffusion did not play a large role, this was associated with the low-temperature growth mode proposed by Gossmann and Feldman.

The second dimer-opening mode observed resulted when a surface atom diffused to the site of a surface dimer and 'bumped' a nearby atom into the center of the dimer. This mechanism resulted in atoms which occupied lattice sites, and produced a surface dimer which remained open for the course of the simulation. Because the final atomic positions corresponded to lattice sites, and because a high rate of surface diffusion was required to produce the 'bump', this mechanism was associated with the high-temperature epitaxial growth mode identified by Gossmann and Feldman.

A related area of crystal growth for which the short timescale required by molecular dynamics is more appropriate is kinetic-energy-enhanced epitaxial growth. In this technique, an energized beam is used to deposit atoms on a surface. For silicon beam deposition of energies 10–65 eV, epitaxial growth has been reported for colder substrate temperatures than is required when thermal beams are used[170]. For energies above about 100 eV, however, enhanced damage of up to 400 Å below the surface has been reported. Based on these experimental observations, there appears to be a limited range of energies which are of use for producing good-quality films. Molecular dynamics simulations have been used to better quantify damage caused by the energetic beam, and to suggest extensions to this technique.

Dodson has used molecular dynamics to study atom–surface dynamics for silicon atom energies ranging from 10 to 100 eV incident on a silicon (111) surface[171]. In this study a modified form of the Tersoff potential was employed[45], and substrates of various sizes were used. For an energy of 10 eV and near-perpendicular angles of incidence, 30% of the deposited atoms were reported to come to rest on the top of the substrate, and the remaining 70% penetrated into subsurface interstitial sites. If the angle of incidence was changed to 60°, however, roughly half of the atoms were reported to have remained on the surface. Dodson's simulations also indicate that kinetic energy transfer between the incoming atoms and the lattice is rapid, with a majority of the initial kinetic energy of the atom being dissipated by phonons within about 0.08 ps.

Dodson also explored the motion of silicon atoms with kinetic energies in the range 20–100 eV that were initially incident on the surface with near grazing angles. In these trajectories the incoming atoms were observed to skip along the surface, traveling with ranges of up to thousands of angstroms. This observation suggests that energized beams could be used to greatly increase surface diffusion, and thereby produce an efficient mode of transporting atoms to steps or defects. This could result in better-quality films, and could offer promise for effectively controlling growth structures.

In another molecular dynamics study, Garrison, Miller and Brenner characterized the chemical dynamics of silicon atoms with energies of 0.026–20 eV which were deposited on a silicon (001) dimer-reconstructed surface[172]. For atomic energies of 0.026 eV and perpendicular incidence, the atoms in the beam remained on the surface, and no significant motion of the dimer-reconstructed surface atoms occurred. For energies in the range of 5–10 eV, however, significant motion of the surface atoms was observed which led to epitaxial atomic configurations. One mechanism which occurred in this energy range was the direct insertion of the incoming atom into the dimerized pair of surface atoms (Fig. 8). In a second mechanism, the incoming atom replaced one of the atoms in a surface dimer pair, which then became the inserted atom. Finally, the incoming atom could also knock open the dimer, and bind to the surface so that the dimer remained open. Each of these mechanisms took place on a timescale of ~ 100 fs, and so they are best characterized as direct dimer-opening processes. This is opposed to a thermal process were many vibrations might be required. Although not all trajectories in this energy range produced open dimer pairs, the results of the simulation indicate that the energy range of up to 10 eV would enhance epitaxial growth without introducing subsurface defects.

In addition to the 5–10 eV energy range, incoming atoms with energies of up to 20 eV were also explored. These atoms were observed to implant into the lattice, and presumably produce damage in a growing film. This conclusion agrees with that of Dodson, where a majority of atoms incident perpendicular to the (111) with energies of 10 eV implanted. These studies indicate that for perpendicular incidence, atoms with energies in the range 5–10 eV would be the most effective for producing low-temperature epitaxial films.

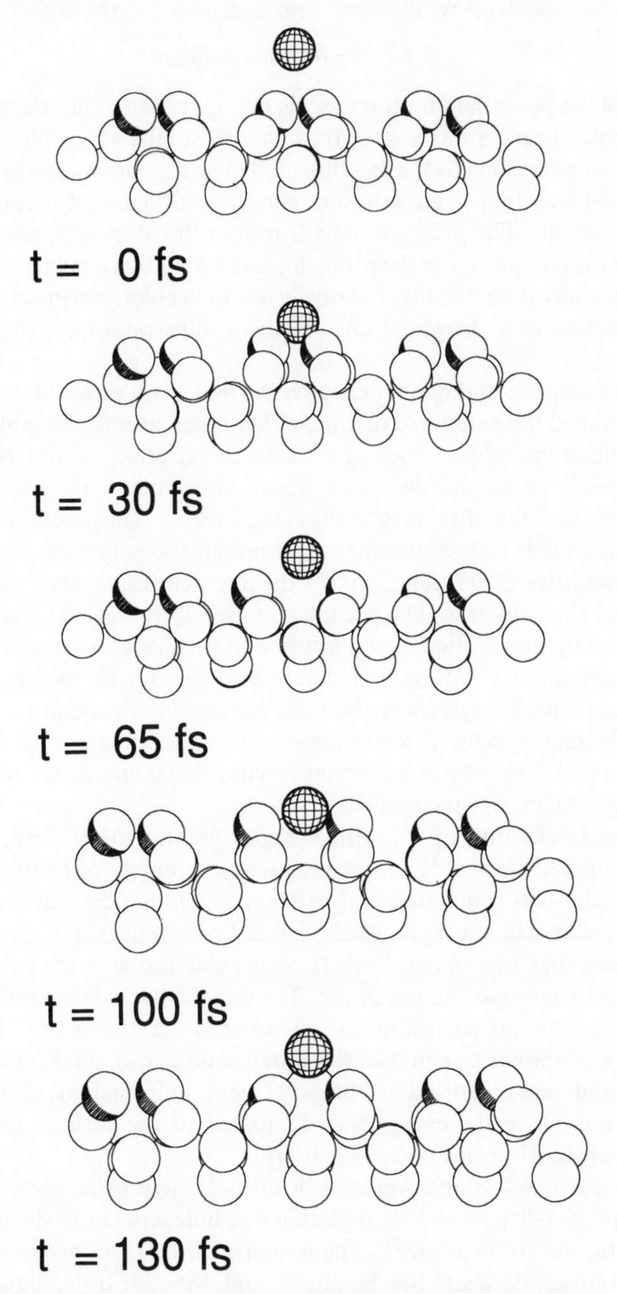

Fig. 8. Insertion mechanism of dimer opening as a function of time in femtoseconds (fs). The adatom started with 7.5 eV of kinetic energy and oriented perpendicular to the surface. The hatched circles represent the adatoms, the shaded circles the original surface dimer atoms and the open circles the substrate atoms. Only four layers of the ten used in the simulation are shown. (*From Ref. 172.*)

3.3.2. Hydrogen on silicon

Despite the potential for atomic-scale manipulation of interfaces displayed by molecular-beam epitaxial growth, a majority of the vapor-phase growth of silicon is accomplished by the reaction of silane with silicon substrates[153]. This is because much higher growth rates can be achieved, and because silane is readily available. The presence of hydrogen in the reacting species, however, increases the complexity of the problem, and makes this a much more difficult system to study theoretically. Relatively few molecular dynamics simulations of the reaction of hydrogen atoms on silicon substrates have therefore been reported.

Raff, Thompson and coworkers have carried out a series of studies which have examined the reaction dynamics of hydrogen atoms and molecules with silicon substrates[173,174]. Rice *et al.* used a variation of the NoorBatcha silicon-on-silicon model described above to simulate the scattering and dissociative sticking of H_2 on the Si(111) surface[174]. The potential used was a valence-force field expression for the atoms in the substrate, and a sum of pairwise-additive Morse functions for the interactions between the hydrogen atoms and the substrate. The parameters used in the Morse functions were determined by the isolated H_2 molecule, and by fitting to various H_2–silicon cluster calculations[173]. Rice *et al.* determined that for H_2 molecules striking the surface with energies less than the barrier for dissociative adsorption (0.18 eV for their potential), scattering was elastic and predominately specular. They also found very little rotational or vibrational energy transfer between the H_2 molecules and the surface.

For the interaction of H_2 with energies greater the 0.18 eV, Rice *et al.* reported that all cases of H_2 adsorption were accompanied by dissociation of the molecule. This is in agreement with experimental observation, where the chemisorption of intact H_2 molecules has not been reported[175]. Furthermore, for all cases they report that both H atoms chemisorb to the surface rather than reflect back into the gas phase. The dissociative adsorption of H_2 was accompanied by an energy release of between 2.5 and 4.3 eV. The energy released was shown to enhance the initial mobility of the H atoms on the surface, and was reported to be somewhat independent of the surface temperature. Once the energy was dissipated to the surface, however, the mobility of the H atoms decreased sharply.

In a subsequent study, Agrawal, Raff and Thompson showed that the sticking probability for the H_2 molecule was independent of the interactions used for the substrate atoms[40]. The mobility of the H atoms and the rate of energy transfer between the H atoms and the substrate, however, were reported to depend somewhat on the lattice. Despite the small dependence on the substrate model, the major results of the initial study remained unchanged.

The reaction of silane in the gas phase and with silicon surfaces is a complex topic to understand on an atomic scale. The molecular dynamics studies described above, however, have shed light on the rates and mechanisms of various reactions involved. Further studies along these lines will undoubtedly prove valuable to understanding the details of this process.

4. THE FUTURE OF MOLECULAR DYNAMICS FOR MODELING GAS–SURFACE REACTIONS

The past few years have been an exciting time for modeling gas–surface reactions. Computational techniques as well as potential energy functions have become sufficiently advanced that dynamics simulations can now described many realistic situations without introducing severe approximations. As computer resources continue to grow, the impact which computer modeling has on science and engineering will also continue to increase.

Significant progress has recently been made in several areas which will have a profound effect on the ability of molecular dynamics to handle more complex problems. In this section we speculate on several areas which appear to hold promise for advancing computer modeling studies. In section 4.1, recent progress in both analytic potential energy expressions and 'first principles' calculations are briefly mentioned. Recent advances in computational techniques are discussed in section 4.2. These include the use of constraints within the classical equations of motion to model thermostats in the surface region, and the incorporation of Monte Carlo techniques into molecular dynamics simulations.

4.1. Further development of potential-energy expressions

A great deal of recent success has been achieved in writing simple analytic potential energy expressions which capture the essence of chemical bonding. Much of the inspiration for these efforts has come from the desire to realistically model reactions in condensed phases and at surfaces. As computer simulations grow in importance, continued progress in the development of new potential energy functions will be needed.

In the near future, the expansion of the covalent-bonding formalisms developed to model silicon to other systems appears promising. Very recently the extension of the Abell–Tersoff covalent-bonding formalism to few-body reactive systems has been demonstrated by the development of an accurate potential energy expression for H_3[176]. In the determination of an analytic potential function for silicon, Tersoff's main objective was to describe the energetics of stationary points on the potential surface[41,42]. This emphasis did not include properties of importance to chemical dynamics such as potential barriers, and so it was not clear if this formalism could be used to describe chemical reactions. In the H_3 study, the potential energy for different atomic configurations which had been calculated by *ab initio* methods were accurately fitted using a Tersoff-type expression. This demonstrated that potential barriers to reaction can be introduced within this formalism, and that it is sufficiently flexible to accurately described few-body potential energy surfaces.

The Abell–Tersoff potential energy expression has also been used to describe reactive collisions in molecular solids[177]. These studies have modeled detonation and energetic-ion induced chemistry, and have demonstrated that the energetics of propagating reactions in solids can be understood using models derived from gas-phase reactivity[178]. The application of this approach to these systems suggests that this formalism will be useful for modeling molecule–surface reactions where the dynamics of both substrate atoms and arrays of molecules are of importance. For example, the incorporation SiH_4 molecules into a silicon substrate could possibly be modeled using this formalism. Such studies could be thought of as incorporating the advantages of simulations which use LEPS potentials, such as adjustable potential energy barriers, with the ability to describe covalent bonding in solids.

In a similar fashion, the introduction of angle-dependent electron densities into the EAM[33] suggests that this formalism may be successfully extended to chemical reactions. This would allow the study, for example, of the reaction of a metal–ligand cluster with a metal surface. This would enhance the applicability of the EAM, and would increase the realm of processes which computer simulations can effectively model.

In addition to analytic potential energy expressions, studies have begun which introduce forces calculated from *ab initio* total energy techniques directly into dynamics simulations. One method which has attracted considerable attention is based on the concept of simulated annealing[179]. Car and Parrinello originally introduced this technique as a way of unifying density-functional theory and molecular dynamics[180]. The idea is that the (classical) nuclear degrees of freedom and the electronic degrees of freedom (which enter through a variational wavefunction) are varied simultaneously. The approach taken is to treat the parameters which enter the wavefunction as classical 'particles', and to write equations of motion for each of the parameters. The forces on the atoms are derived from the electronic wavefunction, while the forces on the wavefunction 'particles' are determined by the condition that the electronic energy be minimized subject to the relative positions of the nuclei. By integrating the electronic degrees of freedom simultaneously with the nuclear degrees of freedom, the minimum energy state of the system can be determined with a smaller amount of effort than would be required if both were minimized separately. Examples where variations of this method have been applied include the calculation of the structure and energetics of various silicon[180–182], germanium[183–185] and silicon oxide compounds[186], and the optimization of basis sets for diatomic molecules[187] and solvated electrons[188,189].

The major drawback for employing the Car–Parrinello approach in dynamics simulations is that since a variational wavefunction is required, the electronic energy should in principle be minimized before the forces on the atoms are calculated. This greatly increases the amount of computer time required at each step of the simulation. Furthermore, the energies calculated with the electronic structure methods currently used in this approach are not exceptionally accurate. For example, it is well established that potential energy barriers, which are of importance to chemical reactivity, often require sophisticated methods to be accurately determined. Nonetheless, the 'first-principles' calculation of the forces during the dynamics is an appealing idea, and will continue to be developed as computer resources expand.

A second marriage between electronic structure techniques and molecular dynamics simulation has been the calculation of atomic forces using semiempirical tight-binding electronic structure methods[190–192]. Menon and Allen, for example, have used this method to model the dynamics of a gas-phase atom interacting with a semiconductor surface[191]. While simulations of this type have proven to be feasible, again finite computer resources restrict the size and length of time of the processes which can be modeled. Also, the energetics calculated by tight-binding methods are not exceptionally accurate for properties such as barriers to reaction, and so the dynamics generated in this way may not always be physically significant.

4.2. Advances in dynamics techniques

In addition to the study of atomic motion during chemical reactions, the molecular dynamics technique has been widely used to study the classical statistical mechanics of well-defined systems. Within this application considerable progress has been made in introducing constraints into the equations of motion so that a variety of ensembles may be studied[2]. For example, classical equations of motion generate constant energy trajectories. By adding additional terms to the forces which arise from properties of the system such as the pressure and temperature, other constants of motion have been introduced.

The addition of constraints to the equations of motion have also been used to produce thermostats at surfaces which control the flux of heat in and out of the substrate. For example, Riley *et al.* have proposed a velocity reset procedure which regulates atomic motion by coupling the current velocity of each atom with a velocity chosen from a Maxwellian distribution[193]. In a similar scheme, Agrawal *et al.* have added a friction term to atomic velocities which depends in part on the difference between the current temperature of the surface region and that desired for the substrate[40]. This approach was originally proposed by Berendsen *et al.* as an alternative to the generalized Langevin equation for bulk phases[194], and it can apparently be applied to surface dynamics. Other similar constrained dynamics schemes, which have arisen from simulations of bulk materials, will also play a role in the dynamics of surface reactions.

One final area which will increase the flexibility of molecular dynamics simulations is the incorporation of Monte Carlo techniques. Because molecular dynamics follows atomic motion in real time, it can be used to study nonequilibrium as well as equilibrium dynamics. Molecular dynamics timescales, however, are currently limited to about the nanosecond range, and so it cannot be used to model processes which require long times. For example, surface diffusion at room temperature often consists of jumps between lattice sites. The time spent at each lattice site, however, can be long compared to the time required to complete a jump, and so the observation of a sufficient number of jumps to estimate surface diffusion properties can require an enormous amount of computer time. In contrast to molecular dynamics, the Monte Carlo technique involves moving atoms in a random fashion with the acceptance of moves depending on the potential energy and a predetermined temperature[195]. The main advantage of the Monte Carlo technique over molecular dynamics is that the configurations sampled by the moves can be restricted to those which are of direct interest to the process being modeled. For example, in the case of surface diffusion the position of the diffusing species may be restricted to regions near lattice sites and saddle points[196–198]. This approach, in conjunction with techniques such as transition-state theory, can lead to accurate estimates of surface diffusion without the need to perform extensive dynamics calculations between jumps.

A marriage between Monte Carlo techniques and molecular dynamics simulations can be envisioned for situations where both short-time nonequilibrium dynamics and long-time equilibrium processes are important. For example, the growth of silicon from silane involves the dissociative chemisorption of molecules on the surface, as well as the long-time diffusion of surface species. If the reactions are studied with molecular dynamics, and surface diffusion is included using Monte Carlo techniques, then a complete picture of the dynamics of growth can in principle be obtained. Although few studies of this type have been undertaken, we believe that such techniques will ultimately prove very useful for modeling technologically important processes.

In this article we have tried to present a general, although somewhat limited overview of molecular dynamics simulations of gas–surface reactions as they pertain to technologically important processes. In the course of this review we have undoubtedly left out a great deal of very important work. We hope, however, that the prospect for significant advances in this area has been conveyed, and that the continued success of these types of studies is apparent.

Acknowledgements

This venture into modeling 'real' reactions at solid surfaces initiated some ten years ago and was greatly encouraged and aided by Nicholas Winograd of the Pennsylvania State University and Don E. Harrison, Jr of the Naval Postgraduate School. Their support and continuous stream of ideas are essential to these studies. Over the years a number of students, postdoctoral associates, and collaborators have contributed to these studies. These include Gregg Buczkowski, Che-Chen Chang, Brian Craig, Alain Diebold, David Deaven, Mohamed El-Maazawi, Karen Foley, Peter Haff, Jian-hui He, Robert E. Johnson, Shukla Kapur, Jung-hui Lin, Davy Lo, Mitch Miller, John Olson, Curt Reimann, Tracy Schoolcraft, Mark Shapiro, R. Srinivasan and Tom Tombrello. The manuscript was critically read by Richard Mowrey, Tracy Schoolcraft and Nicholas Winograd. The preparation of the manuscript was processed with the best of humor by Chris Moyer and Sabrina Glasgow.

Finally we gratefully acknowledge the financial support of the Office of Naval Research, the National Science Foundation, the IBM Program for the Support of the Materials and Processing Sciences, the Shell Development Corporation, the Camille and Henry Dreyfus Foundation, the Sloan Foundation and the Research Corporation. Two people at these agencies, Larry Cooper (ONR) and Henry Blount (NSF) have been particularly supportive.

References

1. Gay, W. L., and Harrison, D. E., Jr., *Phys. Rev.*, **135**, A1780 (1964); Harrison, D. E., Jr., Johnson, J. P. III, and Effron, H. M., *Appl. Phys. Lett.*, **8**, 33 (1966); Harrison, D. E. Jr., and Delaplain, C. B., *J. Appl. Phys.*, **47**, 2252 (1976); Harrison, D. E. Jr., *CRC Critical Reviews in Solid State and Materials Sciences*, **14**, S1 (1988); Garrison, B. J., and Winograd, N., *Science*, **216**, 805 (1982).
2. Hoover, W. G., *Molecular Dynamics*, Springer-Verlag, Berlin, 1986.
3. Heerman, D. W., *Computer Simulation Methods in Theoretical Physics*, Springer-Verlag, Berlin, 1986.
4. Abraham, F. F., *Adv. Phys.*, **35**, 1 (1986).
5. Fincham, D., and Heyes, D. M., *Adv. Chem. Phys.*, **63**, 493 (1985).
6. Klein, M. L., *Ann. Rev. Phys. Chem.*, **36**, 525 (1985).
7. Hockney, R., and Eastwood, J., *Computer Simulations using Particles*, McGraw-Hill, New York, 1981.
8. Hafner, J., and Heine, V., *J. Phys. F: Met.Phys.*, **13**, 2479 (1983).
9. Garrison, B. J., Winograd, N., Deaven, D. M., Reimann, C. T., Lo, D. Y., Tombrello, T. A., Harrison, D. E., and Shapiro, M. H., *Phys. Rev.* B, **37**, 7197 (1988).

10. Torrens, I. M., *Interatomic Potentials*, Academic Press, New York, 1972.
11. Bohr, N., *Kgl. Dansk. Vid. Selsk. Mat.-Fys. Medd.*, **18**, No. 8 (1948).
12. Born, M., and Mayer, J. E., *Z. Phys.*, **75**, 1 (1932).
13. Daw, M. S., and Baskes, M. I., *Phys. Rev.* B, **29**, 6443 (1984).
14. Daw, M. S., *Surf. Sci.*, **166**, L161 (1986).
15. Ercolessi, F., Tosatti, E., and Parrinello, M., *Phys. Rev. Lett.*, **57**, 719 (1986).
16. Ercolessi, F., Parrinello, M., and Tosatti, E., *Surf. Sci.*, **177**, 314 (1986).
17. Garofalo, M., Tosatti, E., and Ercolessi, F., *Surf. Sci.*, **188**, 321 (1987).
18. Tomanek, D., and Bennemann, K. H., *Surf. Sci.*, **163**, 503 (1985).
19. Dodson, B. W., *Phys. Rev.* B, **35**, 880 (1987).
20. Chen, S. P., Voter, A. F., and Srolovitz, D. J., *Phys. Rev. Lett.*, **57**, 1308 (1986).
21. Sathyamurthy, N., *Computer Phys. Reports*, **3**, 1 (1985), and references therein.
22. Murrell, J. N., Carter, S., Farantos, S. C., Huxley, P., and Varandas, A. J. C., *Molecular Potential Energy Functions* Wiley, New York, 1984, and references therein.
23. McCreery, J. H., and Wolken, G., *J. Chem. Phys.*, **63**, 2340 (1975).
24. McCreery, J. H., and Wolken, G., *J. Chem. Phys.*, **67**, 2551 (1977).
25. Wolken, G., and McCreery, J. H., *Chem. Phys. Lett.*, **54**, 35 (1978).
26. Wolken, G., *J. Chem. Phys.*, **68**, 4338 (1978).
27. Purvis, G. D., and Wolken, G., *Chem. Phys. Lett.*, **62**, 42 (1979).
28. Barnett, R. N., Cleveland, C. L., and Landman, U., *Phys. Rev. Lett.*, **54**, 1679 (1985).
29. Foiles, S. M., Basks, M. I., and Daw, M. S., *Phys. Rev.* B, **33**, 7983 (1986).
30. Foiles, S. M., *Phys. Rev.* B, **32**, 7685 (1986).
31. Felter, T. E., Foiles, S. M., Daw, M. S., and Stulen, R. H., *Surf. Sci.*, **171**, L379 (1986).
32. Deaven, D. M., Honors Thesis, The Pennsylvania State University, 1988.
33. Baskes, M. I., *Phys. Rev. Lett.*, **59**, 2666 (1987).
34. Stillinger, F. H., and Weber, T. A., *Phys. Rev.* B, **31**, 5262 (1985).
35. Khor, K. E., and Das Sarma, S., *Phys. Rev.* B, **36**, 7733 (1987).
36. Abraham, F. F., and Batra, I. P., *Surf. Sci.*, **163**, L752 (1985).
37. Pearson, E., Takai, T., Halicioglu, T., and Tiller, W. A., *J. Crystal Growth*, **70**, 33 (1984).
38. Axilrod, B. M., and Teller, E., *J. Chem. Phys.*, **11**, 299 (1943).
39. Brenner, D. W., and Garrison, B. J., *Phys. Rev.* B, **34**, 1304 (1986).
40. Agrawal, P. M., Raff, L. M., and Thompson, D. L., *Surf. Sci.*, **188**, 402 (1987).
41. Tersoff, J., *Phys. Rev. Lett.*, **56**, 632 (1986).
42. Tersoff, J., *Phys. Rev.* B, **37**, 6991 (1988).
43. Abell, G. C., *Phys. Rev.* B, **31**, 6184 (1985).
44. Brenner, D. W., Ph.D. Thesis, The Pennsylvania State University, 1987.
45. Dodson, B. W., *Phys. Rev.* B, **35**, 2795 (1987).
46. Biswas, R., and Hamann, D. R., *Phys. Rev. Lett.*, **55**, 2001 (1985).
47. Khor, K. E., and Das Sarma, S., *Chem. Phys. Lett.*, **134**, 43 (1987).
48. Ding, K., and Anderson, H. C., *Phys. Rev.* B, **34**, 6987 (1986).
49. Grabow, M. H., and Gilmer, G. H., in *initial Stages of Epitaxial Growth* (Eds. J. M. Gibson, R. Hull and D. A. Smith), Materials Research Society, Pittsburgh, 1987, p. 15.
50. Fitz, D. E., Bowagan, A. O., Beard, L. H., Kouri, D. J., and Gerber, R. B., *Chem. Phys. Lett.*, **80**, 537 (1981).
51. Barker, J. A., Kleyn, A. W., and Auerbach, D. J., *Chem. Phys. Lett.*, **97**, 9 (1983)
52. Polanyi, J. C., and Wolf, R. J., *J. Chem. Phys.*, **82**, 1555 (1985).

53. Muhlhausen, C. W., Williams, L. H., and Tully, J. C., *J. Chem. Phys.*, **83**, 2594 (1985).
54. Certain, P. R., and Bruch, L. W., in *Physical Chemistry*, Series One, Vol. I (Ed. W. B. Brown), University Park Press, Baltimore, 1972.
55. Fitts, D. D., *Ann. Rev. Phys. Chem.*, **17**, 59 (1966).
56. Benninghoven, A., Jaspers, D., and Sichterman, W., *Appl. Phys.*, **11**, 35 (1976).
57. Baxter, J. P., Schick, G. A., Singh, J., Kobrin, P. H., and Winograd, N., *J. Vac. Sci. Technol.*, **4**, 1218 (1986).
58. Schick, G. A., Baxter, J. P., Singh, J., Kobrin, P. H., and Winograd, N., in *Secondary Ion Mass Spectrometry-SIMS V*, Vol. 44 of Springer Series in Chemical Physics (Eds A. Benninghoven, R. J. Colton, D. S. Simons, and H. W. Werner), Springer, New York, 1986, p. 90.
59. Winograd, N., Kobrin, P. H., Schick, G. A., Singh, J., Baxter, J. P., and Garrison, B. J., *Surf. Sci. Lett.*, **176**, 1817 (1987).
60. Walder, K. T., and Urbassek, H. M., *Appl. Phys.* A, **45**, 207 (1988).
61. Garrison, B. J., *Nuclear Instru. and Methods*, B, **17**, 305 (1986).
62. Garrison, B. J., Reimann, C. T., Winograd, N., and Harrison, D. E., *Phys. Rev.* B, **36**, 3516 (1987).
63. Adelman, S. A., *Adv. Chem. Phys.*, **44**, 143 (1980).
64. Tully, J. C., *J. Chem. Phys.*, **73**, 1975 (1980).
65. DePristo, A. E., *Surf. Sci.*, **141**, 40 (1984).
66. Huheey, J. E., *Inorganic Chemistry*, Harper and Row, New York, 1978.
67. Bardoli, R. S., Vickerman, J. C., and Wolstenholme, J., *Surf. Sci.*, **85**, 244 (1979).
68. Winograd, N., Garrison, B. J., and Harrison, D. E., *J. Chem. Phys.*, **73**, 3473 (1980).
69. Holland, S. P., Garrison, B. J., and Winograd, N., *Phys. Rev. Lett.*, **44**, 756 (1980).
70. Winograd, N., Garrison, B. J., and Harrison, D. E., *Phys. Rev. Lett.*, **41**, 1120 (1978).
71. Gibbs, R. A., and Winograd, N., *Rev. Sci. Instru.*, **52**, 1148 (1981).
72. Kobrin, P. H., Schick, G. A., Baxter, J. P., and Winograd, N., *Rev. Sci. Instru.*, **57**, 1354 (1986).
73. Garrison, B. J., Reimann, C. T., Winograd, N., and Harrison, D. E., *Phys. Rev.* B, **36**, 3516 (1987).
74. Castner, D. G., Sexton, B. A., and Somorjai, G. A., *Surf. Sci.*, **71**, 519 (1978).
75. Reimann, C. T., El.-Maazawi, M., Walzl, K., Winograd, N., Garrison, B. J., and Deaven, D. M., *J. Chem. Phys.*, **90**, 2027 (1989)
76. Demuth, J. E., Christmann, K., and Sanda, P. N., *Chem. Phys. Lett.*, **76**, 201 (1980).
77. Moon, D. W., Winograd, N., and Garrison, B. J., *Chem. Phys. Lett.*, **114**, 237 (1985).
78. Moon, D. W., Bleiler, R. J., Karwacki, E. J., and Winograd, N., *J. Am. Chem. Soc.*, **105**, 2916 (1983).
79. Garrison, B. J., *J. Am. Chem. Soc.*, **104**, 6211 (1982).
80. Foley, K. E., Winograd, N., Garrison, B. J., and Harrison, D. E., *J. Chem. Phys.*, **80**, 5254 (1984).
81. Erley, W., and Wagner, H., *Surf. Sci.*, **74**, 333 (1978).
82. Erley, W., Ibach, H., Lehwald, S., and Wagner, H., *Surf. Sci.*, **83**, 585 (1979).
83. Garrison, B. J., Diebold, A. C., and Lin, J.-H., *Surf. Sci.*, **124**, 461 (1983).
84. Olson, J. A., and Garrison, B. J., *J. Chem. Phys.*, **83**, 1392 (1985); Olson, J. A., and Garrison, B. J., *J. Vac. Sci. Technol.*, A, **4**, 1222 (1986).
85. Gibbs, R. A., Holland, S. P., Foley, K. E., Garrison, B. J., and Winograd, N., *Phys. Rev.* B, **24**, 6178 (1981).

86. For a review see Johnston, H. S., *Gas Phase Reaction Rate Theory*, Ronald, New York, 1966, and references therein.
87. Polanyi, J. C., *Acc. Chem. Res.*, **5**, 161 (1972).
88. McCreery, J. H., and Wolken, G., *J. Chem. Phys.*, **60**, 2316 (1977).
89. For a review see Truhlar, D., and Wyatt, R., *Adv. Chem. Phys.*, **36**, 141 (1977).
90. McCreery, J. H., and Wolken, G., *J. Chem. Phys.*, **64**, 2845 (1976).
91. Elkowitz, A. B., McCreery, J. H., and Wolken, G., *Chem. Phys.*, **17**, 423 (1976).
92. Light, J. C., *J. Chem. Phys.*, **40**, 3221 (1964).
93. Pechukas, P., and Light, J. C., *J. Chem. Phys.*, **42**, 3281 (1965).
94. McCrerry, J. H., and Wolken, G., *Chem. Phys. Lett.*, **39**, 478 (1976).
95. McCrerry, J. H., and Wolken, G., *J. Chem. Phys.*, **65**, 1310 (1976).
96. Gelb, A., and Cardillo, M., *Surf. Sci.*, **59**, 128 (1976).
97. Gelb, A., and Cardillo, M., *Surf. Sci.*, **64**, 197 (1977).
98. Gelb, A., and Cardillo, M., *Surf. Sci.*, **75**, 199 (1977).
99. Lee, C.-Y., and DePristo, A. E., *J. Chem. Phys.*, **84**, 485 (1986).
100. Lee, C.-Y., and DePristo, A. E., *J. Chem. Phys.*, **85**, 4161 (1986).
101. McCreery, J. H., and Wolken, G., *Chem. Phys. Lett.*, **54**, 35 (1978).
102. Diebold, A. C., and Wolken, G., *Surf. Sci.*, **82**, 245 (1979).
103. Tantardini, G. F., and Simonetta, M., *Chem. Phys. Lett.*, **87**, 420 (1982).
104. Kara, A., and DePristo, A. E., *J. Chem. Phys.*, **88**, 2033 (1988).
105. Pfnur, H. E., Rettner, C. T., Lee, J., Madix, R. J., and Auerbach, D. J., *J. Chem. Phys.*, **85**, 7452 (1986).
106. Sexton, B. A., and Madix, R. J., *Chem. Phys. Lett.*, **76**, 294 (1980).
107. Barteau, M. A., and Madix, R. J., *Surf. Sci.*, **97**, 101 (1980).
108. Backx, C., deGroot, C. P. M., and Biloen, P., *Appl. Surf. Sci.*, **6**, 256 (1980).
109. Backx, C., deGroot, C. P. M., and Biloen, P., *Surf. Sci.*, **104**, 300 (1981).
110. Engelhardt, H. A., and Menzel, D., *Surf. Sci.*, **57**, 591 (1976).
111. Bowker, M., Barteau, M. A., and Madix, R. J., *Surf. Sci.*, **92**, 528 (1980).
112. Rovida, G., and Pratesi, F., *Surf. Sci.*, **52**, 542 (1975).
113. Engelhardt, H. A., Bradshaw, A. M., and Menzel, D., *Surf. Sci.*, **40**, 410 (1973).
114. Upton, T. H., Stevens, P., and Madix, R. J., *J. Chem. Phys.*, **88**, 3988 (1988).
115. Martin, R. L., and Hay, P. J., *Surf. Sci.*, **130**, 1283 (1983).
116. Lin, J.-H., and Garrison, B. J., *J. Chem. Phys.*, **80**, 2904 (1984).
117. Tully, J. C., *J. Chem. Phys.*, **73**, 6333 (1980).
118. Girifalco, L. A., and Weizer, V. G., *Phys. Rev.*, **114**, 687 (1959).
119. Binning, G. K., Rohrer, H., Gerber, Ch., and Stoll, E., *Surf. Sci.*, **144**, 321 (1984).
120. Binning, G., Rohrer, H., Gerber, Ch., and Weibel, E., *Surf. Sci.*, **131**, L379 (1983).
121. Robinson, I. K., *Phys. Rev. Lett.*, **50**, 1145 (1983).
122. Marks, L. D., *Phys. Rev. Lett.*, **51**, 1000 (1983).
123. Moritz, W., and Wolf, D., *Surf. Sci.*, **163**, L655 (1985).
124. Moller, J., Snowdon, K. J., Heiland, W., and Niehaus, H., *Surf. Sci.*, **178**, 475 (1986).
125. Copel, M., and Gustafsson, T., *Phys. Rev. Lett.*, **57**, 723 (1986).
126. Davenport, J. W., and Weinert, M., *Phys. Rev. Lett.*, **58**, 1382 (1987).
127. Brocksch, H.-J., and Bennemann, K. H., *Surf. Sci.*, **161**, 321 (1985).
128. Foiles, S. M., *Phys. Rev.* B, **32**, 7685 (1985).
129. Foiles, S. M., *Surf. Sci.*, **191**, 329 (1987).
130. Foiles, S. M., and Daw, M. S., *J. Mater. Res.*, **2**, 5 (1987).
131. Foiles, S. M., *Surf. Sci.*, **191**, L779 (1987).
132. Dodson, B. W., *Phys. Rev. Lett.*, **60**, 2288 (1988).
133. Marville, L., and Andreoni, W., *J. Phys. Chem.*, **91**, 2645 (1987).
134. Voter, A. F., and Chen, S. P., *Mat. Res. Soc. Symp. Proc.*, **82**, 175 (1987).

135. Thompson, M. W., *Phil. Mag.*, **18**, 377 (1968).
136. Reimann, C. T., El-Maazawi, M., Walzl, K., Winograd, N., Garrison, B. J., and Deaven, D. M., *J. Chem. Phys.*, **89**, 2539 (1988).
137. Lo, D. Y., Tombrello, T. A., Shapiro, M. H., Garrison, B. J., Winograd, N., and Harrison, D. E., *J. Vac. Sci. Technol. A*, **6**, 708 (1988).
138. Schlier, R. E., and Farnsworth, H. E., *J. Chem. Phys.*, **30**, 917 (1959).
139. Redondo, A., and Goddard, W. A., *J. Vac. Sci. Technol.*, **21**, 344 (1982).
140. Chadi, D. J., *Phys. Rev. Lett.*, **43**, 43 (1979).
141. Yin, M. T., and Cohen, M. L., *Phys. Rev.* B, **24**, 2303 (1981).
142. Yang, W. S., Jona, F., and Marcus, P. M., *Phys. Rev.* B, **28**, 2049 (1983).
143. Tromp, R. M., Hamers, R. J., and Demuth, J. E., *Phys. Rev. Lett.*, **55**, 1303 (1985).
144. Hamers, R. J., Tromp, R. M., and Demuth, J. E., *Phys. Rev.* B, **34**, 5343 (1986).
145. Pollman, J., Kalla, R., Kruger, P., Mazur, A., and Wolfgarten, G., *Appl. Phys. A*, **41**, 21 (1986).
146. Eastman, D. E., *J. Vac. Sci. Technol.*, **17**, 492 (1980).
147. Rowe, J. E., and Phillips, J. C., *Phys. Rev. Lett.*, **32**, 1315 (1974).
148. Mark, P., Levine, J. D., and McFarlane, S. H., *Phys. Rev. Lett.*, **38**, 1408 (1977).
149. Levine, J. D., McFarlane, S. H., and Mark, P., *Phys. Rev.* B, **16**, 5415 (1977).
150. Pandey, K. C., *Phys. Rev. Lett.*, **47**, 1913 (1981).
151. Chadi, D. J., *Phys. Rev.* B, **26**, 4762 (1982).
152. Pandey, K. C., *Phys. Rev. Lett.*, **49**, 223 (1982).
153. Joyce, B. A., *Rep. Prog. Phys.*, **37**, 363 (1974).
154. Cho, A. Y., and Arthur, J. R., *Prog. Sol. State Chem.*, **10**, 157 (1975).
155. Panish, M. B., *Science*, **208**, 915 (1980).
156. Dohler, G. H., *Sci. Am.*, **249**, 144 (1983).
157. Joyce, B. A., Bradley, R. R., and Booker, G. R., *Philos. Mag.*, **15**, 1167 (1967).
158. Abbink, H. C., Broudt, R. M., and McCarthy, G. P., *J. Appl. Phys.*, **39**, 4673 (1968).
159. NoorBatcha, I., Raff, L. M., and Thompson, D. L., *J. Chem. Phys.*, **81**, 3715 (1984).
160. NoorBatcha, I., Raff, L. M., and Thompson, D. L., *J. Chem. Phys.*, **82**, 1543 (1985).
161. NoorBatcha, I., Raff, L. M., and Thompson, D. L., *J. Chem. Phys.*, **83**, 6009 (1985).
162. Kasper, E., *Appl. Phys. A*, **28**, 129 (1982).
163. Aizaki, N., and Tatsumi, T., *Surf. Sci.*, **174**, 658 (1986).
164. Khor, K. E., and Das Sarma, S., *Chem. Phys. Lett.*, **134**, 43 (1987).
165. Gawlinski, E. T., and Gunton, J. D., *Phys. Rev.* B, **36**, 4774 (1987).
166. Brenner, D. W., and Garrison, B. J., *Surf. Sci.*, **198**, 151 (1988).
167. Schneider, M., Rahman, A., and Schuller, I. K., *Phys. Rev. Lett.*, **55**, 604 (1985).
168. Gossman, H.-J., and Feldman, L. C., *Phys. Rev.* B, **32**, 6 (1985).
169. See for example, Feldman, L. C., Mayer, J. W., and Picraux, S. T., *Materials Analysis by Ion Channeling*, Academic, New York, 1982.
170. Zuhr, R. A., Alton, G. D., Appleton, B. R., Herbots, N., Noggle, T. S., and Pennycock, S. J., in *Materials Modification and Growth Using Ion Beams* (Eds U. J. Gibson, A. E. White and P. P. Pronko), Materials Research Society, Pittsburgh, 1987, p. 243.
171. Dodson, B. W., *Phys. Rev.* B, **36**, 1068 (1987).
172. Garrison, B. J., Miller, M. T., and Brenner, D. W., *Chem. Phys. Lett.*, **146**, 553 (1988).
173. Raff, L. M., NoorBatcha, I., and Thompson, D. L., *J. Chem. Phys.* **85**, 3081 (1986).
174. Rice, B. M., NoorBatcha, I., Thompson, D. I., and Raff, L. M., *J. Chem. Phys.*, **86**, 1608 (1987).
175. Froitzheim, H., Lammering, H., and Gunter, H. L., *Phys. Rev.*, B, **27**, 2278 (1978).
176. Brenner, D. W., *J. Mat. Res. Soc. Symp. Proc.*, 1988, in press.

177. Brenner, D. W., unpublished.
178. Brenner, D. W., Elert, M. L., Walker, F. E., and White, C. T., unpublished.
179. Kirkpatrick, S., Gelatt, C. D., and Vecchi, M. P., *Science*, **220**, 671 (1983).
180. Car, R., and Parrinello, M., *Phys. Rev. Lett.*, **55**, 2471 (1985).
181. Car, R., and Parrinello, M., *Phys. Rev. Lett.*, **60**, 204 (1988).
182. Ballone, P., Andreoni, W., Car, R., and Parrinello, M., *Phys. Rev. Lett.*, **60**, 271 (1988).
183. Payne, M. C., Joannopoulos, J. D., Allan, D. C., Teter, M. P., and Vanderbilt, D. H., *Phys. Rev. Lett.*, **56**, 2656 (1986).
184. Payne, M. C., Bristowe, P. D., and Joannopoulos, J. D., *Phys. Rev. Lett.*, **58**, 1348 (1987).
185. Needels, M., Payne, M. C., and Joannopoulos, J. D., *Phys. Rev. Lett.*, **58**, 1765 (1987).
186. Allan, D. C., and Teter, M. P., *Phys. Rev. Lett.*, **59**, 1136 (1987).
187. Pederson, M. R., Klein, B. M., and Broughton, J. Q., *Phys. Rev.* B, **38**, 3825 (1988).
188. Sprik, M., and Klein, M. L., *J. Chem. Phys.*, **87**, 5987 (1987).
189. Sprik, M., and Klein, M. L., *J. Chem. Phys.*, **89**, 1592 (1988).
190. Allen, R. E., and Menon, M., *Phys. Rev.* B, **33**, 5611 (1986).
191. Menon, M., and Allen, R. E., *Phys. Rev.* B, **33**, 7099 (1986).
192. Sankey, O. F., and Allen, R. E., *Phys. Rev.* B, **33**, 7164 (1986).
193. Riley, M. E., Coltrin, M. E., and Diestler, D. J., *J. Chem. Phys.*, **88**, 5934 (1988).
194. Berendsen, H. J. C., Postma, J. P. M., van Gunsteren, W. F., DiNola, A., and Haak, *J. R., J. Chem. Phys.*, **81**, 3684 (1984).
195. Metropolis, N., Rosenbluth, A. W., Rosenbluth, M. N., Teller, A. H., and Teller, E., *J. Chem. Phys.*, **21**, 1087 (1953).
196. Voter, A. F., and Doll, J. D., *J. Chem. Phys.*, **80**, 5832 (1984).
197. Voter, A. F., and Doll, J. D., *J. Chem. Phys.*, **82**, 80 (1985).
198. Voter, A. F., *J. Chem. Phys.*, **82**, 1890 (1985).

Advances in Chemical Physics
Edited by K. P. Lawley

THEORY OF RESONANT CHARGE TRANSFER IN ATOM–SURFACE SCATTERING

A. T. AMOS and K. W. SULSTON[†‡]

Department of Mathematics, University of Nottingham, Nottingham NG7 2RD, UK

and

S. G. DAVISON[*†]

Department of Physics, Texas A&M University, College Station, Texas 77843–4242, USA

CONTENTS

[*]On leave from the Departments of Applied Mathematics and Physics, University of Waterloo, Waterloo, Ontario, Canada N2L 3G1.
[†]Work supported by the Natural Sciences and Engineering Research Council of Canada.
[‡]Present address: Department of Physics and Atmospheric Science, Drexel University, Philadelphia, PA 19104, USA.

1. INTRODUCTION

When an atom or ion is scattered from a solid surface, charge transfer from or to the surface can occur under certain conditions. For example, experiments show[1,2] that a neutral sodium atom can be scattered from the tungsten (110) surface as positively ionized Na^+, having lost an electron to the metal. On the other hand, Li^+ can become neutralized when scattered from cesiated tungsten, because it can gain an electron from the 5d band in the metal[3]. During these two processes, when the atom or ion and surface strongly interact, the energy of the occupied (unoccupied) valence orbital of the atom (ion) is aligned with an empty (filled) level in an energy band of the solid during the extremely brief ($\sim 10^{-16}$ s) collision period and, as a consequence, resonant charge transfer between the two can take place. Many atom–surface scattering experiments have been performed which suggest that resonant charge transfer is a fairly pervasive mechanism and this has led, over the past ten years, to an increasing theoretical interest in the topic[4–7]. It is the purpose of this article to describe the theoretical model most often used to characterize resonant charge transfer and to review some of the results obtained from that model.

We shall be almost exclusively concerned with the resonant process, although we should mention that other mechanisms for charge exchange are possible, as illustrated in Fig. 1. The quasi-resonant transition[8–10] is somewhat similar to the resonant one and proceeds by direct transfer of an electron from a core state of the solid to an ion. The processes differ in that the quasi-resonant transition is non-adiabatic, with an energy difference of up to 5–10 eV between the two energy levels, producing either a slight change in the ion's kinetic energy or the emission of a phonon into the solid. Auger neutralization[11], on the other hand, is a two-electron process, whereby one band-state electron transfers to the ion while a second is excited so as to ensure energy conservation.

Returning to resonant charge transfer, Fig. 2 shows, schematically, resonance between a single valence orbital $|0\rangle$, associated with an atom outside a solid, and a set of wavefunctions $|i\rangle$ associated with an energy band in the solid. The orbital and band states can be coupled together through the matrix elements $\langle 0|\mathscr{H}|i\rangle$, where $\mathscr{H}$ is the Hamiltonian, and those band states, with energies close to the valence-orbital energy ε_0, become mixed in with $|0\rangle$ – the so-called 'broadening' of the valence level. If ε_0 lies in the empty part of the band, sufficiently large matrix elements will lead to a high probability for the transfer of an electron from $|0\rangle$ (if there is one in $|0\rangle$) into the empty band orbitals. Conversely, if ε_0 lies in the filled part of the band, there can be a high probability of transfer from the band to $|0\rangle$ (if $|0\rangle$ is empty).

While such a description of resonant charge transfer is broadly correct, it is, of course, an entirely static account which requires modifying before it can be

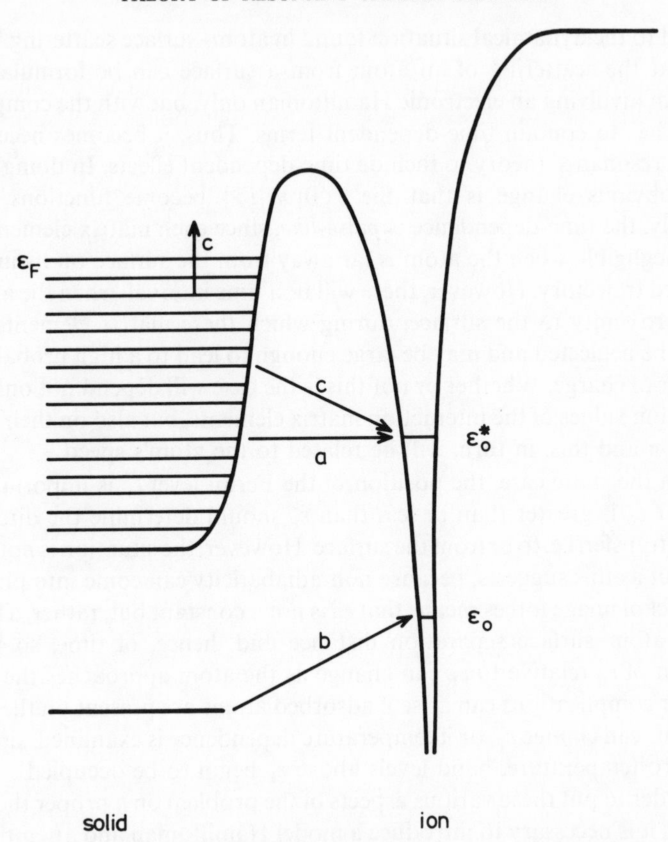

Fig. 1. Schematic illustration of electron transfer via (a) resonant, (b) quasi-resonant, (c) Auger transition.

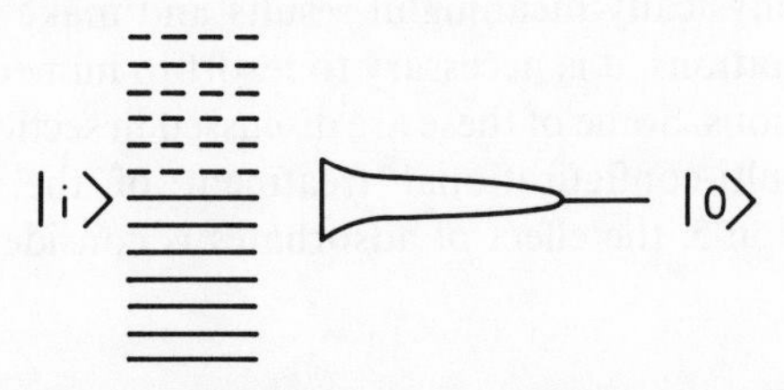

Fig. 2. Resonant coupling between atom state $|0\rangle$ and band states $|i\rangle$ in a solid. Full (broken) lines represent the filled (unfilled) part of the band.

applied to the dynamical situation found in atom–surface scattering.† It turns out that the scattering of an atom from a surface can be formulated as a problem involving an electronic Hamiltonian only, but with the complication that it has to contain time-dependent terms. Thus, it becomes necessary to extend resonance theory to include time-dependent effects. In doing this, the most obvious change is that the $\{\langle 0|\mathscr{H}|i\rangle\}$ become functions of time. Actually, the time-dependence is *pulse-like*, since each matrix element will be quite negligible when the atom is far away from the surface on its inward or outward trajectory. However, there will be a time interval, when the atom is in close proximity to the surface, during which these matrix elements can no longer be neglected and may be large enough to lead to a high probability for transfer of charge. Whether or not this is the case will depend not only on the maximum values of the interaction matrix elements, but also on their effective duration and this, in turn, will be related to the atom's speed.

As in the static case, the position of the Fermi level ε_F is important, since whether ε_0 is greater than or less than ε_F should determine the direction of charge transfer, i.e. to or from the surface. However, the situation is not quite as clear-cut as this suggests, because non-adiabaticity can come into play. Also, the effect of image forces means that ε_0 is not a constant but, rather, a function of the atom–surface separation distance and, hence, of time, so that the position of ε_0 relative to ε_F can change as the atom approaches the surface. Further complications can arise if adsorbed atoms are present on the surface, since this can change ε_F, or if temperature dependence is examined, since, with non-zero temperature, band levels above ε_F begin to be occupied.

In order to put these various aspects of the problem on a proper theoretical footing, it is necessary to introduce a model Hamiltonian and attempt to find solutions to the time-dependent Schrödinger equation apropriate to it. It has been found that the simplest Hamiltonian which contains most of the physically important aspects of the problem is the Anderson–Newns Hamiltonian[12,13]. Therefore, in section 2, we begin by showing how resonant charge transfer can be formulated in terms of the time-dependent Anderson–Newns (TDAN) model. Even though this model leads to formally simple equations, it appears to be impossible to obtain analytical solutions. Thus, in order to extract physically meaningful results and make interesting predictions and interpretations, it is necessary to resort to numercial methods or to apply approximations. Some of these are discussed in section 3. In section 4, a many-electron multi-configurational treatment of the TDAN model is presented. In section 5, the effect of adsorbates is considered.

†We shall use the term atom to apply to either an ion or neutral atom.

2. A MODEL FOR THE THEORY OF RESONANT CHARGE TRANSFER

Tully[9] has discussed how the classical-path method, used originally for gas-phase collisions, can be applied to the study of atom–surface collisions. It is assumed that the motion of the atomic nucleus is associated with an effective potential energy surface and can be treated classically, thus leading to a classical trajectory $\mathbf{R}(t)$. The total Hamiltonian for the system can then be reduced to one for electronic motion only, associated with an electronic Hamiltonian $\mathscr{H}(\mathbf{R}) = \mathscr{H}(t)$ which, as indicated, depends parametrically on the nuclear position and through that on time. Therefore, the problem becomes one of solving a time-dependent Schrödinger equation†

$$\mathscr{H}(t)|\Phi(t)\rangle = i\partial|\Phi(t)\rangle/\partial t \tag{1}$$

for the electronic wavefunction $|\Phi(t)\rangle$.

It is hardly possible, and probably not very useful, to find the exact form for $\mathscr{H}(t)$. Instead, as frequently happens, much progress can be made by adopting a model Hamiltonian. The work of Blandin *et al.*[14], Bloss and Hone[15] and the earlier study of sputtering by Sroubek[16] shows that a suitable one is the TDAN Hamiltonian, which is a generalization of the time-independent one originally introduced to discuss impurities in metals[12] and later applied to hydrogen chemisorption on metals[13].

The first simplification in the TDAN model is to consider only a few electronic orbitals on the scattered atom. For many applications, it is sufficient to consider one only, that from which, or into which, an electron is transferred. Let the ket $|0\rangle$ denote the spatial part of the orbital. When far from the surface, suppose its energy is ε_0, and let U_0 be the Coulomb repulsion integral associated with the energy change when it is occupied by two electrons of opposite spin. In terms of creation and annihilation operators $c_{0\sigma}^{\dagger}$ and $c_{0\sigma}$ for $|0\rangle$, with σ ($=\alpha$ or β) a spin index, that part of $\mathscr{H}$ which refers to the free atom is

$$\mathscr{H}_{\mathrm{A}} = \varepsilon_0 \sum_{\sigma} c_{0\sigma}^{\dagger} c_{0\sigma} + U_0 c_{0\alpha}^{\dagger} c_{0\alpha} c_{0\beta}^{\dagger} c_{0\beta} \tag{2}$$

In a similar fashion, it usually suffices to consider a single band of the solid. If the band orbitals are $|\chi_k\rangle$ ($k = 1, \ldots, N$) with energies ε_k, that part of $\mathscr{H}$ which refers to the band before it interacts with the atom is

$$\mathscr{H}_{\mathrm{B}} = \sum_{\sigma} \sum_{k=1}^{N} \varepsilon_k c_{k\sigma}^{\dagger} c_{k\sigma} \tag{3}$$

†Except where otherwise stated, we use atomic units.

Since the c-operators in (2) and (3) operate on space and spin functions, the kets $|0\rangle$ and $|\chi_k\rangle$, which correspond to spatial components of the orbitals only, have to be multiplied by spin kets $|\alpha\rangle$ and $|\beta\rangle$ to give the set of $2N+2$ space-spin functions of the problem. Occasionally we shall use the notation $|\psi, \sigma\rangle$ for the product $|\psi\rangle|\sigma\rangle$ when we need to make explicit which spin (α or β) is associated with the spatial ket $|\psi\rangle$.

To complete the specification of $\mathscr{H}$, it is necessary to introduce time-dependent terms. These are of two types. The first allows for many-body effects which, to a first approximation, correspond to an interaction of the atom with its image in the solid. As a result, the ionization level and the Coulomb repulsion integral for $|0\rangle$ become functions, $E_0(z)$ and $U(z)$, of the perpendicular distance z of the atom from the surface. The classical electrostatic forms[17] for these functions are

$$E_0(z) = \varepsilon_0 + F(z), \qquad U(z) = U_0 - 2F(z) \tag{4}$$

with $F(z) = (4z)^{-1}$. However, more complicated, albeit more accurate forms, have been suggested[18]. Of course, the distance z, via the trajectory $\mathbf{R}(t)$, will be a function of t so that we can write $F(t) = [4z(t)]^{-1}$. As a rule, it is assumed that the atom moves with constant perpendicular velocity v, so that $z = z_0 + v|t|$, taking z_0 as the distance of closest approach at time $t = 0$. The time-dependent part of $\mathscr{H}$, due to this image effect, is

$$\mathscr{H}_{\mathrm{I}}(t) = F(t)\left[\sum_{\sigma} c_{0\sigma}^{\dagger} c_{0\sigma} - 2c_{0\alpha}^{\dagger} c_{0\alpha} c_{0\beta}^{\dagger} c_{0\beta}\right] \tag{5}$$

The second type of time-dependent term, which couples $|0\rangle$ to the band orbitals and is therefore responsible for charge transfer, is

$$\mathscr{H}_{\mathrm{C}} = \sum_{\sigma}\sum_{k=1}^{N} V_{k\sigma}(t)\{c_{k\sigma}^{\dagger} c_{0\sigma} + c_{0\sigma}^{\dagger} c_{k\sigma}\} \tag{6}$$

so that $V_{k\alpha}(t)$, for example, will represent the time-dependent coupling between $|0\rangle$ and $|\chi_k\rangle$ when both are associated with α-spin. As is implied in (6), in some cases it can be more satisfactory to make this coupling spin-dependent, i.e., $V_{k\alpha} \neq V_{k\beta}$.

A variety of forms have been suggested[5,19] for the time-dependent functions in (6). A useful simplification is to assume that $V_k(t)$ (we drop the spin subscript to indicate the same form of the equation can be applied to both $V_{k\alpha}$ and $V_{k\beta}$) can be factored as a scalar, dependent on k, and a time-dependent function independent of k, i.e.

$$V_k(t) = v_k V(t) \tag{7}$$

Frequently, v_k is assumed to be $\langle \chi_k | s \rangle$, where $|s\rangle$ is an atomic orbital on the target atom, i.e. the surface atom with which the scattered atom most strongly interacts. $V(t)$ should go to zero, as $t \to \pm\infty$, when the atom is far away from the surface, and rise to a peak around $t = 0$, the time of closest approach. Therefore, $V(t)$ is likely to have a pulse-like form. Very widely used is

$$V(t) = V_0 e^{-\lambda|t|} \tag{8}$$

where the parameter V_0 represents the maximum strength of the interaction, while λ^{-1} is a measure of the length of time that the interaction is effective. It is to be expected that λ is related to the perpendicular speed of the atom, e.g. a fast atom will interact with the surface for a shorter time than will a slow atom. Equation (8) can be rationalized on the basis that interactions between orbitals usually vary like overlaps, leading to an exponential dependence of the form $A\exp(-Cz)$, which becomes (8) on putting $z = z_0 + v|t|$ with λ proportional to the velocity v. Other possible forms for $V(t)$, with similar properties, are V_0 sech(λt) and Gaussians[19].

Combining $\mathscr{H}_A$, $\mathscr{H}_B$, $\mathscr{H}_I$ and $\mathscr{H}_C$ together gives the TDAN Hamiltonian $\mathscr{H}(t)$ which we shall use in this chapter. We must mention, however, that somewhat more complicated forms have been introduced in order to consider some special aspects of the problem. Some of the most notable include that used by Bloss and Hone[15] (which has a core level in the solid and two orbitals on the atom), those which take into account the scattering of atoms with large velocity components parallel to the surface[20,21] and that recently considered by Kawai *et al.*[22] in order to suggest a possible new experimental technique for studying energy levels of atoms in collision with surfaces.

The $\mathscr{H}(t)$ to be considered here can be rewritten as

$$\mathscr{H}(t) = H_\alpha(t) + H_\beta(t) + [U_0 - 2F(t)]c_{0\alpha}^\dagger c_{0\alpha} c_{0\beta}^\dagger c_{0\beta} \tag{9}$$

where

$$H_\sigma = [\varepsilon_0 + F(t)]c_{0\sigma}^\dagger c_{0\sigma} + \sum_k \varepsilon_k c_{k\sigma}^\dagger c_{k\sigma} + \sum_k V_{k\sigma}(t)[c_{k\sigma}^\dagger c_{0\sigma} + c_{0\sigma}^\dagger c_{k\sigma}] \tag{10}$$

with $\sigma = \alpha$ or β. It will be noticed that H_α and H_β are the sum of one-electron operators which are, in principle, easy to handle. The difficult term in $\mathscr{H}(t)$ is the two-electron Coulomb repulsion term and the methods used to solve the TDAN model can be divided into two classes according to the way chosen to deal with this term. By far the largest class consists of those methods which avoid the complications arising from the Coulomb term by the simple expedient of neglecting it entirely, or by using the Hartree–Fock approximation, where the two-electron term is averaged to give an effective one-electron term. The great advantage in each of these is that the equations reduce to one-electron equations. In the next section we shall discuss these equations, their solutions and the conclusions to be drawn from them. If the two-electron term in $\mathscr{H}(t)$ is not neglected or averaged, a many-electron approach must be used. This will be considered in section 4.

3. ONE-ELECTRON THEORY

3.1. One-electron method and TDAN model

When $\mathscr{H}$ is approximated as the sum of one-electron operators, considerable simplifications in the theory can be made. Following the important pioneering work of Bloss and Hone[15], this has been exploited, most often, by using the Heisenberg representation and solving the equations of motion for the c-operators. In a slightly different approach, Sebastian *et al.*[23,24] wrote the Hartree–Fock wavefunction as a single Slater determinant of one-electron functions and then considered the evolution operator, which generates the time variation of these functions. For the present chapter, we believe it is more transparent to deal with the one-electron functions themselves, rather than the c-operators or the evolution operator. Consequently, we work entirely in the Schrödinger representation.

Suppose there are $2n$ electrons in the system, half with α-spin and half with β-spin, and we form a single-determinant wavefunction with the spatial functions $\{|\psi_{j\alpha}(t)\rangle\}$ and $\{|\psi_{j\beta}(t)\rangle\}$, so that

$$|\Phi(t)\rangle = ||\psi_{1\alpha}, \alpha\rangle|\psi_{1\beta}, \beta\rangle \ldots |\psi_{n\alpha}, \alpha\rangle|\psi_{n\beta}, \beta\rangle| \tag{11}$$

When the Coulomb term is entirely neglected, substitution into (1) shows that $|\Phi(t)\rangle$ will be a solution provided the spatial functions satisfy

$$\mathrm{i}\partial|\psi_{j\sigma}\rangle/\partial t = h_\sigma|\psi_{j\sigma}\rangle \tag{12}$$

where the spatial one-electron operator is

$$h_\sigma = \{\varepsilon_0 + F(t)\}|0\rangle\langle 0| + \sum_k \varepsilon_k|\chi_k\rangle\langle\chi_k| + \sum_k V_{k\sigma}\{|0\rangle\langle\chi_k| + |\chi_k\rangle\langle 0|\} \tag{13}$$

If the Hartree–Fock approximation is used, the additional term

$$[U_0 - 2F(t)]\langle P_\beta(t)\rangle|0\rangle\langle 0| \tag{14}$$

involving the β-spin occupation number P_β of the scattered atom, must be added to h_α, with a similar one to supplement h_β. Note that, in this case, the equations become non-linear and have to be solved iteratively, which is such a difficult computational problem that only a few calculations have been attempted[25–27].

To solve (12) uniquely, we must impose the initial conditions which arise from the original state of the system $|\Phi(t = -\infty)\rangle$. Considering a case where, initially, the valence orbital $|0\rangle$ of the atom is empty, then $|\Phi(-\infty)\rangle$ must represent a situation where the n band orbitals of lowest energy are doubly occupied by electrons, i.e.

$$|\psi_{j\alpha}(-\infty)\rangle = |\psi_{j\beta}(-\infty)\rangle = |\chi_j\rangle, \qquad j = 1, \ldots, n \tag{15}$$

If we treat α and β-spin electrons equivalently by choosing $V_\alpha = V_\beta$, so that

$h_\alpha = h_\beta$, then the system evolves with $|\psi_{j\alpha}\rangle$ and $|\psi_{j\beta}\rangle$ equal at all times. Consequently, the probability that, at time $t = \infty$, the atom has captured an α-spin electron from the band, will be

$$P_\alpha = \sum_{j=1}^{n} |\langle 0|\psi_{j\alpha}(\infty)\rangle|^2 \tag{16}$$

which is the same as the probability that a β-spin electron has been transferred. The probabilities, P_0, P_1 and P_2, of the atom capturing no electrons, one electron and two electrons (one of each spin) from the surface are

$$P_0 = (1 - P_\alpha)^2, \qquad P_1 = 2(P_\alpha - P_\alpha^2), \qquad P_2 = P_\alpha^2 \tag{17a}$$

Sebastian[27] has emphasized that (17a) implies $P_1 \leqslant 0.5$ (since $0 \leqslant P_\alpha \leqslant 1$), which is contradicted by many experiments where values close to unity have been found. Therefore, the approach outlined above cannot be used in situations where $P_1 \approx 1$, $P_2 \approx 0$. We must stress that the difficulty arises here from the restricted form the wavefunction (11) has when the α and β-spin orbitals are constrained to be equal. It can be circumvented by removing this constraint and using different spatial orbitals for electrons with different spin, which is accomplished by making different choices for the coupling functions, i.e., $V_{k\alpha}(t) \neq V_{k\beta}(t)$.

If it is believed that P_2 is very close to zero, the simplest form of (11), which necessarily gives the result $P_2 = 0$, is that where one set of orbitals (the β-set, say) is held fixed at their $t = -\infty$ values, so that only transfer of an α-spin electron is allowed. To do this, one chooses $V_{k\beta}(t) \equiv 0$ and then,

$$P_0 = 1 - P_\alpha, \qquad P_1 = P_\alpha, \qquad P_2 = 0 \tag{17b}$$

Rather loosely, this may be thought of as taking $U \approx \infty$, so that double occupancy of $|0\rangle$ is totally forbidden. Because the wavefunction (11) has different α and β-spin spatial orbitals, it is not a pure spin state (it will be a mixture of singlet, triplet, quintet, etc.). However, for the problem in hand, this does not seem to be much of a disadvantage.

Similar *ad hoc* methods must be applied in other circumstances, for example, when $|0\rangle$ is doubly or singly occupied. In the latter case, suppose there are $2n$ electrons in the band with a single α-spin electron $|0\rangle$. Then we must add to the determinant (11) the additional orbital $|\psi_{n+1,\alpha}, \alpha\rangle$, which at $t = -\infty$ equals $|0\rangle$. The probabilities for finding electrons of α-spin and β-spin on the atom will be

$$P_\alpha = \sum_{j=1}^{n+1} |\langle 0|\psi_{j\alpha}(\infty)\rangle|^2, \qquad P_\beta = \sum_{j=1}^{n} |\langle 0|\psi_{j\beta}(\infty)\rangle|^2 \tag{18}$$

so that

$$P_0 = 1 - P_\alpha - P_\beta + P_\alpha P_\beta, \qquad P_1 = P_\alpha + P_\beta - 2P_\alpha P_\beta, \qquad P_2 = P_\alpha P_\beta \tag{19}$$

Cases[1,2] such as the scattering of Na off clean W, where $P_2 \approx 0$, correspond to $P_\beta \approx 0$ and can be treated by taking $V_{k\beta}(t) \equiv 0$, thus keeping the β-spin orbitals fixed, and allowing only the α-spin orbitals to evolve. The appropriate probabilities are then $P_1 = P_\alpha$ and $P_0 = 1 - P_\alpha$.

Although the methods suggested above are not, by any means, completely satisfactory, they are sufficient to describe the main qualitative aspects of the problem. In the remainder of this section, therefore, we shall discuss methods of solving the one-electron equations (12), with h_σ defined by (13). We shall assume that the Coulomb repulsion term can be neglected and, therefore, will not consider Hartree–Fock solutions. Details of the latter can be found in Refs. 25–27.

3.2. The one-electron TDAN equations of motion

In this section, we write down the explicit forms for the one-electron TDAN equations of motion for the situation where the Coulomb repulsion integral is set equal to zero. Also, we consider only the case where the band energy levels contain $2n$ electrons and the valence orbital of the scattered atom is initially empty. It will be obvious how to modify the equations to deal with other possible cases; mostly this will involve just a change in the initial conditions.

For our purposes, the valence orbital $|0\rangle$ plus the band orbitals $\{|\chi_k\rangle\}$ ($k = 1, 2, \ldots, N$; occupied and unoccupied) can be regarded as a complete set of functions, and so we can write for each of the n occupied α-spin orbitals ($j = 1, 2, \ldots, n$)

$$|\psi_{j\alpha}(t)\rangle = a_{0j}(t)\mathrm{e}^{-\mathrm{i}\varepsilon_0 t}|0\rangle + \sum_{k=1}^{N} a_{kj}(t)\mathrm{e}^{-\mathrm{i}\varepsilon_k t}|\chi_k\rangle \tag{20}$$

Substituting into (12) gives the equations of motion[28] for the complex coefficients in (20):

$$\dot{a}_{0j}(t) = -\mathrm{i}F(t)a_{0j} - \mathrm{i}\sum_{k=1}^{N} v_k V(t)\mathrm{e}^{\mathrm{i}\omega_k t}a_{kj}(t) \tag{21a}$$

$$\dot{a}_{kj}(t) = -iv_k V(t)\mathrm{e}^{-\mathrm{i}\omega_k t}a_{0j}(t) \tag{21b}$$

where $\omega_k = \varepsilon_0 - \varepsilon_k$. The initial conditions are

$$a_{0j}(-\infty) = 0, \qquad a_{kj}(-\infty) = \delta_{kj} \tag{22}$$

For some purposes, it is better to work with a scaled a_{0j} defined by

$$f_j = v_j^{-1}a_{0j} \tag{23}$$

For each occupied orbital $j = 1, \ldots, n$, we have to solve the set of $N + 1$ linear first-order differential equations (21). Unfortunately, it does not seem possible to obtain analytical solutions for any realistic choice of $V(t)$ and, therefore, it is necessary to resort to finding approximate or numerical solutions. Some of these can be developed directly from (21) while, for others, it is better to convert these equations into an integro-differential equation for f_j. This is obtained by formally integrating (21b) and substituting into (21a) to give

$$\dot{f}_j = -\mathrm{i}F(t)f_j - \mathrm{i}V(t)\mathrm{e}^{\mathrm{i}\omega_j t} - V(t)\int_{-\infty}^{t} G_1(t-u)\mathrm{e}^{\mathrm{i}\varepsilon_0(t-u)}V(u)f_j(u)\,\mathrm{d}u \quad (24)$$

where

$$G_1(t-u) = \sum_{k=1}^{N} v_k^2 \mathrm{e}^{-\mathrm{i}\varepsilon_k(t-u)} \quad (25)$$

A further integration, followed by a change in the order of integration, converts this into a Volterra integral equation, which can be solved formally using the Picard method. However, this does not appear to be very practical or efficient, except in those rare cases where V is small and the Picard series converges rapidly.

Once the $f_j(t)$ have been obtained – either approximately or numerically – the function $P_\alpha(t)$, required to evaluate the probability of charge transfer through (17a) or (17b), can be found. Since $|\langle 0|\psi_j^\alpha\rangle| = |v_j f_j(t)|$, it follows from (16) that

$$P_\alpha(t) = \sum_{j=1}^{n} |v_j f_j(t)|^2 \quad (26a)$$

If the v_j are set equal to $\langle \chi_j | s\rangle$, where s is the target atom in the surface, then

$$P_\alpha(t) = \sum_{j=1}^{n} |\langle \chi_j | s\rangle|^2 |f_j(t)|^2 \quad (26b)$$

As shown, for example, in Ref. 28, the surface density of states $\rho_S(\varepsilon)$ at the target atom can be expressed as

$$\rho_S(\varepsilon) = \sum_{j=1}^{N} |\langle \chi_j | s\rangle|^2 \delta(\varepsilon - \varepsilon_j) \quad (27)$$

Consequently, if, in (26b), we replace the discrete energy variable ε_j by a continuous one ε, so that $f_j(t)$ becomes a function $f(\varepsilon, t)$, we can write

$$P_\alpha(t) = \int_{-\infty}^{\varepsilon_F} \rho_S(\varepsilon)|f(\varepsilon, t)|^2\,\mathrm{d}\varepsilon \quad (28)$$

Thus, the charge transfer probability can be related to the surface density of states.

At this point, it is interesting to extend the theory to include temperature effects, as was first done by Brako and Newns[29]. When the temperature is non-zero, the Fermi–Dirac function

$$\mathscr{F}(\varepsilon, T) = \{1 + \exp[(\varepsilon - \varepsilon_F)/kT]\}^{-1} \quad (29)$$

must be included in the integrated of (28) and the factor $\mathscr{F}(\varepsilon_j, T)$ in the summand of (26) in order to give the value of $P_\alpha(t)$ at temperature T. Note, also, that ε_F will be a function of T in (28) and the summation in (26) must be extended to allow for this.

Before proceeding, in later sections, to review some of the numerical and approximate methods used to handle equations (21) and (24), we think it helpful to consider a simple model to illustrate the nature and properties of these equations and their solutions.

3.3. Tight-binding linear chain

The simplest model of a solid is a linear chain of N atoms, with one end of the chain corresponding to the surface. If an atomic orbital $|r\rangle$ ($r = 1, \ldots, N$) is associated with the rth atom, then, in the *tight-binding approximation*[30], the matrix elements of the Hamiltonian for the solid, $\mathscr{H}_B$, can be expressed in terms of the site energy α of an atomic orbital and the hopping integral β between nearest-neighbor orbitals, i.e.

$$\langle r|\mathscr{H}_B|s\rangle = \alpha\delta_{rs} + \beta(\delta_{r,s-1} + \delta_{r,s+1}) \tag{30}$$

In the notation of section 2, the eigenvalues and eigenfunctions of $\mathscr{H}_B$ will be

$$\varepsilon_j = \alpha + 2\beta\cos\left(\frac{j\pi}{N+1}\right) \tag{31}$$

$$|\chi_j\rangle = \sum_{r=1}^{N}\left(\frac{2}{N+1}\right)^{1/2}\sin\left(\frac{rj\pi}{N+1}\right)|r\rangle \tag{32}$$

From (31), it is clear that the bandwidth is $4|\beta|$, while the Fermi level ε_F depends on the number of electrons. Two cases are of interest: (1) where each atom in the solid contributes two electrons to the band so that the band is completely full, with $\varepsilon_F = \alpha + 2|\beta|$, and (2) where each atom contributes one electron so that the band is half-full, with $\varepsilon_F = \alpha$.

The parameter v_k in the interaction potential will be $v_k = \langle\chi_k|1\rangle$, since the target atom is $s = 1$, and for $V(t)$ we use (8), which involves the two parameters λ and V_0. We write ε_0 in terms of α and β as

$$\varepsilon_0 = \alpha + \delta|\beta| \tag{33}$$

In this section, we ignore image effects and assume that this orbital energy remains fixed throughout the motion. Therefore, in (21) and (24), we set $F(t) = 0$.

Now, in order to illustrate how the values of P_α change when the parameters are varied, we present some results obtained by solving Eq. (21) numerically with $N = 20$ (for details see section 3.5) and using (26b) to find P_α. For similar, but more extensive results, see Refs. 23, 24, 31, 32, and particularly the paper by

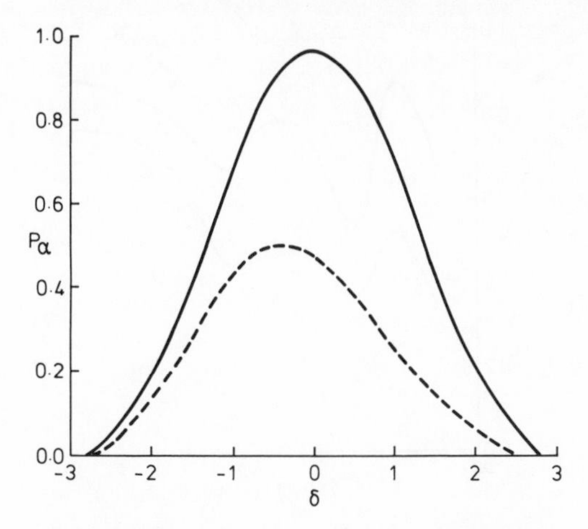

Fig. 3. P_α versus δ for filled (solid curve) and half-filled (dashed curve) bands, where $\alpha = 0.0$, $\beta = -0.125$, $\lambda = 0.1$.

McDowell[33]. In all of the present calculations we have taken $V_0 = 0.5$, which is perhaps a little high but not untypical of the values found in the literature. For β we used either -0.125, corresponding to a broad band of width 13.6 eV, or -0.0125 for a very narrow band, while λ was given values ranging from $\lambda = 0.1$ (slow atom) to $\lambda = 1.0$ (fast atom).

As would surely be expected on physical grounds, a crucial factor is the value of δ which determines the position of ε_0 relative to the band. With a half-filled band, δ in the range $[-2, 0]$ places ε_0 in the occupied part of the band and, on energy grounds, charge transfer should be likely. With a full band, the condition is $\delta \in [-2, 2]$. Figure 3 shows plots of P_α as a function of δ, for the wide-band case. The figure includes graphs both for the filled and half-filled bands. As expected, P_α reaches a maximum when ε_0 is in the occupied part of the band, but it goes to zero much less rapidly than might have been expected as ε_0 moves out of this region. It is particularly noticeable that, for the half-filled band, P_α has quite large values when ε_0 is above the Fermi level but with an energy equivalent to unoccupied band orbitals. Of course, P_α is larger for the full band than the half-filled band, simply because there are twice as many electrons available for transfer in the former case than the latter.

In Fig. 4, we examine the variation of P_α with λ. Figure 4a plots P_α against λ^{-1} for the wide band and demonstrates an oscillatory behavior for small values of λ^{-1}, i.e. for fast atoms, which is lost for larger values corresponding to slow atoms. Generally speaking, the values of P_α are larger for slow atoms, indicating a higher probability for charge transfer, but P_α can be high for some fast atoms with specific values of λ. Figure 4b gives an equivalent graph for the narrow band, and displays a very different behavior, which is oscillatory throughout, being damped to zero for large λ^{-1}.

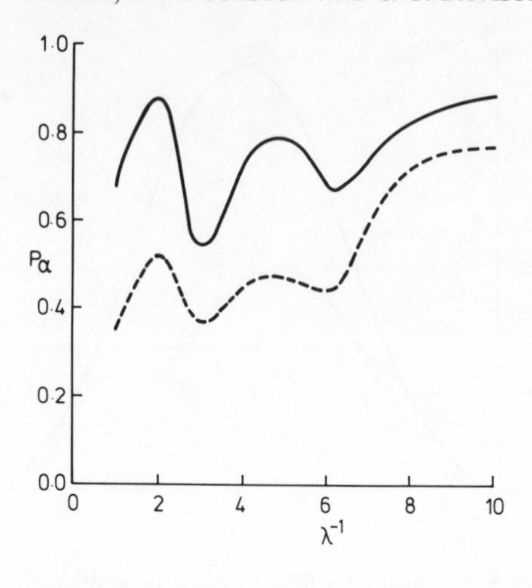

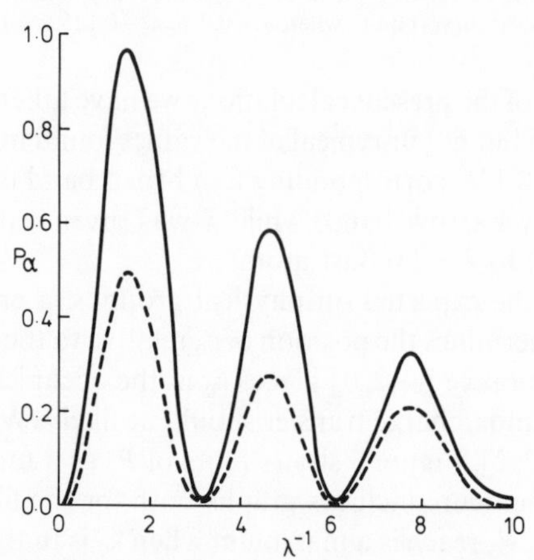

Fig. 4. P_α versus λ^{-1} where $\beta = $ (a) -0.125, (b) -0.0125, for filled (solid curve) and half-filled (dashed curve) bands. $\varepsilon_0 = -0.2$, $\alpha = -0.1$.

We examine the time dependence of the charge transfer $P_\alpha(t)$ in Fig. 5a for the wide band and Fig. 5b for the narrow band. Both curves show some oscillation, indicating there can be a series of transfers between the atom and the band. We have used $\lambda = 0.1$, so the atom is moving slowly and is in the vicinity of the surface for a sufficiently long time for these multiple transfers to occur. The graphs show that the effective interaction time, when transfers take place, is about $[-4\lambda, 4\lambda]$ for the narrow band and $[-2\lambda, 2\lambda]$ for the wide band. It has been suggested that it is only the outward ($t > 0$) part of the atom's trajectory which is important for the charge-transfer process, but that is certainly not true for the cases shown in Fig. 5.

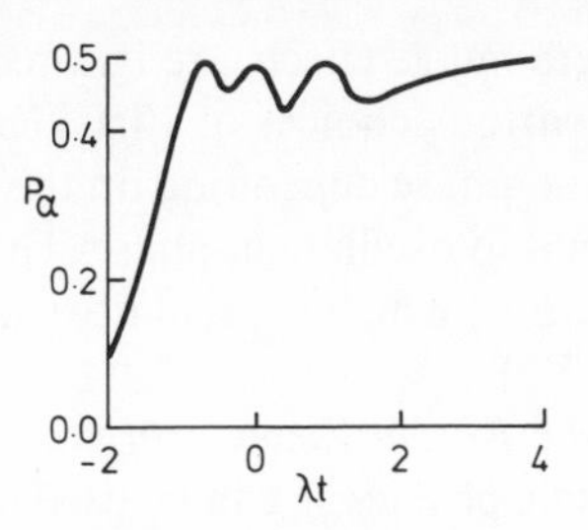

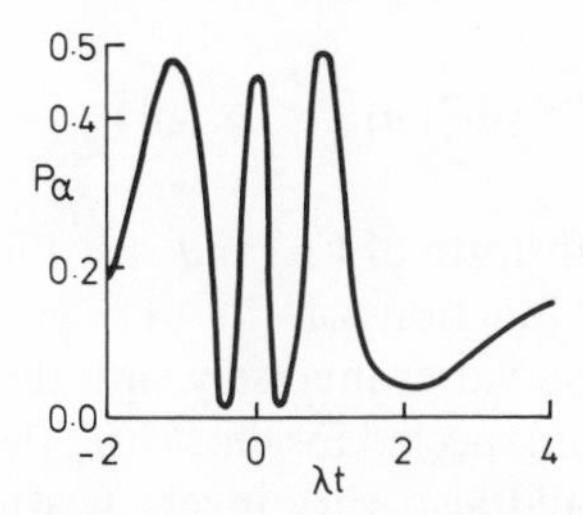

Fig. 5. P_α versus λt for a half-filled band where $\beta =$ (a) -0.125 (b) -0.0125, with $\lambda = 0.1$, $\varepsilon_0 = -0.2$, $\alpha = -0.1$.

Comparing the two graphs in Fig. 5, it is obvious that the behavior, in the case of the wide band, where the oscillations are relatively small in magnitude, differs qualitatively from the rapid and large variations found for the narrow band. Approximate methods can be used to explain these and other differences.

3.4. Approximate methods

In view of the results discussed in the previous section, it is not surprising that, in finding suitable approximations[31,32], a critical factor is bandwidth. We consider first an approximation suitable for a very narrow band. Since the difference between energy levels in the band must be small, $G_1(t-u)$ in (25) can be approximated by setting $\varepsilon_k = \varepsilon_j$ (for all k), so that

$$G_1(t-u) = \mathrm{e}^{-\mathrm{i}\varepsilon_j(t-u)} \tag{34}$$

Substituting (34) into (24) gives an equation equivalent to that of the atom scattering off a single atom of energy ε_j in the surface, and for this reason the approximation has been called the *single-orbital approximation* (SOA)[34]. We note, in passing, that a second approximation, the *narrow-band approximation*[32], can be obtained in a very similar way and gives almost identical results to the SOA in the region where both are valid.

In the special case where image effects are ignored, it is possible to use the SOA to solve for f_j, for various choices of $V(t)$. Generally, the solutions are oscillatory, with period and phase depending on the ε_j, but, since all the ε_j are effectively equal, the f_j tend to oscillate in phase. Therefore, when substituted into (26b) they produce a $P_\alpha(t)$ which is oscillatory also, and this explains the behavior of $P_\alpha(t)$ in Fig. 5(b).

In calculating $P_\alpha(\infty)$, the *Rosen–Zener approximation*[35], familiar from the two-level problem in atomic physics[36], can be used for $f_j(\infty)$ and this leads to the compact expression, for any pulse-like $V(t)$,

$$P_\alpha = A^{-2}\sin^2(A)\int_{-\infty}^{\varepsilon_F} \rho_S(\varepsilon)|\bar{V}(\varepsilon_0 - \varepsilon)|^2\,\mathrm{d}\varepsilon \tag{35}$$

where $\bar{V}$ is the Fourier transform of $V(t)$ and $A = \bar{V}(0)$ is the pulse area. When $V(t)$ is the exponential function of (8), $A = 2V_0/\lambda$, which implies A is proportional to λ^{-1} and so varies inversely with the atom's speed. Because a similar dependence is to be expected for most $V(t)$, the oscillations in P_α, due to the sine term in (35), should also vary inversely with the atom's speed. The remaining terms in (35) form an envelope to these oscillations, causing their amplitude to decrease as A increases, so that $P_\alpha \to 0$ for large A, i.e. low speeds. These general conclusions are clearly in accord with the results illustrated in Fig. 4b, as is the high-velocity result (small A) that $P_\alpha \sim A^2 \to 0$. Thus, charge transfer becomes improbable both for high- and low-velocity atoms and this is also the case if ε_0 differs considerably from the band energies, since the integral in (35) will then be small.

A full analysis of the SOA suggests that it should be restricted to narrow bands or to high-velocity atoms, but that, within these regimes, it works well. The complementary approximation for broad bands and low velocity atoms is the *wide-band approximation* (WBA)[14,15,29]. The essence of the WBA lies in the recognition that, for wide bands, the function $G_1(t-u)$ in (25) behaves like a delta function. This is because the sinusoidal terms in G_1 oscillate out of phase when the band is broad, leading to a destructive cancellation of terms, which is avoided only when $t = u$. For example, in the one-dimensional problem of section 3.3,

$$G_1(t-u) = 2|\beta|^{-1}\mathrm{e}^{-\mathrm{i}\alpha(t-u)}\{J_1[2|\beta|(t-u)]/2(t-u)\} \tag{36}$$

where J_1 is the first-order Bessel function. As $|\beta| \to \infty$, the term in braces becomes $\delta(t-u)$. Therefore, for wide bands, we make the approximation

$$G_1(t-u) = 2B^{-1}\mathrm{e}^{-\mathrm{i}\alpha(t-u)}\delta(t-u) \tag{37}$$

where B is a constant, which is large in the WBA, since it is, in general, related to the bandwidth; usually B is proportional to the bandwidth. In some treatments, an equivalent approximation identifies B^{-1} with the broadening of the valence level. Substituting into (24) leads to

$$\dot{f}_j = -\mathrm{i}F(t)f_j - \mathrm{i}V\mathrm{e}^{\mathrm{i}\omega_j t} - B^{-1}V^2 f_j \tag{38}$$

which can be solved to give

$$f_j = -\mathrm{ie}^{-Q(t)}\int_{-\infty}^{t} V(u)\exp\{\mathrm{i}\omega_j u + Q(u)\}\,\mathrm{d}u \tag{39}$$

where

$$Q(t) = \int_{-\infty}^{t} \{\mathrm{i}F(u) + B^{-1}V^2(u)\}\,\mathrm{d}u \tag{40}$$

Letting $t \to \infty$ and substituting into (26) will give P_α. This, however, is based on the assumption that the valence orbital of the scattered atom is originally unoccupied. If the contrary is true, an additional term must be included. A composite expression, which includes both cases, is[29]

$$P_\alpha(\infty) = \mathrm{e}^{-2R(\infty)}\{P_\alpha(-\infty) + \sum_{j\mathrm{occ}} |v_j \mathscr{F}(\omega_j)|^2\} \tag{41}$$

where $\mathscr{F}$ is the Fourier transform of $V(t)\exp Q(t)$ and $R(t)$ is the real part of $Q(t)$.

Brako and Newns[29] have analysed an expression equivalent to (41) and have determined the behavior of P_α in various important circumstances. A good discussion can be found, also, in the article by Yoshimori and Makoshi[5], and we ourselves[32] have examined the range of validity of the WBA by using the one-dimensional model of section 3.3. It is possible to develop along these lines a theory of the sputtering of atoms from surfaces but that topic lies outside the scope of this chapter. (For a recent review see the paper by Yu and Lang[37].) However, Lang's paper[38] on sputtering contains interesting results which can, with minor modification, be used in the WBA.

Clearly, an important quantity in (41) is $R(\infty)$, the area under the curve $B^{-1}V^2(t)$. With $V^2(t)$ of exponential form (Eq. (8)),

$$R(\infty) = V_0^2/\lambda B \tag{42}$$

leading to two extreme cases: (1) $V_0^2 \ll \lambda B$, giving a small value to $R(\infty)$ and corresponding to weak interactions and fast speeds; (2) $V_0^2 \gg \lambda B$, giving a large value to $R(\infty)$ and corresponding to strong interactions and slow speeds. Similar results will hold with other appropriate expressions for $V(t)$.

The first term in (41) has been called the *memory* term since it contains the memory of the initial occupancy of the valence orbital. For slow atoms with large interactions, $\exp[-2R(\infty)]$ will be small and the memory term insignificant. For fast atoms with weak interactions, $\exp[-2R(\infty)]$ will be close to unity, so the memory of the initial state is retained.

The second term in (41) is more difficult to deal with in general terms, although it is easy to evaluate in particular cases. One example is when the band is completely full (or half-full if it is symmetrical), ε_0 is level with the band centre and image effects are ignored ($F(t) = 0$). After some manipulation, it follows that

$$P_\alpha = b + \{P_\alpha(-\infty) - b\}e^{-2R(\infty)} \tag{43}$$

with $b = 1$ or $\frac{1}{2}$ according to whether the band is full or half-full. The properties of $R(\infty)$ imply that charge transfer is favoured for slow atoms and is less likely for fast atoms.

Equation (43) assumes a specific value for ε_0. A more general result follows from the fact that the Fourier transforms in (41) will be small, except for small values of ω_j, so that, as expected, to obtain high probabilities of charge transfer, ε_0 should either be in the solid band or, at least, not lie very far from the band. An interesting result, showing the dependence of P_α on ε_0, can be obtained for the one-dimensional model of section 3.3 with a half-filled band, $V(t)$ of exponential form, $P_\alpha(-\infty) = 0$ and $V_0^2 \gg \lambda B$. This is

$$P_\alpha = \tfrac{1}{2} + \pi^{-1} \tan^{-1} \sinh \pi(\varepsilon_F - \varepsilon_0)/2\lambda \tag{44}$$

It shows, as expected, that P_α increases as ε_0 drops below the Fermi level. Note, however, that P_α can be appreciable even when ε_0 is above the Fermi level (cf. Fig. 3). The condition $V_0^2 \gg \lambda B$ will usually be due to small λ, in which case (44) reduces asymptotically to the result that $P_\alpha = 1$, if $\varepsilon_0 < \varepsilon_F$ and $P_\alpha = 0$, if $\varepsilon_0 > \varepsilon_F$.

If $F(t)$ is not equal to zero, so that image effects are included, it does not seem possible to obtain any simple analytical expression for P_α, although there is no difficulty in obtaining results by numerical quadrature. For the special case where $F(z)$ is approximated as a linear function of z, however, an asymptotic result has been found[29,38].

Another method suitable for wide bands is the *local-time approximation* (LTA) discussed and used by Kawai[39–41] and exploited by several authors[21,42]. The fundamental idea of the LTA is that the time variation of f_j in the integral in (24) can be ignored, so that $f_j(u)$ can be replaced by $f_j(t)$, which can then be removed from the integral. Therefore, (24) takes the form

$$\dot{f}_j = -iF(t)f_j - iV(t)e^{i\omega_j t} - f(t)V(t)\int_{-\infty}^{t} G_1(t-u)e^{i\varepsilon_0(t-u)}V(u)\,du \tag{45}$$

A comparison of this approximation with numerical solutions suggests that its range of validity is for wide bands and high-velocity ions.

Finally in this section, we note that the use of perturbation theory, particularly for quasi-resonant charge transfer, has been developed by Battaglia, George and Lanaro[43,44]. They show that first-order perturbation theory is satisfactory for high-velocity atoms, with ε_0 lying outside the solid band, and they have examined in detail protons scattered from alkali-halide surfaces, where the bands are narrow. This is where the SOA should be valid and, indeed, as $A \to 0$ in Eq. (35), the first-order perturbation result is obtained.

3.5. Numerical techniques

Accurate numerical techniques are available, which allow quantitative investigation of the atom–surface scattering process to be made, and which provide a standard for assessing the validity of the various approximations[31,32]. Despite the different approaches adopted by the numerical methods reviewed here, they are, nonetheless, interrelated and do lead to substantially the same results.

The approach introduced by the present authors (SAD)[45] is to integrate directly the $N+1$ differential equations of motion (21), for each value of j in (26b) corresponding to an energy ε_j below the Fermi level ε_F. The first step is the evaluation of the energies ε_k and coefficients v_k, which requires that the time-independent Hamiltonian $\mathscr{H}_B$ of the isolated finite solid be specified. Knowing v_k and ε_k, it is straightforward to integrate Eq. (21) using a Runge–Kutta method; in our own case, the Merson variation. Runge–Kutta methods have the advantage that they tend to be very accurate, but they have the drawback of potentially being much more time-consuming than other numerical procedures. However, this has not been a problem for the systems that we have studied. The size of the model solid required to ensure satisfactory convergence of the results is parameter-dependent. In the case of a fast-moving atom interacting with a narrow-band one-dimensional substrate, as few as 8 atoms in the chain can be sufficient, while the requisite number can be much larger for the contrasting situation of a slow atom and wide-band solid. There is no major difficulty in using the procedure to investigate three-dimensional systems, although the necessary number of atoms in the solid increases, but not to the point of intractability.

Muda and Hanawa (MH)[46] have approached the problem by considering the time variation of the quantities $\eta_{mn}(t) = \langle \Phi(t) | c_m^\dagger c_n | \Phi(t) \rangle$, which are expectation values of products of creation and annihilation operators for site-centered orbitals $|n\rangle$. The Schrödinger equation then leads to a set of first-order differential equations, viz.,

$$\mathrm{i}\dot{\eta}_{mn}(t) = \sum_l \left[\langle n | h_\sigma(t) | l \rangle \eta_{ml}(t) - \langle l | h_\sigma(t) | m \rangle \eta_{ln}(t) \right] \tag{46}$$

This system of equations can be integrated numerically by a Runge–Kutta method, and the charge transfer probability is then obtained as $P_\alpha(\infty) = \lim_{t \to \infty} \eta_{00}(t)$. The quantities $\eta_{mn}(t)$ can be related to $a_{kj}(t)$ in (21) by writing

$$\eta_{mn}(t) = \sum_{j \in \mathrm{occ}} \sum_{kk'} A_{km}^* A_{k'n} a_{kj}^*(t) a_{k'j}(t) \mathrm{e}^{\mathrm{i}(\varepsilon_k - \varepsilon_{k'})t} \tag{47}$$

where $A_{km} = \langle m|\chi_k\rangle$. It is clear from (47) that, whereas SAD solve directly for a_{kj}, MH basically calculate the squares of those quantities.

Although the MH method yields $P_\alpha = \eta_{00}$ directly, it is not without its difficulties. Foremost is the large number of atoms in the substrate, which are necessary for convergence of the results. The one-dimensional system studied by MH required a chain length of 100 atoms. This suggests that application of their procedure to three-dimensional substrates could lead to an intractably large number of equations, of type (46), to be solved. MH suggested that the convergence problem is due to some long-range property of the charge-transfer process, but it seems more likely that it is partially a result of calculating the products $a^*_{kj}a_{k'j}$, rather than $a_{k'j}$, as well as of modeling relatively wide bands within the next-nearest neighbour approximation.

Sebastian *et al.* (SJG)[23,24] concerned themselves with the time-evolution operator of the system, which satisfies the equation

$$i\partial U(t,t_0)/\partial t = h_\sigma(t)U(t,t_0), \qquad U(t_0,t_0) = 1 \tag{48}$$

In the interaction representation, the occupation number of atomic orbital $|0\rangle$, at time t, is

$$P_\alpha(t) = \sum_{j\text{occ}} |\langle 0|\hat{U}(t,t_0)|\chi_j\rangle|^2 \tag{49}$$

where the matrix element $\hat{U}(0,j,t) = \langle 0|\hat{U}(t,t_0)|\chi_j\rangle$ can be shown[23] to be the solution of a particular Volterra integro-differential equation. In this formulation, $\hat{U}(0,j,t)$ plays the same role as $a_{0j}(t)$ does in that of SAD and, indeed, the Volterra equation for $\hat{U}(0,j,t)$ is equivalent to that satisfied by $a_{0j}(t)$, namely (24) with $a_{0j} = v_j f_j$. In the large-N limit used by SJG, (25) becomes

$$G_1(\tau) = \int_{-\infty}^{\infty} \rho_S(\varepsilon)e^{-i\varepsilon\tau}\,d\varepsilon \tag{50}$$

which is just the Fourier time transform of the surface density of states $\rho_S(\varepsilon)$. When $\rho_S(\varepsilon)$ has semi-elliptical form, as in SJG, it is possible to obtain the exact analytical form of $G_1(\tau)$. For more complicated systems, it is not, in general, possible to do so, and $G_1(\tau)$ must be either approximated analytically or evaluated numerically. The former situation results in a loss of accuracy, while the latter causes an increase in computation time.

Using Simpson's rule to evaluate the integral on the right-hand side of (24), SJG solved the equation by means of the Euler method. Although the technique is straightforward and efficient to apply for simple systems, it could prove more cumbersome for complicated three-dimensional systems and require the use of a more accurate method than the Euler one.

McDowell introduced two numerical approaches, the first[33,47] of which is closely related to that of SJG. His molecular-orbital treatment started with the large-N limit of (24), with $a_{0j}/v_j \to f(\varepsilon, t)$ and $F(t) = 0$, so that

$$\mathrm{i}\dot{f}(\varepsilon, t) = V(t)\mathrm{e}^{\mathrm{i}(\varepsilon_0 - \varepsilon)t} - \mathrm{i}V(t)\int_{-\infty}^{t} \mathrm{d}u\, V(u)\mathrm{e}^{\mathrm{i}\varepsilon_0(t-u)}G_1(t-u)f(\varepsilon, u) \tag{51}$$

with G_1 given by (50), and the atom occupation number $P_\alpha(t)$ found from (28). At this point, McDowell parts company from SJG by defining an auxiliary function

$$I_{\varepsilon,\varepsilon'}(t) = \mathrm{e}^{-\mathrm{i}(\varepsilon' - \varepsilon_0)t}\int_{-\infty}^{t} \mathrm{d}u\mathrm{e}^{\mathrm{i}(\varepsilon' - \varepsilon_0)u}V(u)f(\varepsilon, u) \tag{52}$$

whose derivative is

$$\dot{I}_{\varepsilon,\varepsilon'}(t) = -\mathrm{i}(\varepsilon' - \varepsilon_0)I_{\varepsilon,\varepsilon'}(t) + V(t)f(\varepsilon, t) \tag{53}$$

so that (51) becomes

$$\dot{f}(\varepsilon, t) = -\mathrm{i}V(t)\mathrm{e}^{\mathrm{i}(\varepsilon_0 - \varepsilon)t} - V(t)\int_{-\infty}^{\infty} \rho_S(\varepsilon')I_{\varepsilon,\varepsilon'}(t)\,\mathrm{d}\varepsilon' \tag{54}$$

The integrals in (28) and (54) are approximated by q Gaussian quadrature points, so that, for each quadrature energy point in (28), there is a set of $q + 1$ first-order differential equations to be solved, since

$$\int_{-\infty}^{\infty} \rho_S(\varepsilon')I_{\varepsilon,\varepsilon'}(t)\mathrm{d}\varepsilon' \simeq \sum_{n=1}^{q} w_n\rho_S(\varepsilon_n)I_{\varepsilon,\varepsilon_n}(t) \tag{55}$$

where ε_n and w_n are the quadrature points and weights, respectively. The system of differential equations can then be integrated by a Runge–Kutta technique, with about the same computational effort as that for the SAD equations, since the number of quadrature points required is roughly equal to the number of substrate atoms needed by SAD. The finite-N form of (53) and (54), obtained by replacing the integral $\int \mathrm{d}\varepsilon'\rho_S(\varepsilon')$ by the sum $\sum_k v_k^2$, can, in fact, be transformed back into the SAD equations of motion by means of the relationship

$$a_{0j} = v_j f_j, \qquad a_{kj} = \delta_{kj} - \mathrm{i}v_j v_k \mathrm{e}^{\mathrm{i}(\varepsilon_k - \varepsilon_0)t}I_{jk} \tag{56}$$

For simple systems, the McDowell molecular-orbital technique would seem to be more time-consuming than that of SJG. In more complicated situations, however, this approach should lead to more accurate results, not only by using a Runge–Kutta rather than Euler method, but also by employing directly $\rho_S(\varepsilon)$, rather than its Fourier transform $G_1(\tau)$, whose explicit form may not be known.

McDowell's second method[33,48] was a Langevin treatment, based upon the work of Adelman and Doll[49], the basic idea of which was the division of the system into a primary zone, consisting only of the incoming atom and target atom, and a 'background' bath, characterizing the substrate, which was represented by the Fourier transform (25) and a driving term. Using these notions, the equations of motion for the annihilation operators $c_n(t)$ of the atomic orbitals of the system can be solved in a manner similar to the molecular-orbital treatment, with comparable accuracy and computation time.

Inglesfield[50] used a Green's function technique to write the one-electron wavefunctions as

$$|\psi(\mathbf{r},t)\rangle = |\psi_0(\mathbf{r},t)\rangle - \int_{-\infty}^{t} \mathrm{d}u \int \mathrm{d}\mathbf{r}' G_0(\mathbf{r},\mathbf{r}';t-u)V(u)|\psi(\mathbf{r}',u)\rangle \tag{57}$$

where $|\psi_0\rangle$ is the wavefunction for the separated substrate and scatterer, without the interaction potential $V(t)$. G_0 is the corresponding unperturbed Green's function satisfying

$$(H_0 - \mathrm{i}\partial/\partial t)G_0(\mathbf{r},\mathbf{r}';t-u) = \delta(\mathbf{r}-\mathbf{r}')\delta(t-u) \tag{58}$$

where

$$H_0 = \mathscr{H}_{\mathrm{B}} + \varepsilon_0|0\rangle\langle 0| \tag{59}$$

Notice that G_0 is the Fourier time transform of the one-electron energy-dependent Green's function $G_0(\mathbf{r},\mathbf{r}';E)$. Expansion of a particular wavefunction, in terms of site atomic orbitals, as

$$|\psi_j(\mathbf{r},t)\rangle = \sum_m b_{jm}(t)|m\rangle \tag{60}$$

and similarly for $|\psi_0\rangle$, and subsequent substitution in (57), leads to a coupled pair of integral equations for the coefficients of the projectile orbital $|0\rangle$ and the target-site orbital $|1\rangle$:

$$b_{j0}(t) = b_{j0}^{0}(t) - \int_{-\infty}^{t} \mathrm{d}u \langle 0|G_0(t-u)|0\rangle V(u) b_{j1}(u) \tag{61}$$

$$b_{j1}(t) = b_{j1}^{0}(t) - \int_{-\infty}^{t} \mathrm{d}u \langle 1|G_0(t-u)|1\rangle V(u) b_{j0}(u) \tag{62}$$

The unperturbed coefficients are

$$b_{j0}^{0}(t) = \delta_{j0}\mathrm{e}^{-\mathrm{i}\varepsilon_0 t}, \qquad b_{j1}^{0}(t) = v_j \mathrm{e}^{-\mathrm{i}\varepsilon_j t}(1-\delta_{j0}) \tag{63}$$

while the quantity $\langle n|G_0(t-u)|n\rangle$ is the local Green's function on the site $|n\rangle$. The occupation number $P_\alpha(t)$ of the orbital on the scatterer is then given by

$$P_\alpha(t) = \sum_j |b_{j0}(t)|^2 \tag{64}$$

where the summation includes all initially occupied states. It should be noted that the coefficients, calculated via (61) and (62), are related to those of the SAD approach by

$$b_{j0}(t) = \mathrm{e}^{-\mathrm{i}\varepsilon_0 t} a_{0j}(t), \qquad b_{j1}(t) = \sum_k v_k \mathrm{e}^{-\mathrm{i}\varepsilon_k t} a_{kj}(t) \tag{65}$$

Thus, the SAD equations of motion can be transformed into (61) and (62), by formally integrating (21) to obtain a system of integral equations, which can then be put into Inglesfield's form, by making the change of variables (65), followed by some rearrangement of terms.

The two complex equations (61) and (62) give rise to a set of four real integral equations, which must be solved numerically for each value of j corresponding to an initially occupied state. This can be done by discretizing the time variable in the integrals (using Simpson's rule, for example), which then allows $b_{j0}(t)$ and $b_{j1}(t)$ to be calculated, with about the same computational effort as the SAD technique.

4. MANY-ELECTRON THEORY

Although the one-electron theory of resonant charge transfer has been quite successful in explaining much of the experimental data, it is not without its failings. Firstly, Auger transitions play a role in many particle–surface scattering systems, and therefore require rigorous theoretical treatment[51,52]. However, because Auger transfer is a two-electron process, it cannot be properly described within the framework of a one-electron theory, so many-electron basis functions must be used. Secondly, as was discussed in section 3.1, the two-electron term $Uc^{\dagger}_{0\alpha}c_{0\alpha}c^{\dagger}_{0\beta}c_{0\beta}$ in the TDAN Hamiltonian is difficult to treat satisfactorily using a one-electron approximation, although Sebastian[53] has shown how to improve upon the Hartree–Fock approximation by writing the two-electron time-evolution operator as a functional average of a one-electron operator. In any event, rigorous treatment of the two-electron term requires a many-electron method. This is related to the fact that a single-configuration Hartree or Hartree–Fock wavefunction is not suitable for discussing charge transfer because the separating of the particle from the solid is, effectively, a dissociation process, for which multi-configuration wavefunctions must be used.

The conclusion that the theory of surface-charge transfer should be formulated using many-electron wavefunctions was reached by Tully[9], and formed the foundation for his theory which, in its full generality, can encompass the resonant, quasi-resonant and Auger mechanisms. Kasai and Okiji[54] have introduced a method which goes beyond Hartree–Fock and which utilizes the Heisenberg equations of motion. Brako and Newns[55] have devised a many-body treatment of the TDAN Hamiltonian for the particular case of multiply degenerate atomic orbitals and infinitely large Coulomb repulsion. Sebastian[27] attacked the same problem from quite a different angle, using the time-dependent version of the coupled-cluster technique[56], based on the wavefunction $\exp[T_0(t) + T_1(t) + T_2(t) + \cdots]|\Phi_0\rangle$, where $|\Phi_0\rangle$ is a Slater determinant and $T_n(t)$ creates n particle–hole excitations in it. This approach allowed the two-electron term in the TDAN Hamiltonian to be included in the theory, without the need to resort to the one-electron approximation. In calculations on Li^+ scattered from a Ni surface, Sebastian obtained very different results from those of the one-electron theory. This must cause a little concern regarding the application of the one-electron theory to some systems.

In this section, we shall outline a many-electron treatment of charge transfer[42], similar in spirit to that of Tully, which enables different charge-exchange mechanisms to be incorporated in the formalism simultaneously. Although we shall concentrate on the TDAN model of resonant neutralization and negative ionization, we shall indicate how other neutralization processes can be included, and the approach for the reverse process of positive ionization will be fairly apparent.

In considering the resonant neutralization of a positive ion, there are two possible cases to examine, depending on whether the incoming ion has a completely empty orbital (e.g. H^+) or one containing a single electron (e.g. He^+). We begin by discussing the former case where, henceforth, the ket notation is generally omitted, for compactness.

If the solution $\Phi(t)$ to the time-dependent Schrödinger equation is expanded as a linear combination of time-independent orthonormal many-electron basis functions Φ_i, i.e.

$$\Phi(t) = \sum_i a_i(t)\Phi_i \tag{66}$$

then the coefficients satisfy

$$i\dot{a}_i = \sum_j \langle \Phi_i|\mathscr{H}(t)|\Phi_j\rangle a_j \tag{67}$$

If the original state $\Phi(-\infty)$ of the system is designated as Φ_0, then the initial condition for (67) is $a_i(-\infty) = \delta_{i0}$. In theory, $\{\Phi_i\}$ should be a complete set, but, in practice, it must be truncated to include only the more important elements present.

Now, if the impinging ion's empty valence orbital, denoted by ω_0, interacts with a surface, one electron (or two of opposite spins) can be transferred into that orbital. Let the solid band consist of n doubly occupied one-electron orbitals $\{\chi_i\}_{i=1,\ldots,n}$ and v unoccupied orbitals $\{\chi_\mu\}_{\mu=n+1,\ldots,n+v}$.† Hence, the wavefunction Φ_0 for the non-interacting system (at $t = -\infty$) will be the Slater determinant for the occupied band, namely,

$$\Phi_0 = |\chi_1\bar{\chi}_1 \cdots \chi_n\bar{\chi}_n| \tag{68}$$

†Roman and Greek subscripts shall refer to occupied and unoccupied solid orbitals, respectively.

where each unbarred (barred) orbital is occupied by an $\alpha(\beta)$-spin electron. Corresponding to the resonant transition of an electron from the ith solid orbital into the ion's vacant orbital is the wavefunction

$$\Phi_{i0} = 2^{-1/2}\{|\chi_1\bar{\chi}_1\cdots\chi_i\bar{\omega}_0\cdots| + |\chi_1\bar{\chi}_1\cdots\omega_0\bar{\chi}_i\cdots|\}, \qquad i = 1, \cdots, n \tag{69}$$

the form of which is such that the total spin remains zero.

When two electrons are transferred to the ion, there are two forms of wavefunction to be considered, depending upon whether the electrons originate from the same solid orbital or different ones. These functions are, respectively,

$$\Phi_{i0i0} = |\cdots\chi_{i-1}\bar{\chi}_{i-1}\omega_0\bar{\omega}_0\chi_{i+1}\bar{\chi}_{i+1}\cdots| \tag{70a}$$

$$\Phi_{i0k0} = 2^{-1/2}\{|\cdots\chi_i\bar{\omega}_0\cdots\omega_0\bar{\chi}_k\cdots| + |\cdots\omega_0\bar{\chi}_i\cdots\chi_k\bar{\omega}_0\cdots|\} \quad (i < k) \tag{70b}$$

There are n functions of type (70a) and $\frac{1}{2}(n^2 - n)$ of type (70b).

After an electron has been transferred from an occupied substrate orbital χ_i to the ionic one ω_0, it is possible that the electron can then be transferred back to the solid again, either into χ_i, yielding the original wavefunction Φ_0, or into an unoccupied orbital χ_μ, in which case the system is represented by the wavefunction

$$\Phi_{i\mu} = 2^{-1/2}\{|\cdots\chi_i\bar{\chi}_\mu\cdots| + |\cdots\chi_\mu\bar{\chi}_i\cdots|\} \tag{71}$$

The process $\Phi_0 \rightarrow \Phi_{i0} \rightarrow \Phi_{i\mu}$ is, in effect, an excitation of an electron from χ_i to χ_μ. In a similar way, the negatively ionized species, represented by (70), can lose one of its electrons (or possibly both, though we shall ignore this) into an unoccupied substrate state, giving a wavefunction of the form either of (69) or

$$\Phi_{i0i\mu} = 2^{-1/2}\{|\cdots\chi_{i-1}\bar{\chi}_{i-1}\omega_0\bar{\chi}_\mu\chi_{i+1}\bar{\chi}_{i+1}\cdots| + |\cdots\chi_{i-1}\bar{\chi}_{i-1}\chi_\mu\bar{\omega}_0\chi_{i+1}\bar{\chi}_{i+1}\cdots|\} \tag{72a}$$

$$\Phi_{i0k\mu} = 2^{-1/2}\{|\cdots\chi_i\bar{\omega}_0\cdots\chi_\mu\bar{\chi}_k\cdots| + |\cdots\omega_0\bar{\chi}_i\cdots\chi_k\bar{\chi}_\mu\cdots|\} \quad (i \neq k) \tag{72b}$$

Equation (72) corresponds to a neutralized atom, accompanied by an excitation in the solid.

We shall, for a moment, discuss how to expand the above set of basis functions to include quasi-resonant and Auger phenomena. For quasi-resonant neutralization, the TDAN Hamiltonian (9) must be modified[15] so that the ion orbital ω_0 is coupled to a solid core orbital χ_d (of energy ε_d); i.e. to $\mathscr{H}(t)$ must be added the terms

$$\mathscr{H}' = \sum_\sigma \varepsilon_d c_{d\sigma}^\dagger c_{d\sigma} + V_{d0}(t) \sum_\sigma [c_{d\sigma}^\dagger c_{0\sigma} + c_{0\sigma}^\dagger c_{d\sigma}] \tag{73}$$

Correspondingly, this process is represented by the basis function

$$\Phi_{d0} = 2^{-1/2}\{|\cdots\chi_d\bar{\omega}_0\cdots| + |\cdots\omega_0\bar{\chi}_d\cdots|\} \tag{74}$$

Auger transitions are due to the transfer of a solid electron to the ion, coupled with the excitation of a second electron into an Auger state χ_A of energy E_A in the solid. Thus, the TDAN Hamiltonian is augmented by the terms[51,57]

$$\mathscr{H}'' = \sum_A E_A c_A^\dagger c_A + \sum_{i,k,A,\sigma} V_{0Aik}(t)[c_{0\sigma}^\dagger c_{A-\sigma}^\dagger c_{i\sigma} c_{k-\sigma} + c_{k-\sigma}^\dagger c_{i\sigma}^\dagger c_{A-\sigma} c_{0\sigma}] \quad (75)$$

subject to the energy-conserving condition

$$E_0 + E_A = \varepsilon_i + \varepsilon_k \quad (76)$$

The relevant basis functions are

$$\Phi_{i0iA} = 2^{-1/2}\{|\cdots\chi_{i-1}\bar{\chi}_{i-1}\omega_0\bar{\chi}_A\chi_{i+1}\bar{\chi}_{i+1}\cdots| + |\cdots\chi_{i-1}\bar{\chi}_{i-1}\chi_A\bar{\omega}_0\chi_{i+1}\bar{\chi}_{i+1}\cdots|\} \quad (77a)$$

$$\Phi_{i0kA} = 2^{-1/2}\{|\cdots\chi_i\bar{\omega}_0\cdots\chi_A\bar{\chi}_k\cdots| + |\cdots\omega_0\bar{\chi}_i\cdots\chi_k\bar{\chi}_A\cdots|\}, \qquad (i \neq k) \quad (77b)$$

The functions discussed above will play dominant roles in most situations. However, when necessary, additional basis functions can be introduced, possibly requiring more ion or metal orbitals to be used, as would be the case, for instance, if excited states of the ion were playing a part in the scattering process[22].

For current purposes, it is sufficient to consider a truncated basis set consisting only of Eqs. (68)–(71), i.e. one or two resonant transitions and a single excitation in the solid. Rewriting (66) in terms of this set gives

$$\phi(t) = a_0(t)\phi_0 + \sum_{i=1}^{n} b_i(t)\phi_{i0} + \sum_{j=1}^{m}\sum_{i=1}^{j} c_{ij}(t)\phi_{i0j0} + \sum_{i=1}^{n}\sum_{\mu=n+1}^{n+v} d_{i\mu}(t)\phi_{i\mu} \quad (78)$$

To a good approximation, the basis set can be assumed to be time-independent and orthonormal, so that the coefficients in (78) are given by the differential equations (67). Using the TDAN Hamiltonian (9), the matrix elements of $\mathscr{H}(t)$, in the above basis, can be calculated in a straightforward manner, and substituting them into (67) produces the equations of motion:

$$\begin{aligned}
\dot{a}_0 &= -\mathrm{i}E_0 a_0 - \mathrm{i}2^{1/2}V(t)\sum_j v_j b_j \\
\dot{b}_k &= -\mathrm{i}2^{1/2}v_k V(t)a_0 - \mathrm{i}[E_0 + \varepsilon_0(t) - \varepsilon_k]b_k \\
&\quad - \mathrm{i}2^{1/2}v_k V(t)c_{kk} - \mathrm{i}V(t)\left[\sum_{j<k} v_j c_{jk} + \sum_{j>k} v_j c_{kj}\right] - \mathrm{i}V(t)\sum_\mu v_\mu d_{k\mu} \\
\dot{c}_{kk} &= -\mathrm{i}2^{1/2}v_k V(t)b_k - \mathrm{i}[E_0 - 2\varepsilon_k + 2\varepsilon_0(t) + U(t)]c_{kk} \\
\dot{c}_{kl} &= -\mathrm{i}v_l V(t)b_k - \mathrm{i}v_k V(t)b_l - \mathrm{i}[E_0 - \varepsilon_k - \varepsilon_l + 2\varepsilon_0(t) + U(t)]c_{kl} \quad (k < l) \\
\dot{d}_{k\mu} &= -\mathrm{i}v_\mu V(t)b_k - \mathrm{i}(E_0 - \varepsilon_k + \varepsilon_\mu)d_{k\mu}
\end{aligned} \quad (79)$$

This set of first-order differential equations can be solved, approximately or numerically, for a specific system. The theory has been applied[42] to Li^+ scattered from Cs/W, and gives more satisfactory agreement with experiment than does the one-electron approach.

Turning now to the situation where the ion orbital contains a single electron, assumed to be of α-spin, the formulation is much the same, except that the basis functions must be modified to take into account the extra electron in the system. Thus, the wavefunction representing the initial state of the system is

$$\Phi_0 = |\chi_1\bar{\chi}_1 \cdots \chi_n\bar{\chi}_n\omega_0| \tag{80}$$

The wavefunction corresponding to the transfer of an electron from the ith solid orbital to the vacant ion orbital $\bar{\omega}_0$ is

$$\Phi_{i0} = |\chi_1\bar{\chi}_1 \cdots \chi_i\bar{\omega}_0 \cdots \chi_n\bar{\chi}_n\omega_0|, \qquad (i = 1, \ldots, n) \tag{81}$$

It is also possible for an electron to transfer from (or via) the ion into a vacant solid orbital (producing an excitation in the substrate), the associated wavefunctions being

$$\Phi_{0\mu} = |\chi_1\bar{\chi}_1 \cdots \chi_n\bar{\chi}_n\chi_\mu| \tag{82}$$

and

$$\Phi_{i\mu} = 2^{-1/2}\{|\chi_1\bar{\chi}_1 \cdots \chi_i\bar{\omega}_0 \cdots \chi_n\bar{\chi}_n\chi_\mu| + |\chi_1\bar{\chi}_1 \cdots \chi_i\bar{\chi}_\mu \cdots \chi_n\bar{\chi}_n\omega_0|\} \tag{83}$$

Other processes are, of course, possible, but since the form of their wavefunctions is of similar structure to the ones above, we shall omit them here. With (80)–(83) as the basis set, an expansion like (78) can be made, leading to equations of motion analogous to (79), which can be solved in a similar way.

The many-electron theory of charge transfer discussed here possesses the versatility needed in order to treat different mechanisms within the same quantum-mechanical framework. However, it remains for future work to decide how successful the present formalism will be in providing a comprehensive many-electron theory of surface charge transfer.

5. EFFECTS OF ADSORBATES

The theory of the previous sections, in its various formulations, has been used to investigate the scattering properties of a number of systems. The majority of these investigations have been concerned with substrates possessing clean surfaces, and the associated results have been reviewed elsewhere[5,7]. However, the presence of adsorbates on the surface gives rise to some interesting features, apart from producing considerable changes in the charge-transfer probability from its value for the clean surface. This is to be expected, since the appearance of the surface density of states in (28) shows its strong influence on the charge-transfer process.

Recently, we have studied[34,45] the effect of the surface density of states on the charge-transfer probability, in the case where the surface possesses localized states created by surface perturbations or the presence of adatoms. For the tight-binding linear chain these perturbations or adatoms are taken into account by changing the electronic energy of the end atom of the chain to α', which differs from the energy α of the other atoms in the chain. This difference can lead to the formation of a localized surface state, whose energy is

$$E_s = \alpha + 2\beta[z + (4z)^{-1}] \tag{84}$$

in terms of the dimensionless surface perturbation parameter

$$z = (\alpha' - \alpha)/2\beta \tag{85}$$

with corresponding intensity $I_s = 1 - (2z)^{-2}$, which is subject to the existence condition $|z| > \frac{1}{2}$. The single-orbital approximation produced virtually exact values of $P_\alpha(\infty)$, as a function of ε_0, as shown in Fig. 6[34]. It is clear from Fig. 6 that, for any particular choice of ε_0, $P_\alpha(\infty)$ depends strongly on the value of α', and that the range of values of ε_0, which give high $P_\alpha(\infty)$, varies dramatically with α'. In the case of $\alpha' = 0.0$ (a clean surface), significant amounts of neutralization can occur only if ε_0 is aligned with the occupied portion of the band (-1 to 0 a.u.). The maximum of the curve occurs when ε_0 is close to the

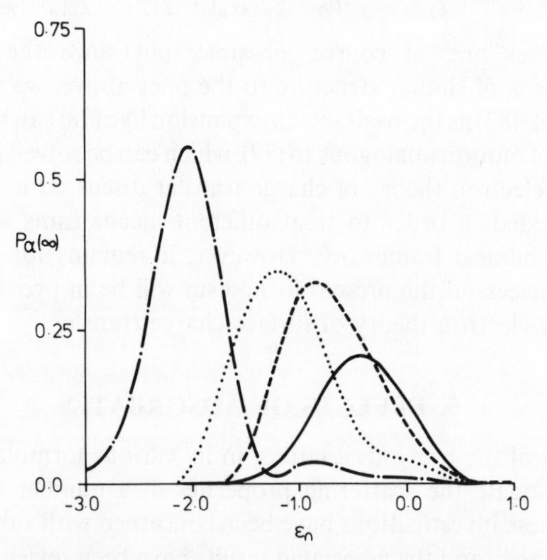

Fig. 6. Atom occupation number $P_\alpha(\infty)$ versus ε_0 for $\alpha = 0.0$, $\beta = -0.5$, $\lambda = 0.4$ and $V_0 = 0.2$. $\alpha' = 0.0$ (solid curve), -0.5 (dashed), -1.0 (dotted) and -2.0 (dash-dotted). $E_S = -1.0$ (dashed), -1.25 (dotted) and -2.125 (dash-dotted). There is no localized state for $\alpha' = 0.0$. (*Reprinted with permission from Ref. 34, Copyright 1988, Pergamon Journals Ltd.*)

centre of the region. As $|\alpha'|$ increases, a localized state emerges from the lower band edge and moves to lower energies, with the peak in the curve simultaneously shifting to correspond to this energy. At the extreme of $\alpha' = -2.0$, charge transfer is likely to occur only if ε_0 is more or less aligned with E_s, rather than the band. This situation is reminiscent of the quasi-resonant neutralization process associated with core-level states. It is important to notice that, even when the perturbation on the end atom is not large enough to create a localized state ($\alpha' = -0.5$), the change in the surface density of states is still sufficient to produce considerable differences in the charge-transfer probabilities from those for the clean surface. It must be concluded that the presence of adatoms on a surface will generally give rise to charge-transfer behavior quite distinct from that produced by a clean surface.

It is also significant, in Fig. 6, that the maximum value of $P_\alpha(\infty)$ increases with the creation and greater intensity of a localized state (i.e. larger $|\alpha'|$), due to the fact that electrons in these states are more localized at the surface, making them more likely to be transferred to the ion than bulk-state electrons, which are delocalized throughout the solid.

The influence of adsorbates on the resonant exchange mechanism goes beyond the perturbation of the surface potential and the creation of localized states. The formation of a monolayer of adatoms on the surface can increase the number of electrons available to a partially filled energy band, having the effect of raising the Fermi level ε_F substantially (i.e. lowering the work function ϕ). For example, the accumulation of Na on W(110) lowers ϕ from about 5 eV to 2 eV[2]. This is important because, as we have seen, the position with respect to the Fermi level of an energy level on the atom, at closest approach, is crucial in determining the probable occupancy of that level as the atom recedes from the surface; only if it is below ε_F is the level likely to be occupied.

During the scattering of neutral Na from W(110)[2], Na adatoms can accumulate on the surface, thereby lowering ϕ. The products Na^+, Na^0 and Na^- are all observed in proportions strongly dependent on the value of ϕ. Na^+ is formed by way of the transfer of the electron in the ionization level to the substrate, while Na^- results from the affinity level picking up an electron from the occupied part of the band. In their theoretical investigation of this system, Grimley *et al.*[24] modelled the substrate as a one-dimensional chain of W atoms, and the effect of Na adatoms was represented solely by varying ε_F to simulate different degrees of coverage. At large separations from the surface, the initially occupied ionization level has energy -5.14 eV, but as the atom approaches the surface, the level is shifted upwards by $F(t)$, so that at closest approach, it has energy -2.84 eV. Thus, when ε_F is substantially below -2.84 (i.e. relatively clean surfaces), the ionization level almost certainly loses its electron into one of the unoccupied orbitals of the solid, so there is virtually total conversion of Na^0 to Na^+. Conversely, for $\varepsilon_F > -2.84$ (surfaces with heavy coverage by adatoms), the ionization level is always aligned with occupied substrate states, strongly inhibiting any charge transfer to the solid, and Na^+ is a minority product. An analogous argument can be applied to the initially empty affinity level. That level has a large-separation energy of -0.549 eV, which the image interaction shifts downwards by $F(t)$, to a minimum value of -2.84 eV (equal to the closest-approach energy of the ionization level). For low values of ε_F (i.e. high ϕ), the affinity level is aligned with unoccupied levels of the solid, so essentially no charge transfer is possible. When $\varepsilon_F > -2.84$ (low ϕ), the circumstances are conducive to an electron transferring to the affinity level, so that, assuming the ionization level keeps its electron, Na^- is the final product.

The Grimley approach, discussed above, is able to account for the main experimental features of the Na–W system, despite its oversimplified treatment of adatom effects, which was restricted to a variation of the surface work function. One possible way to improve the model would be to adopt a suggestion by Brako and Newns[29] that, at low temperatures and coverage by adatoms, individual Na adatoms can act as 'active sites' which inhibit ionization in the following way. Being essentially ionic, an adsorbed Na atom has an effective dipole moment, normal to the surface, which can counterbalance the image interaction shift in Na atoms several angstroms away, thereby lowering the ionization level so that it becomes more closely aligned with occupied band states. Thus, the process of positive ionization becomes more unlikely, and the neutral fraction is increased from what might otherwise be expected.

The explicit inclusion of Na adatoms in the surface density of states, along the lines discussed earlier[34,45], has been considered by us[58], enabling the different situations of Na and W target atoms to be studied. In addition, variation of ϕ was used to model the degree of coverage of the surface by adatoms. Typical results are shown in Fig. 7[58], where only the positive fraction is considered. Although the general conclusions of Grimley *et al.*[24]

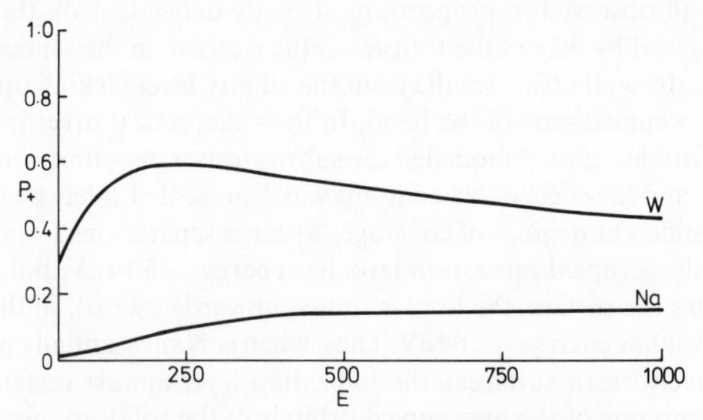

Fig. 7. Positive fraction P_+ versus incident Na energy E in eV for W and Na target atoms, with $\phi = 3.5$ eV. (*Reprinted with permission from Ref. 58, Copyright 1988, Pergamon Journals Ltd.*)

are confirmed, it is clearly seen that the most likely product (Na^+ or Na^0) can be determined by the specific target (W or Na) with which an incoming atom interacts, i.e. because collision with a Na adatom reduces the ionization probability, it is possible for the neutral fraction to be the majority product, rather than the minority, as tends to happen when W is the target atom. The primary reason for this reduction in P_+ is an intrinsic decrease in the image interaction shift for Na adatoms, which keeps $E_0(t)$ more closely aligned with the occupied portion of the band, thereby deterring charge transfer into the band.

Another system of great theoretical and experimental interest is W(110) with a Cs adlayer. The best-known experimental work is that of Geerlings *et al.*[3,18,59–65] who bombarded this surface with positive hydrogen and alkali (Li, K, Cs) ions, with the aim of producing intense negative beams, the motivation being their potential application to heating of fusion plasmas. The parameters of most interest are the surface work function and the electron affinity of the ion. As in the Na-W system, the largest negative-ion yields are generally obtained for the lowest work function surfaces.

During H^+ scattering[59,60], there are two independent charge-transfer events which occur. The first of these is the total conversion of H^+ into H^0, either by direct Auger neutralization into the ground state, or by resonant neutralization into an excited state followed by Auger de-excitation. Negative ionization can then occur via resonant transfer. Theoretical models[18] ignore the two-electron neutralization process, essentially treating H^0 as the initial state. Because decreasing ϕ increases the overlap between occupied band states and the affinity level, it is not surprising that the final H^- fraction varies strongly with ϕ, which is, of course, dependent on the degree of surface coverage by Cs adatoms. Lowering ϕ increases the negative fraction, up to a maximum of as much as 67% with $E = 100\,\text{eV}$, for the minimum work function $\phi = 1.45\,\text{eV}$, which corresponds to a Cs coverage of 0.6 times the saturation value.

The high H^- yield, obtained from collisions with cesiated tungsten, can be markedly decreased, by as much as a factor of 5, by the co-adsorption of hydrogen. However, this reduction cannot be attributed to any change in the surface work function, because hydrogen is adsorbed neutrally, which means that there will be no variation in the work function. The experimental observation[61], that the reduction in H^- production is proportional to the number of hydrogen adatoms, implies that each adatom acts as a local 'sink' for H^- formation, within a range of about 5 Å. A proposed mechanism[61,66] is that the electron in the affinity level of a newly formed H^- ion, acquired during a collision with a Cs or W atom, transfers to the unoccupied affinity level of a nearby H adatom, and hence into one of the unoccupied metal levels above ε_F. Consequently, the H atom is unable to retrieve the lost electron, and rebounds from the surface neutralized.

The scattering of alkali (Li, K, Cs) ions (and atoms) from cesiated tungsten[3,62–64] differs from the previous system in that the positioning of the ionization and affinity levels is such that all electron transfers to (or from) them occur by resonant transitions. In the case of Li^+ scattering, the products Li^+, Li^0, Li^- and $Li^0(2p)$ have all been observed under different conditions, depending on the degree of surface coverage by Cs adatoms. For high coverages (over 0.25 monolayer), this dependence manifests itself through the macroscopic work function, while for low coverages, individual Cs adatoms act as neutralization sites, so that charge exchange is controlled by the local electrostatic potential, creating a 'local' work function. The situation for K^+ and Cs^+ projectiles is similar[63]. For all three types of alkali scattering, the basic results are as for the H^+ case; namely, with decreasing work function, the negative fraction increases at the expense of the positive fraction.

6. CONCLUSIONS

In this chapter we have shown that the TDAN model gives a good description of the resonant charge-transfer process in atom–surface scattering. While it is unfortunate that exact solutions for the TDAN wavefunction cannot be obtained, the one-electron method can be used to find approximate solutions which allow qualitative predictions to be made. On the whole, these predictions are in reasonable accord with experimental findings.

Nevertheless, the one-electron approach does have its deficiencies, and we believe that a major theoretical effort must now be devoted to improving on it. This is not only in order to obtain better quantitative results but, perhaps more importantly, to develop a framework which can encompass all types of charge-transfer processes, including Auger and quasi-resonant ones. To do so is likely to require the use of many-electron multi-configurational wavefunctions. There have been some attempts along these lines and we have indicated, in detail, how such a theory might be developed. The few many-electron calculations which have been made do differ qualitatively from the one-electron results for the same systems and, clearly, further calculations on other systems are required.

There are some indications that interesting new effects, which it should be possible to treat by a modification of the TDAN model, are beginning to be discussed theoretically or observed experimentally. We have in mind especially those which involve excitation of the atom during the scattering process[67] and those in which a coupling to the medium[68], represented as a heat bath, is important. It seems likely that the theory we have described can be readily extended to handle these, and similar problems, and so we can expect it to be used increasingly in future theoretical work on charge transfer in atom–surface scattering.

References

1. Overbosch, E. G., Rasser, B., Tenner, A. D., and Los, J., *Surf. Sci.*, **92**, 310 (1980).
2. Overbosch, E. G., and Los, J., *Surf. Sci.*, **108**, 99 (1981); **108**, 117 (1981).
3. Geerlings, J. J. C., and Los, J., *Phys. Letts. A*, **102**, 204 (1984).
4. Newns, D. M., Makoshi, K., Brako, R., and van Wunnik, J. N. M., *Physica Scripta*, **T6**, 5 (1983).
5. Yoshimori, A., and Makoshi, K., *Prog. Surf. Sci.*, **21**, 251 (1986).
6. Davison, S. G., Sulston, K. W., and Amos, A. T., in *Asia Pacific Symposium on Surface Physics* (Ed. X. Xide), World Scientific Publishers, 1987, p. 80.
7. Modinos, A., *Prog. Surf. Sci.*, **26**, 19 (1987).
8. Erickson, R. L., and Smith, D. P., *Phys. Rev. Letts.*, **34**, 297 (1975).
9. Tully, J. C., *Phys. Rev.* B, **16**, 4324 (1977).
10. Easa, S. I., and Modinos, A., *Surf. Sci.*, **161**, 129 (1985).
11. Hagstrum, H. D., *Phys. Rev.*, **96**, 336 (1954); **122**, 83 (1961).
12. Anderson, P. W., *Phys. Rev.*, **124**, 41 (1961).
13. Newns, D. M., *Phys. Rev.*, **178**, 1123 (1969).
14. Blandin, A., Nourtier, A., and Hone, D. W., *J. de Phys.*, **37**, 369 (1976).
15. Bloss, W., and Hone, D., *Surf. Sci.*, **72**, 277 (1978).
16. Sroubek, Z., *Surf. Sci.*, **44**, 47 (1974).
17. Gomer, R., and Swanson, L. W., *J. Chem. Phys.*, **38**, 1613 (1963).
18. Rasser, B., van Wunnik, J. N. M., and Los, J., *Surf. Sci.*, **118**, 697 (1982).
19. Sroubek, Z., in *Inelastic Particle–Surface Collisions*, Springer Series in Chemical Physics, vol. 17 (Eds. E. Taglauer and W. Heiland), Springer-Verlag, 1981, p. 277.
20. van Wunnik, J. N. M., Brako, R., Makoshi, K., and Newns, D. M., *Surf. Sci.*, **126**, 618 (1983).
21. Easa, S. I., and Modinos, A., *Surf. Sci.*, **183**, 531 (1987).
22. Kawai, R., Liu, K. C., Newns, D. M., and Burnett, K., *Surf. Sci.*, **183**, 161 (1987).
23. Sebastian, K. L., Jyothi Bhasu, V. C., and Grimley, T. B., *Surf. Sci.*, **110**, L571 (1981).
24. Grimley, T. B., Jyothi Bhasu, V. C., and Sebastian, K. L., *Surf. Sci.*, **124**, 305 (1983).
25. Yoshimori, A., Kawai, H., and Makoshi, K., *Prog. Theoret. Phys. Suppl.*, **80**, 203 (1984).
26. Makoshi, K., Kawai, H., and Yoshimori, A., *J. Phys. Soc. Japan*, **53**, 2441 (1984).
27. Sebastian, K. L., *Phys. Rev.* B, **31**, 6976 (1985).
28. Davison, S. G., Sulston, K. W., and Amos, A. T., *J. Electroanal. Chem.*, **204**, 173 (1986).
29. Brako, R., and Newns, D. M., *Surf. Sci.*, **108**, 253 (1981).
30. Ziman, J. M., *Principles of the Theory of Solids*, Cambridge University Press, 1965, p. 80.
31. Shindo, S., and Kawai, R., *Surf. Sci.*, **165**, 477 (1986).
32. Sulston, K. W., Amos, A. T., and Davison, S. G., *Surf. Sci.*, **197**, 555 (1988).
33. McDowell, H. K., *J. Chem. Phys.*, **77**, 3263 (1982).
34. Sulston, K. W., Davison, S. G., and Amos, A. T., *Sol. St. Comm.*, **62**, 781 (1987).
35. Amos, A. T., Davison, S. G., and Sulston, K. W., *Phys. Lett. A*, **118**, 471 (1986).
36. Rosen, N., and Zener, C., *Phys. Rev.*, **40**, 502 (1932).
37. Yu, M. L., and Lang, N. D., *Nucl. Instr. and Meth. Phys. Res. B*, **14**, 403 (1986).
38. Lang, N. D., *Phys. Rev. B*, **27**, 2019 (1983).
39. Kawai, R., in *Dynamical Processes and Ordering on Solid Surfaces*, Springer Series in Solid State Sciences, vol. 59 (Eds. A. Yoshimori and M. Tsukada), Springer-Verlag, 1984, p. 51.

40. Kawai, R., and Ohtsuki, Y. H., *Nucl. Instr. and Meth. Phys. Res. B*, **2**, 414 (1984).
41. Kawai, R., *Phys. Rev. B*, **32**, 1013 (1985).
42. Sulston, K. W., Amos, A. T., and Davison, S. G., *Phys. Rev. B*, **37**, 9121 (1988).
43. Battaglia, F., George, T. F., and Lanaro, A., *Surf. Sci.*, **161**, 163 (1985).
44. Battaglia, F., and George, T. F., *J. Chem. Phys.*, **82**, 3847 (1985).
45. Sulston, K. W., Amos, A. T., and Davison, S. G., *Chem. Phys.*, **124**, 411 (1988).
46. Muda, Y., and Hanawa, T., *Surf. Sci.*, **97**, 283 (1980).
47. McDowell, H. K., *Chem. Phys.*, **72**, 451 (1982).
48. McDowell, H. K., *J. Chem. Phys.*, **83**, 772 (1985).
49. Adelman, S. A., and Doll, J. D., *J. Chem. Phys.*, **61**, 4242 (1974); **63**, 4908 (1975); **64**, 2375 (1976).
50. Inglesfield, J. E., *Surf. Sci.*, **127**, 555 (1983).
51. Hentschke, R., Snowdon, K. J., Hertel, P., and Heiland, W., *Surf. Sci.*, **173**, 565 (1986).
52. Snowdon, K. J., Hentschke, R., Narmann, A., and Heiland, W., *Surf. Sci.*, **173**, 581 (1986).
53. Sebastian, K. L., *Phys. Letts. A*, **98**, 39 (1983).
54. Kasai, H., and Okiji, A., *Surf. Sci.*, **183**, 147 (1987).
55. Brako, R., and Newns, D. M., *Sol. St. Comm.*, **55**, 633 (1985); *Physica Scripta*, **32**, 451 (1985).
56. Cizek, J., *Adv. Chem. Phys.*, **14**, 35 (1969).
57. Ishii, A., *Surf. Sci.*, **192**, 172 (1987).
58. Sulston, K. W., Amos, A. T., and Davison, S. G., *Prog. Surf. Sci.*, **26**, 63 (1987).
59. van Wunnik, J. N. M. and Los, J., *Physica Scripta*, **T6**, 27 (1983).
60. Granneman, E. H. A., Geerlings, J. J. C., van Wunnik, J. N. M., van Bommel, P. J., Hopman, H. J., and Los, J., in *Production and Neutralization of Negative Ions and Beams* (AIP Conf. Proc. 111). (Ed. K. Prelec), 1984, p. 206.
61. van Amersfoort, P. W., Geerlings, J. J. C., Rodink, R., Granneman, E. H. A., and Los, J., *J. Appl. Phys.*, **59**, 241 (1985).
62. Hermann, J., Welle, B., Gehring, J., Schall, H., and Kempter, V., *Surf. Sci.*, **138**, 570 (1984).
63. Geerlings, J. J. C., Kwakman, L. F. Tz., and Los, J., *Surf. Sci.*, **184**, 305 (1987).
64. Geerlings, J. J. C., Rodink, R., Los, J., and Gauyacq, J. P., *Surf. Sci.*, **186**, 15 (1987).
65. Geerlings, J. J. C., van Amersfoort, P. W., Kwakman, L. F. Tz., Granneman, E. H. A., Los, J., and Gauyacq, J. P., *Surf, Sci.*, **157**, 151 (1985).
66. Gauyacq, J. P., and Geerlings, J. J. C., *Surf. Sci.*, **182**, 245 (1987).
67. Brako, R., *Phys. Rev. B*, **30**, 5629 (1984).
68. Tsukada, M., Shima, N., and Tsuneyuki, S., *Prog. Surf. Sci.*, **26**, 47 (1987).

AUTHOR INDEX

COMPOUND INDEX FOR ADSORBATE/ SURFACE SYSTEMS

SUBJECT INDEX